In-Band Full-Duplex Wireless Systems Handbook

For a listing of recent titles in the
Artech House Mobile Communications Series,
turn to the back of this book.

In-Band Full-Duplex Wireless Systems Handbook

Kenneth E. Kolodziej
Editor

ARTECH HOUSE
BOSTON | LONDON
artechhouse.com

Library of Congress Cataloging-in-Publication Data
A catalog record for this book is available from the U.S. Library of Congress.

British Library Cataloguing in Publication Data
A catalog record for this book is available from the British Library.

ISBN-13: 978-1-63081-789-3

Cover design by Kalyani Kastor

685 Canton Street
Norwood, MA 02062

10 9 8 7 6 5 4 3 2 1

To my wife, Maria-Christina, for encouraging and supporting me since the day I met her, and to my daughter, Briana, for making me smile and helping me realize that being a dad is my true calling in life. Vi amo tantissimo!

Contents

Foreword

How hard can it be? *It* here refers to in-band full-duplex wireless, where a node can send and receive independent data simultaneously in the same frequency band. After all, the interfering signal is transmitted by the node itself, and hence it can cancel this (self-)interference in the receiver side processing. However, the self-interference overwhelms the receiver analog-to-digital converters and is time-varying due to propagation over backscatter paths in wireless channels. As a result, in-band full-duplex had remained confined to radars and did not transition to communication systems.

The situation changed in 2010, when two research groups, one at Stanford University and another being my research group at Rice University, independently demonstrated the feasibility of in-band full-duplex. Both groups presented different techniques to reduce only the self-interference power before the point where the combined signal and self-interference signal are digitized by the analog-to-digital converter.

The results were met with a healthy bout of skepticism by the wireless community. Before publishing our paper, we checked every part of our implementation and experimental data analysis many times, as we knew we were about to publish an uncomfortable result. We were going against the wireless community's firm belief that in-band full-duplex is impossible. That firm belief influences all communications standards, wireless or wireline, as all of them are either time-division or frequency-division duplex.

The tide turned as more groups confirmed the results and improved on our original reported results. Even more hearteningly, the research community launched into innovations in many domains: circuits, algorithms, network protocols, and new applications, to name a few.

Considering that in-band full-duplex requires rethinking of network architectures, my expectation was that the technology adoption (e.g., adoption in standards) will take longer than other communications innovations. Surprisingly, the cable modem industry moved rapidly to adopt full-duplex communications, as the fundamental concepts apply to wireless and wired networks. By 2017, DOCSIS 4.0 had been ratified to replace DOCSIS 3.1 (DOCSIS stands for Data Over Cable Service Interface Specification and defines the cable modem standards). The new standard uses in-band full-duplex, replacing the frequency-division duplex used in the previous standard. The use of in-band full-duplex allows the 4.0 standard to support 10 Gbps downstream and 6 Gbps upstream, doubling the downstream and quadrupling the upstream speeds compared to the previous generation.

On the wireless front, full duplex is fortuitously aligned with the adoption of the massive multiple-input multiple output (MIMO) for fifth-generation (5G) networks. Currently, most massive MIMO use time-division duplex, that is, they transmit and receive in the same band. Thus, there is a natural pathway to upgrade these networks into a full-duplex operation. Some vendors have already developed full-duplex product demonstrators for 5G massive MIMO base stations, and standardization proposals continue to be discussed.

As I noted above, in-band full-duplex has inspired research in many communities over the last decade. One reason is that, at its very core, in-band full-duplex requires algorithmic and hardware codesign. As a result, significant ongoing research is developing self-interference cancellation methods for next-generation systems that may have very large bandwidths, tens or hundreds of antennas, potentially packed in small form-factor devices. These developments mean pushing the boundaries in circuits, antennas, algorithms, and, most importantly, their codesign. Complementing these innovations are innovations to translate physical-layer gains to network-scale metrics, such as throughput gains or reduction in latency. Last but not least, an important emerging research area is to leverage full duplex for new network functions, like combining radar and communication in one subsystem. These new applications are becoming highly relevant in the wake of self-driving cars and unmanned aerial vehicles, all of which need methods to communicate and sense the environment.

This book provides both a look back and a look ahead, covering all major research thrusts in one cohesive narrative. Kenneth Kolodziej has assembled a team of leading researchers who have contributed chapters on the fundamentals, practical considerations, and state-of-the-art results for in-band full-duplex wireless systems. Together, this unique collection covers self-interference cancellation techniques that span all transceiver domains and their deployment within various systems and related applications. These distinctive aspects make the *In-Band Full-Duplex Wireless Systems Handbook* an excellent reference for both beginners and professionals to help to propel this technology area for years to come.

Ashutosh Sabharwal
Chair and Professor
Department of Electrical and Computer Engineering
Rice University
Houston, Texas
March 2021

Preface

Although the simplicity of time-division and frequency-division duplex schemes will prevent them from becoming obsolete, in-band full-duplex (IBFD) is the technology of the future. It was considered to be one of the five pillars that was going to enhance our mobile connectivity within fifth-generation (5G) wireless networks. Although IBFD functionality has not yet been incorporated into the 5G specification, it seemed like the right time to organize a book that uniquely focused on the physical-layer aspects of IBFD self-interference cancellation (SIC) techniques, systems, and applications that will help the technology to become widely adopted.

There are many different approaches to mitigating the self-interference (SI) that exists in IBFD systems, combinations of which produce an endless possibility of system-level architectures. Countless university groups, research entities, and corporations around the world are actively investigating these SIC areas as well as their integration to enable novel devices and applications. It is for this reason that I asked the individual experts in these areas for their help in assembling a collection of chapters by authors from different backgrounds. Although this was not a straightforward project, I felt that this approach would produce the best handbook reference for the research community and help the field to progress in a meaningful way.

One of the novel aspects of this book is that it focuses on IBFD topics that have been implemented and demonstrated in the real world. This allows the chapter authors to not only present the theoretical foundation of their work, but also validate their assumptions with measured results that can be used to set expectations for their associated SIC technique, system, or application. Additionally, this uniquely allows them to share insights into the lessons learned during the prototyping process, which should help others to avoid past mistakes as well as highlight problematic areas that require careful attention.

This book was designed to support bouncing around among the chapters (orientation diagrams are provided at the beginning of every chapter). I assumed that not everyone wants to read the details of each different SIC method and instead may focus on his or her research area or current interest. With that said, I recommend you start with Chapter 1. This is not just because I wrote it, but also because it sets the stage by providing a brief overview of IBFD, presents a survey of measured systems, and explains the book organization methodology. I hope that you find this handbook to be a valuable resource, and I wish you the best in your future research efforts that will ensure that IBFD is the technology of the future.

This book would not have been possible without help from a wide range of people. I want to start with my family back in New Jersey: Tim, Dad, Mom, Mr. Joe, and Gianna; thank you for always taking an interest in my work activities and supporting me during every step of my career path. My curiosity in radio frequency (RF) and microwave engineering would not have existed if not for my advisor, Professor Harry Lenzing, to whom I am grateful for his dedication and passion to teaching. At the MIT Lincoln Laboratory, I had the tremendous fortune of working with Dr. Bradley Perry and Dr. Jeffrey Herd, both of whom started my research on IBFD systems and have been excellent mentors along the way. I especially thank Professor Ashutosh Sabharwal from Rice University for not only writing the book's Foreword, but also inspiring me with his seminal paper on IBFD.

This book really would not have happened without the IBFD experts who graciously agreed to write the chapters; you are all tremendous visionaries in this field, and I am honored to have worked with you on this. Finally, and I saved the most important for last, I am forever grateful to my wife, Maria-Christina , for putting up with all of my work on nights and weekends that made this book a reality, as well as for putting her English degree to the test and proofreading every single chapter multiple times; I do not know how you did it, but I am forever grateful. Thank you all for reading!

Kenneth E. Kolodziej
March 2021

CHAPTER 1

In-Band Full-Duplex Overview

Kenneth E. Kolodziej

1.1 Introduction

In-band full-duplex (IBFD) is a technology that allows devices to transmit and receive signals on the same frequency at the same time [1]. Historically, this concept has intentionally been avoided in a wide range of systems, including basic telephony, internet data traffic, Wi-Fi communications, cellular infrastructures, and many radar architectures [2]. This is due to the significant self-interference (SI) that is generated by a transmitter colocated with a receiver, and cannot be suppressed by traditional filters because the interfering signal is within the desired reception band [3–8]. As mentioned in the Foreword, IBFD can be applied to both wireline and wireless systems, the latter of which will be the focus of this book.

Some of the early IBFD research focused on continuous-wave (CW) radar systems that provide simple but effective architectures, which take advantage of continuous transmissions to improve their signal returns [9, 10]. It later became apparent that many other wireless systems can benefit from the incorporation of IBFD technology as well [11]. Chief among them are wireless communications networks that need to contend with scarce spectral resource allocations, for which governments have garnered large sums of money through spectral auctions, often awarded to large telecom companies [12]. By utilizing the same channel for transmit and receive, IBFD can double spectral efficiency for future wireless networks, such as fifth-generation (5G) New Radio (NR), its sixth-generation (6G) successor, and beyond [13, 14]. Additionally, the networking overhead can significantly be reduced for time-division duplex (TDD) systems that restrict only one user to transmit during a given timeslot. Similarly, IBFD can be integrated into military systems to provide novel multifunction capabilities, such as adding radar and/or electronic warfare (EW) techniques on top of normal tactical communications [15, 16]. While

DISTRIBUTION STATEMENT A. Approved for public release. Distribution is unlimited. This material is based upon work supported by the Under Secretary of Defense for Research and Engineering under Air Force Contract No. FA8702-15-D-0001. Any opinions, findings, conclusions or recommendations expressed in this material are those of the author(s) and do not necessarily reflect the views of the Under Secretary of Defense for Research and Engineering.

these examples provide just some insight into the benefits of IBFD being employed in the field today, this technology can be adapted to enable future systems and applications that were previously considered impossible as well.

It should be noted that the term IBFD grew out of the communications community, and that there are other names applied to the same concept. The analogous term, simultaneous transmit and receive (STAR), seems to have been adopted by defense organizations, possibly to encompass the asymmetric data traffic of military systems that support applications other than communications. Additionally, while same-frequency full-duplex (SFFD) can also be used to describe the technology, many publications simply refer to it as full-duplex (FD). Similar to STAR, the FD term traditionally includes systems that transmit and receive at the same time on different frequencies, known as frequency-division duplex (FDD) or out-of-band full-duplex, and should thus be avoided when specifically discussing IBFD systems. With that said, these terms will be used interchangeably throughout the book to all refer to the concept of concurrent transmission and reception within the same frequency band.

In this book, experts from academic, commercial, and federal research institutions will discuss the various approaches that can be taken to suppress the inherent SI that is generated in IBFD systems. Chapter authors will provide details on both static and adaptive techniques that span the various transceiver domains discussed in the following section as well as present IBFD systems that utilize multiple techniques for different applications.

1.2 System Overview

1.2.1 Transceiver Diagram

Many wireless systems concern themselves with interference suppression and protection against external signals to ensure desired operation. However, IBFD systems have the additional challenge of mitigating strong interference signals generated within the node/device itself. While this difference means that many possible coupling paths exist, it also offers designers the benefit of being able to influence the coupling mechanisms and sometimes even the SI signal itself. IBFD technology only becomes viable when these benefits are exploited to provide self-interference cancellation (SIC) that results in effectively isolating the receiver from the transmitter.

Potential locations for SIC can be considered by studying the generic IBFD transceiver diagram of Figure 1.1. This diagram illustrates a transceiver broken down into three different domains: propagation, analog, and digital [17]. The propagation domain applies to systems that share a single antenna through the use of an antenna interface device as well as designs that utilize multiple antennas for transmit and receive capabilities. Additionally, multiples of either antenna approach can be used to create arrays that introduce other SIC options. Both configurations include direct path (DP) coupling between the transmitter and receiver, which often represents the first SI component to arrive at the receiver. For multiple antennas, this is typically the line-of-sight path, and for the shared case, the DP captures the finite amount of leakage provided by the antenna interface. While these DPs are fairly static, IBFD systems must also contend with dynamic SI reflection paths

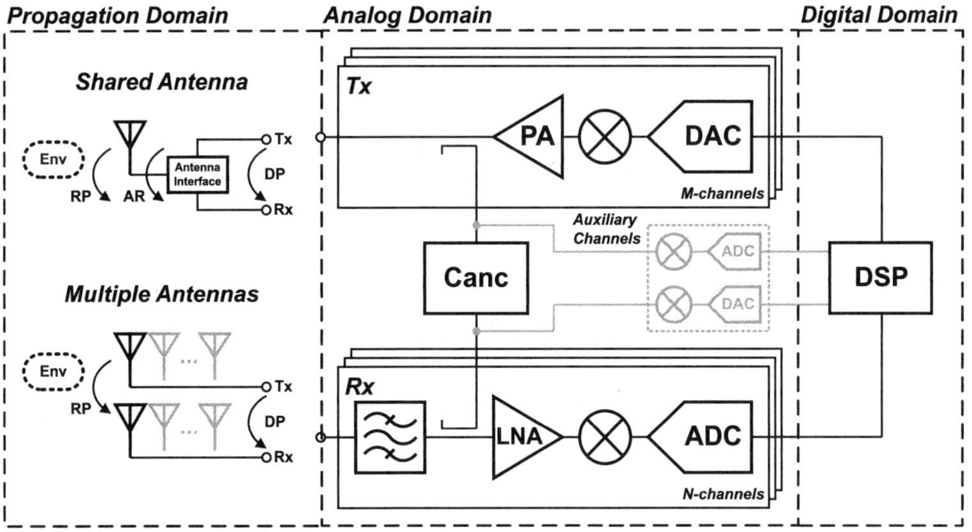

Figure 1.1 Generic IBFD transceiver, highlighting the propagation, analog, and digital domains.

(RPs) that are influenced by the environment surrounding the node. Local multipath scattering can reflect the transmit signal back towards the receiving antenna and effectively produce delayed echoes of the SI signal that become increasingly problematic as the transmit power increases. Finally, for the shared antenna setup, an antenna reflection (AR) is created since the antenna impedance is not perfectly matched to that of the connected antenna interface. Similar to the RPs, this AR is also influenced by the immediate surroundings that can alter the perceived impedance of the antenna and effectively change the reflection signal. Due to this AR and the tightly coupled relationship of the antenna and its interfacing device, SIC techniques related to the antenna interface are included in the propagation domain instead of the analog domain discussed next.

Within the analog domain of the transceiver resides the radio hardware responsible for conditioning and converting the propagation, or radio frequency (RF), signal to and from the digital domain. The diagram depicts M transmit (Tx) channels, each with a digital-to-analog converter (DAC) and a power amplifier (PA). Additionally, a single frequency conversion mixer is also represented, but only depicts a common configuration. Realistic designs may contain more or less frequency conversion stages as well as associated out-of-band filtering and additional gain to appropriately adjust the signal before transmission. Similarly, N receive (Rx) channels are illustrated with a unique same-frequency in-band filter at the input to provide SIC that discriminates the SI by a signal parameter other than frequency. This filter is followed by a low-noise amplifier (LNA), a single frequency conversion mixer, and an analog-to-digital converter (ADC). Both the linearity and saturation level of these active receiver components must be carefully considered to determine how much propagation and analog domain suppression is required before the digital domain. Analogous to the transmit side, realistic receivers will include different amounts of gain and frequency conversion as well as out-of-band filtering to suppress the external interference encountered by traditional wireless systems.

The figure also illustrates a cancellation (Canc) element between the transmit and receive channels. This specific depiction shows the canceller operating at RF (the propagation frequency), and coupling off the transmit signal path after the PA. While this represents a common configuration, both the operating frequency and location within the transceiver can vary, along with the canceller feeding mechanism. Alternative designs employ auxiliary channels to aid in SIC in either the analog and/or digital domains. A representative subset of the required components in these auxiliary channels and their connection paths are also captured within this generic transceiver.

Finally, the digital domain contains all SIC techniques that are controlled through digital signal processing (DSP). The diagram intentionally maintains an abstract representation of these capabilities as they are numerous and ever-growing. As discussed later in this chapter, this generic transceiver diagram will be utilized throughout the book to help orient the reader to which part of the transceiver and associated SIC domain is being highlighted within each individual chapter.

1.2.2 Self-Interference Cancellation Techniques

The remaining chapters in this book discuss SIC techniques and their application within different locations of the generic transceiver shown in Figure 1.1. The previously discussed breakdown among the propagation, analog, and digital domains can be utilized to group these methods as depicted in Figure 1.2. This tree diagram is organized based on the location where cancellation occurs, and provides a convenient list of SIC techniques that have previously been published and initially summarized in [17]. The propagation domain includes various passive and active antenna techniques as well as the several different antenna interfaces. The analog domain contains in-band filtering, assorted cancellation methods and analog receive beamforming, while the digital domain encompasses numerous channel-modeling and digital receive beamforming approaches.

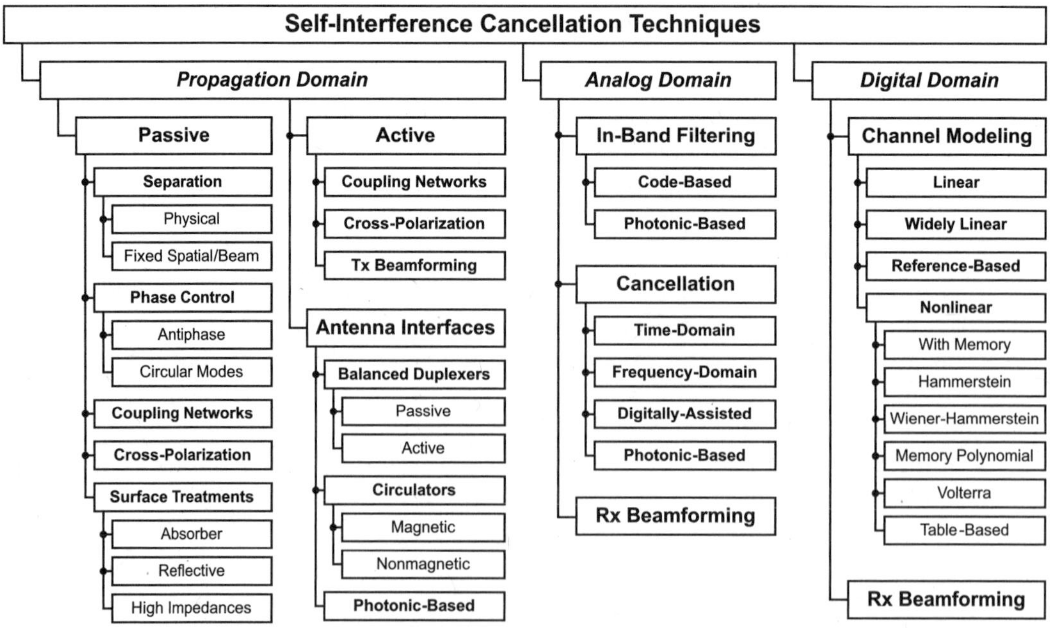

Figure 1.2 SIC techniques tree.

Subsequent chapters will walk readers through the basics of these individual methods as well as illustrate their performance capabilities with measured results. Similar to the generic transceiver diagram, this SIC tree will be referenced at the beginning of every chapter to indicate the chapter focus and techniques utilized.

1.3 Literature Review

1.3.1 Publications

Research on the various aspects of IBFD technology has increased significantly since the initial reports on communications systems in 2010 [18, 19]. This includes investigations of not only all of the SIC techniques previously discussed, but also simulation studies that move beyond the physical layer and emulate IBFD network interactions. A representative plot of this trend is shown in Figure 1.3, where the number of Institute of Electrical and Electronics Engineers (IEEE) publications is depicted between 2010 and 2020. This data was generated for articles that include IBFD and STAR as keywords, and is meant to provide an overview of the research growth, rather than an all-encompassing account. These publications are further segregated based on their type: conferences, journals, and magazines.

The figure illustrates that the number of publications has been steadily increasing since 2010 with the exception of a drop in 2020. This decrease is due to two reasons: first, the results only capture through the end of the third quarter of 2020 when this book was sent to print. Second, many conferences were either cancelled or postponed due to the worldwide pandemic and is clearly evident in the significant drop of that category. If not for those two reasons, one would expect 2020 to produce at least 100 publications, and more than likely surpass the numbers of 2018 and 2019. Overall, this increased interest in IBFD research should continue and help to transition this technology from simulations and laboratory experimentations into adopted network specifications and practical system deployments.

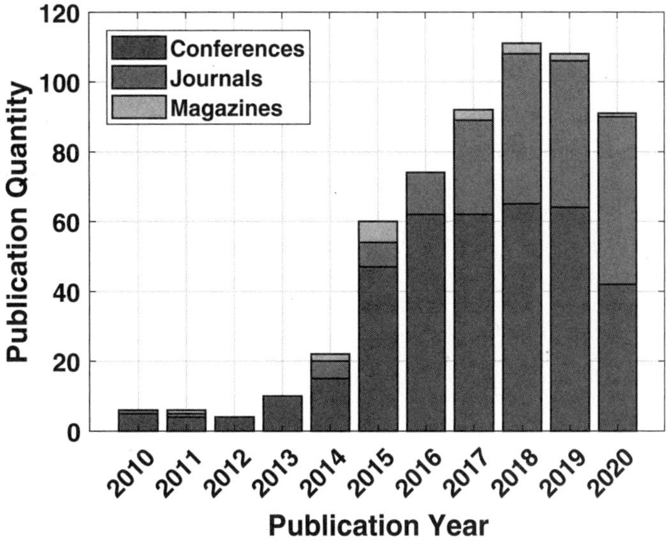

Figure 1.3 IBFD conference, journal, and magazine articles versus publication year.

1.3.2 System Survey

After considering the generic transceiver and SIC techniques shown in Figures 1.1 and 1.2, respectively, it should become apparent that a system-level design approach is required for many IBFD applications. When high amounts of isolation are desired, techniques across the multiple domains discussed are utilized in conjunction. Careful attention must be paid when combining different SIC methods because cancellation performance does not necessarily add linearly. For example, analog circuits can suppress the SI signal, but at the same time introduce nonlinear distortion and noise components that restrict any subsequent digital filtering. Successful system designers should analyze the dynamic range of the individual transceiver components and associated SIC techniques to maximize the overall IBFD performance.

A survey of measured IBFD systems that have combined SIC techniques in more than one domain has been conducted for publications between 2010 and 2020. The investigation analyzed more than 90 total systems with nearly 130 unique measurement scenarios, which encompassed the influence of changing different system parameters. It should be noted that simulation-only results have been excluded from this survey as many real-world implementation issues and practical design tolerances cannot be captured in modeling environments.

Figure 1.4 depicts the total isolation reported for the measured prototypes versus the date of their publication. The data points are differentiated based on the combinations of SIC domains utilized, where "P+A+D" represents a design that leveraged cancellation in each of the propagation (P), analog (A), and digital (D) domains. While this plot only captures measured IBFD systems, it still highlights the trend of increased research over the years, as illustrated in Figure 1.3. Additionally, it can be seen that systems that have incorporated techniques in the digital domain are among the highest-performing designs, specifically considering isolation numbers greater than 100 dB. The highest-isolation system was reported in late 2016 with an isolation of 115.3 dB [20]. That prototype utilized a relay antenna

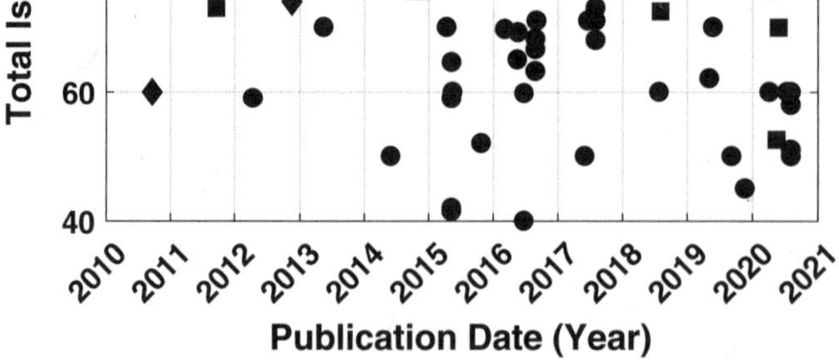

Figure 1.4 Survey of measured isolation versus publication date for IBFD systems that utilized a combination of SIC techniques in the propagation (P), analog (A), and digital (D) domains.

with passive fixed spatial/beam separation, a time-domain-based RF canceller and nonlinear digital filtering, which together represent techniques in all three domains.

These systems have been demonstrated at various center frequencies as shown in Figure 1.5, which illustrates their isolation performance between a frequency range of 0 to 5,000 MHz. The plot clearly indicates the significant number of systems that have been demonstrated within the 2.4-GHz industrial, scientific, and medical (ISM) band, which has most likely been propelled by its unlicensed nature and the vast selection of wireless parts that support commercial applications within that range. Additionally, there is a collection of research around 900 MHz that offers favorable propagation characteristics for cellular networks. While this survey focused on sub-6-GHz designs and did not include millimeter-wave prototypes, the highest reported center frequency was 4.5 GHz for a system that employed transmit and receive beamforming in the propagation and digital domains, respectively [21].

At the previously discussed center frequencies, these systems operated over instantaneous bandwidths that ranged from 25 kHz to 150 MHz, as depicted in Figure 1.6. A concentration of data points with varying levels of isolation are located along the 20-MHz bandwidth line, which corresponds to the standard channel width of Wi-Fi systems and illustrates a common testing configuration over the bandwidth centered at 2.4 GHz, as supported by Figure 1.5. Additionally, groups of measurements are observed at larger standard bandwidths, including 40 and 80 MHz. It is interesting to see that as the instantaneous bandwidth increases beyond 80 MHz, the isolation performance steadily degrades before stopping at 60 dB over a 150-MHz bandwidth for a system that utilized passive antenna phase control and cross-polarization techniques along with analog cancellation [22].

By combining the results of Figures 1.5 and 1.6, the percent bandwidth for these systems can be derived simply by dividing the instantaneous bandwidth by the center frequency. The resulting numbers are plotted in Figure 1.7, and illustrate that systems have ranged from close to 0% up to nearly 12%, which was for a prototype that employed passive antenna separation, analog cancellation and digital

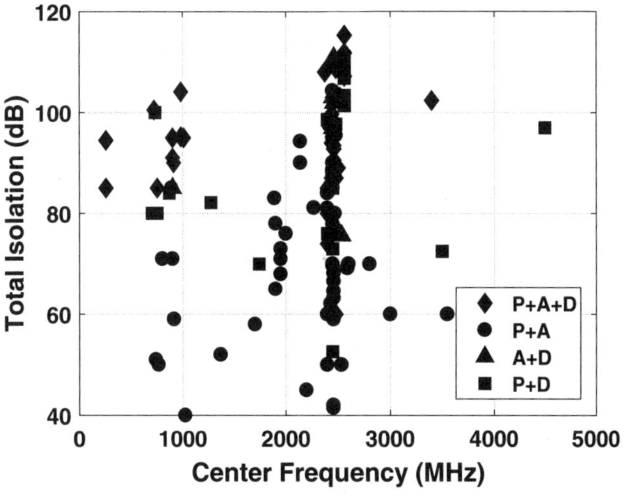

Figure 1.5 Survey of measured isolation versus center frequency for IBFD systems that utilized a combination of SIC techniques in the propagation (P), analog (A), and digital (D) domains.

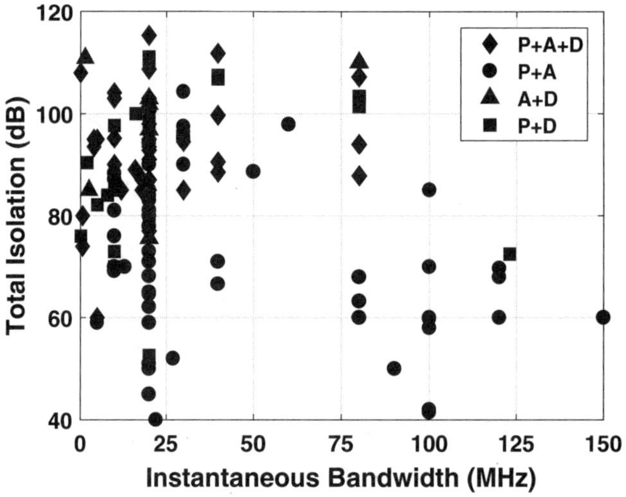

Figure 1.6 Survey of measured isolation versus instantaneous bandwidth for IBFD systems that utilized a combination of SIC techniques in the propagation (P), analog (A), and digital (D) domains.

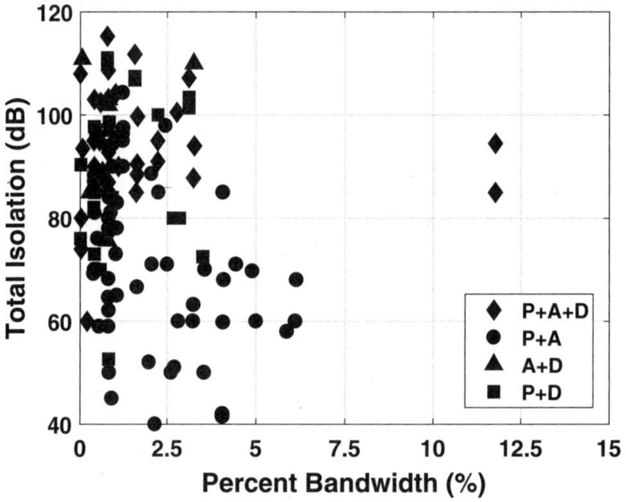

Figure 1.7 Survey of measured isolation versus percent bandwidth for IBFD systems that utilized a combination of SIC techniques in the propagation (P), analog (A), and digital (D) domains.

filtering over a 30-MHz bandwidth centered at 255 MHz [23, 24]. Most of these results indicate that the systems were designed to operate over fairly narrow percent bandwidths, typically less than 5%, with an average of 1.76%.

Another key metric of any prototype evaluation is the transmit output power that was utilized during testing. Higher transmit powers allow nodes to communicate over longer distances, and are not only more difficult to generate, but also typically increase both the nonlinear and noise distortion within the transceiver, which often require additional SIC to avoid degrading the receiver performance. Figure 1.8 illustrates the isolation performance of these systems as a function of their transmit power. This graph shows that many of the highest-isolation demonstrations have occurred at high output powers, close to 30 dBm, or 1W. While the average test case utilized 14.5 dBm of power, the highest reported reached 35 dBm

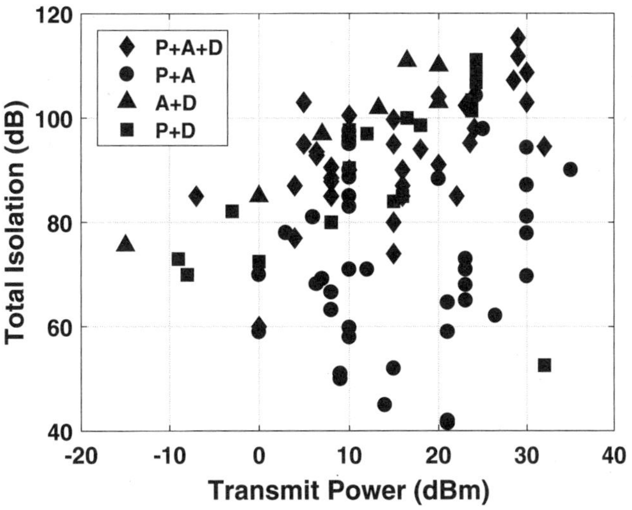

Figure 1.8 Survey of measured isolation versus transmit power for IBFD systems that utilized a combination of SIC techniques in the propagation (P), analog (A), and digital (D) domains.

for a magnetic-circulator-based system augmented with digitally-assisted analog cancellation and nonlinear digital filtering [25].

While Figures 1.4 through 1.8 depicted the measured isolation of the surveyed IBFD systems in relation to several critical design metrics, the combination of these parameters and overall performance was not captured. For this purpose, a unique figure of merit (FOM) was introduced that aggregates the previously discussed design variables, such that

$$\text{FOM} = \frac{\text{ISO}}{\text{RSD}} \bullet \text{IBW} \bullet P_{Tx} \tag{1.1}$$

where ISO is the total system isolation in linear units, RSD is the receiver sensitivity degradation in linear units, IBW is the instantaneous bandwidth in megahertz, and P_{Tx} is the transmit output power in milliwatts [17]. This receiver sensitivity degradation number represents the elevation in the nominal receiver noise floor that comes from implementing IBFD functionality. The degradation can be comprised of residual SI signals that were not mitigated and/or nonlinear and noise distortion components introduced by the various SIC techniques. For the systems surveyed, the receiver sensitivity degradation ranged from 0 to 31 dB, which illustrates that this parameter can drastically impact the noise level in these receivers. It should be noted that this FOM does not capture several additional IBFD design parameters, many of which have not been quantified by the previously surveyed systems in their prototyping stages. Ideally, additional metrics should be integrated into the FOM, including the DC power consumption for the various SIC techniques, the adaptation tuning time, and the system size impact.

The application of this FOM to the surveyed systems is plotted in Figure 1.9 as a function of the article publication dates. While the data points are somewhat scattered, a trend of increasing FOM results with relation to time can be visualized,

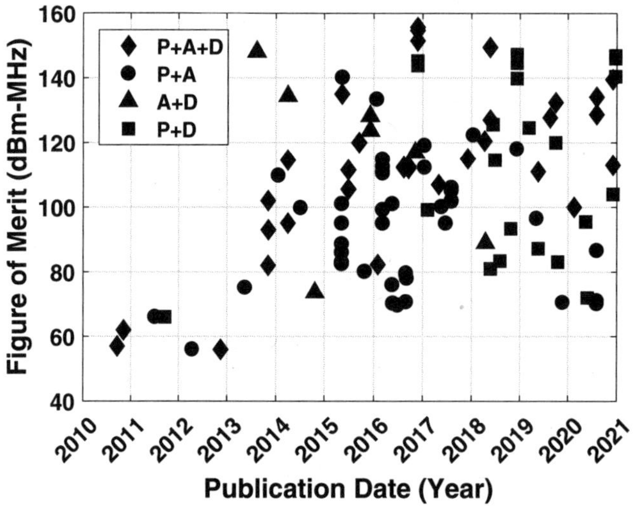

Figure 1.9 Survey of system FOM versus publication date for IBFD systems that utilized a combination of SIC techniques in the propagation (P), analog (A), and digital (D) domains.

which indicates that as IBFD research progresses, advances are being made in regards to not only the total isolation, but also the receiver sensitivity degradation, instantaneous bandwidth, and transmit output power. The system with the highest-reported isolation is also the one with the highest FOM, which speaks to the fact that it was evaluated with advanced specifications. More specifically, that prototype delivered a FOM of 155.6 dBm-MHz and was measured over a 20-MHz bandwidth centered at 2.56 GHz with an output power of 29 dBm and receiver sensitivity degradation of 1.7 dB [20]. Overall, as IBFD wireless systems research continues, this FOM should help to guide designers towards advancing the various system parameters that will propel the technology forward for both communications handheld devices and base stations as well as other novel IBFD applications.

1.4 Book Organization

As previously mentioned, the remaining chapters of this book focus on SIC techniques in the propagation, analog and digital domains as well as IBFD systems and applications that employ methods across multiple domains, as represented in the system survey of the preceding section. The book is divided into four sections centered around these three domains as well as one illustrating how they can be combined to create different IBFD systems and applications.

Figures 1.10 and 1.11 provide a mapping of the book chapters within the generic transceiver and SIC techniques diagrams, respectively. Chapters 2 through 12 are indicated in circles next to the focus of their respective chapters, while Chapters 13 through 17 are intentionally grouped together at the bottom of the figures because they all discuss IBFD systems and applications that utilize multiple techniques in several of the domains.

Within the propagation domain section, Chapter 2 discusses many of the various passive techniques that can be combined with antenna interfaces using

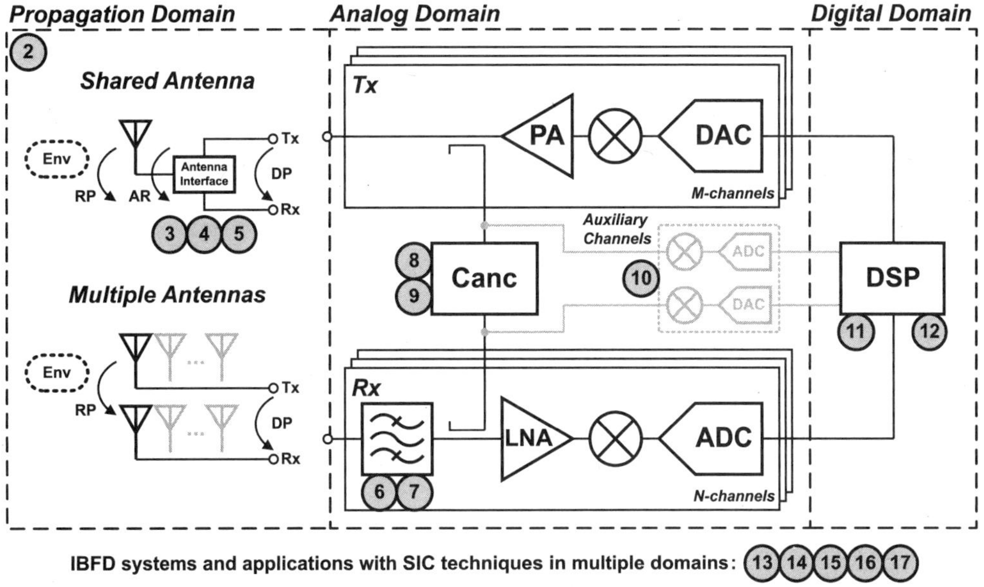

Figure 1.10 Generic IBFD transceiver indicating the focus of Chapters 2 through 17 with numbered circles.

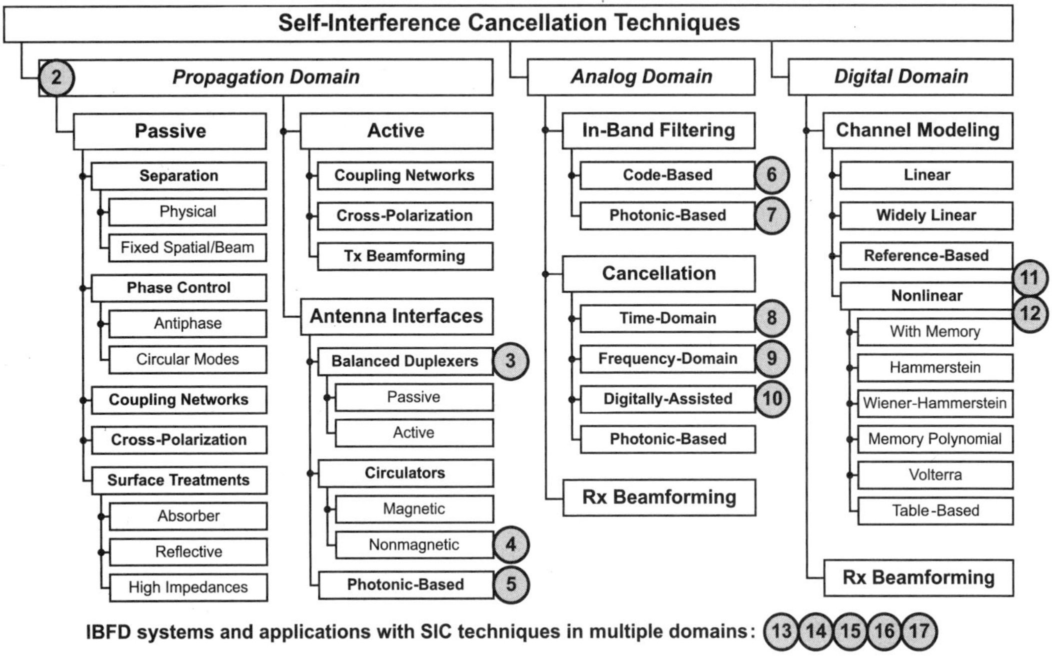

Figure 1.11 SIC techniques tree indicating the focus of Chapters 2 through 17 with numbered circles.

traditional magnetic circulators. Chapters 3, 4, and 5 highlight the design principles and practical performance expectations of antenna interfaces that rely on electrical balanced duplexers, integrated nonmagnetic circulators, and wideband photonic-based devices.

Within the analog domain sections, Chapters 6 and 7 focus on two different types of same-frequency in-band filtering that uniquely utilize codes and photonics to provide interference suppression, respectively. Chapters 8, 9, and 10 subsequently discuss the various signal cancellation methods that can be implemented throughout a transceiver and cover time-domain, frequency-domain, and digitally-assisted architectures, respectively.

Within the digital domain, basic linear channel models are greatly enhanced with the use of nonlinear techniques, which is showcased through comparisons of memory polynomial models and table-based interpolations in Chapter 11. Alternatively, the benefits of reference-based channel modeling to capture nonlinear effects along with memory models are discussed in Chapter 12.

Within the systems and applications section, Chapter 13 provides a comprehensive analysis of different approaches to construct IBFD phased arrays by leveraging passive propagation domain methods along with active transmit and digital receive beamforming, in addition to incorporating the reference-based channel modeling discussed in Chapter 12. Chapter 14 builds upon the array discussion by focusing on multiple-input multiple-output (MIMO) communications applications that employ the nonmagnetic circulators covered in Chapter 4 along with active transmit and analog receive beamforming, in addition to nonlinear channel modeling based on the Volterra series expansion technique.

Following the discussion of time-domain cancellation in Chapter 8, Chapter 15 presents several photonic-based methods for optical-based communications systems that have been combined with traditional magnetic circulators. Chapter 16 discusses the application of IBFD to backscatter modulation and the need for SIC techniques in each of the domains for successful system implementation. Finally, Chapter 17 highlights the potential for IBFD to revolutionize military operations with devices that pull from a range of the previously discussed SIC techniques in every domain.

1.5 Practical Considerations

One of the unique aspects of this book is that it focuses on IBFD techniques, systems, and applications that have been demonstrated with measured results. Each chapter will include a short section like this one that discusses some lessons learned by the authors over the years and should provide some very valuable insights into their area of expertise for all.

The major practical takeaway from this introductory chapter should be that IBFD wireless systems require detailed systems engineering for real-world implementations. Many applications necessitate high amounts of transmit-receive isolation, and both the transceiver design in terms of dynamic range and SIC technique integration should be carefully planned and analyzed. With that said, nothing will help more to build confidence in future system models than creating and measuring prototype subsystems, which can then be iterated and improved to produce the desired IBFD performance for a given application.

1.6 Conclusion

1.6.1 Summary

This chapter provided an overview of IBFD wireless systems, and introduced a generic transceiver architecture and SIC technique diagram that will be utilized in the book to help to guide readers through the various chapters focusing on different aspects of IBFD systems. A literature review was presented to illustrate the increasing interest in this research topic as well as provide some background on demonstrated IBFD system performance. The survey highlighted the use of SIC techniques within multiple domains of IBFD systems and studied the influence of various parameters, including center frequency, bandwidth and transmit power. Additionally, a novel FOM was discussed to capture the interdependency of these system specifications to help to propel future research to investigate advanced designs in each of these aspects. Finally, an overview of the book organizational structure was introduced to include major sections that focus on SIC techniques within the propagation, analog, and digital domains as well as a section dedicated to IBFD systems and applications that utilize methods across multiple domains.

1.6.2 Future Directions

There are many open questions in the field of IBFD technology that are related to not only the integration and maturity of the physical-layer hardware, firmware, and software techniques discussed in this book, but also involve their incorporations into the higher layers that form a complete network. In addition to advancing the development of research through practical demonstrations, the system survey presented in this chapter provides some insight into future research areas. One such direction relates to the center frequency of IBFD designs, and the forthcoming applications that benefit from the technology in the near term. Shifting focus towards research at 3.5 GHz for newly released 5G NR bands as well as 5.9 GHz for vehicle-to-everything (V2X) applications could accelerate the adoption of IBFD into these standards where spectral efficiencies are critical. Additionally, increasing to wider instantaneous bandwidths could not only enable the support of the wider channel allocations of 5G and 6G, but also allow for the option of carrier aggregation that can drastically increase the device bandwidth by combining multiple carriers under certain network constrains. Systems that can also mitigate the increased nonlinear and noise components of higher transmit output powers need to be designed and demonstrated so that IBFD research can be applied to base stations and wireless backhaul links, which are often required to transmit over long distances. The success of IBFD technology is left as a challenge to the reader that may be either learning about a technique for the first time or actively progressing the state-of-the-art systems within the field. It is my hope that this book contributes to that advancement in some way.

References

[1] Sabharwal, A., et al., "In-Band Full-Duplex Wireless: Challenges and Opportunities," *IEEE Journal on Selected Areas in Communications*, Vol. 32, No. 9, September 2014, pp. 1637–1652.

[2] Hong, S., et al., "Applications of Self-Interference Cancellation in 5G and Beyond," *IEEE Communications Magazine*, Vol. 52, No. 2, February 2014, pp. 114–121.

[3] Kim, D., H. Lee and D. Hong, "A Survey of In-Band Full-Duplex Transmission: From the Perspective of PHY and MAC Layers," *IEEE Communications Surveys & Tutorials*, Vol. 17, No. 4, Fourth Quarter 2015, pp. 2017–2046.

[4] Liu, G., et al., "In-Band Full-Duplex Relaying: A Survey, Research Issues and Challenges," *IEEE Communications Surveys & Tutorials*, Vol. 17, No. 2, Second Quarter 2015, pp. 500–524.

[5] Krishnaswamy, H., and G. Zussman, "1 Chip 2x the Bandwidth," *IEEE Spectrum*, Vol. 53, No. 7, July 2016, pp. 38–54.

[6] Zhou, J., et al., "Integrated Full Duplex Radios," *IEEE Communications Magazine*, Vol. 55, No. 4, April 2017, pp. 142–151.

[7] Reiskarimian, N., et al., "One-Way Ramp to a Two-Way Highway: Integrated Magnetic-Free Nonreciprocal Antenna Interfaces for Full-Duplex Wireless," *IEEE Microwave Magazine*, Vol. 20, No. 2, February 2019, pp. 56–75.

[8] Urick, V. J., M. E. Godinez, and D. C. Mikeska, "Photonic Assisted Radio-Frequency Interference Mitigation," *Journal of Lightwave Technology*, Vol. 38, No. 6, March 15, 2020, pp. 1268–1274.

[9] O'Hara, F., and G. Moore, "A High Performance CW Receiver Using Feedthru Nulling," *Microwave Journal*, Vol. 6, September 1963, pp. 63–71.

[10] Lin, L., et al., "A *Ka*-Band FMCW Radar Front-End with Adaptive Leakage Cancellation," *IEEE Transactions on Microwave Theory and Techniques*, Vol. 54, No. 12, December 2006, pp. 4041–4048.

[11] Zhang, Z., et al., "Full-Duplex Wireless Communications: Challenges, Solutions, and Future Research Directions," *Proceedings of the IEEE*, Vol. 104, No. 7, July 2016, pp. 1369–1409.

[12] Kelley, M. R., "The Spectrum Auction: Big Money and Lots of Unanswered Questions," *IEEE Internet Computing*, Vol. 12, No. 1, January-February 2008, pp. 66–70.

[13] Zhang, Z., et al., "Full Duplex Techniques for 5G Networks: Self-Interference Cancellation, Protocol Design, and Relay Selection," *IEEE Communications Magazine*, Vol. 53, No. 5, May 2015, pp. 128–137.

[14] Sharma, S. K., et al., "Dynamic Spectrum Sharing in 5G Wireless Networks with Full-Duplex Technology: Recent Advances and Research Challenges," *IEEE Communications Surveys & Tutorials*, Vol. 20, No. 1, First Quarter 2018, pp. 674–707.

[15] Riihonen, T., et al., "Inband Full-Duplex Radio Transceivers: A Paradigm Shift in Tactical Communications and Electronic Warfare?" *IEEE Communications Magazine*, Vol. 55, No. 10, October 2017, pp. 30–36.

[16] Adrat, M., et al., "Full-Duplex Radio – Increasing the Spectral Efficiency for Military Applications," *2019 International Conference on Military Communications and Information Systems (ICMCIS)*, Budva, Montenegro, 2019, pp. 1–5.

[17] Kolodziej, K. E., B. T. Perry, and J. S. Herd, "In-Band Full-Duplex Technology: Techniques and Systems Survey," *IEEE Transactions on Microwave Theory and Techniques*, Vol. 67, No. 7, July 2019, pp. 3025–3041.

[18] Choi, J. I., et al., "Achieving Single Channel, Full Duplex Wireless Communication," *Proceedings of the 16th Annual International Conference on Mobile Computing and Networking (MobiCom)*, New York, September 2010, pp. 1–12.

[19] Duarte, M., and A. Sabharwal, "Full-Duplex Wireless Communications Using Off-The-Shelf Radios: Feasibility and First Results," *2010 Conference Record of the 44th Asilomar Conference on Signals, Systems and Computers*, Pacific Grove, CA, 2010, pp. 1558–1562.

[20] Korpi, D., et al., "Compact Inband Full-Duplex Relays with Beyond 100 dB Self-Interference Suppression: Enabling Techniques and Field Measurements," *IEEE Transactions on Antennas and Propagation*, Vol. 65, No. 2, February 2017, pp. 960–965.

[21] Lei, L., N. Saba, and S. G. Razul, "A Multichannel Self-Interference Cancellation Prototyping System," *2019 IEEE 2nd 5G World Forum (5GWF)*, Dresden, Germany, 2019, pp. 427–432.

[22] Khaledian, S., et al., "Robust Self-Interference Cancellation for Microstrip Antennas by Means of Phase Reconfigurable Coupler," *IEEE Transactions on Antennas and Propagation*, Vol. 66, No. 10, October 2018, pp. 5574–5579.

[23] Enserink, S., et al., "Joint Analog and Digital Interference Cancellation," *2014 IEEE International Symposium on Dynamic Spectrum Access Networks (DYSPAN)*, McLean, VA, 2014, pp. 378–379.

[24] Fitz, M. P., et al., "Towards Simultaneous Radar and Spectral Sensing," *2014 IEEE International Symposium on Dynamic Spectrum Access Networks (DYSPAN)*, McLean, VA, 2014, pp. 15–19.

[25] Kiayani, A., et al., "Adaptive Nonlinear RF Cancellation for Improved Isolation in Simultaneous Transmit–Receive Systems," *IEEE Transactions on Microwave Theory and Techniques*, Vol. 66, No. 5, May 2018, pp. 2299–2312.

CHAPTER 2

Antenna Subsystems

Prathap Valale Prasannakumar, Carlos A. Mulero Hernández, and Dejan S. Filipovic

2.1 Introduction

As discussed in Chapter 1, the propagation domain is the first in the chain of cancellation domains and contains various antenna techniques. High SIC from antennas may lead to simpler realizations of analog and digital subsystems comprising respective cancellation domains. Whereas approximately 15 dB of isolation in the propagation level may be sufficient for low-power applications, a SIC above 30 dB and often 50 to 60 dB over the entire bandwidth is typically needed. Antenna near-field, spurious feed network radiation, surface currents, and scattering from nearby objects are all major contributors to the reduction in SIC between the Tx and Rx antenna ports. Hence, various IBFD antenna topologies can be deployed at the propagation level of a STAR system, as highlighted within the IBFD transceiver diagram of Figure 2.1(a) and the SIC techniques tree of Figure 2.1(b).

In this chapter, we group antennas based on their realization imprint into monostatic, bistatic, quasi-monostatic, and hybrid IBFD antenna subsystems. Each of these configurations has its own merits and challenges. A comprehensive trade study generally needs to consider application, bandwidth, required performance, available space, complexity, and system cost. A monostatic antenna subsystem uses a single antenna or aperture for Tx and Rx. SIC is achieved by exploiting: (1) inherent antenna features to cancel the induced currents at the receiving port (such as symmetry); (2) careful routing of coupled and leaked signals through a beamforming network (BFN) with or without circulators; and (3) employing mode orthogonality. Some monostatic antennas can achieve very high isolation, theoretically infinite; however, various imperfections limit the achievable SIC. Nonetheless, with careful fabrication and assembly, isolation on the order of 50 dB or more can be achieved over a wide bandwidth. Radiation efficiency and pattern imperfections are typical drawbacks.

Contrarily, a bistatic IBFD antenna subsystem, composed of physically separated Tx and Rx apertures, may achieve isolation that is robust against fabrication anomalies and tolerances. Moreover, it can be more immune to the multipath

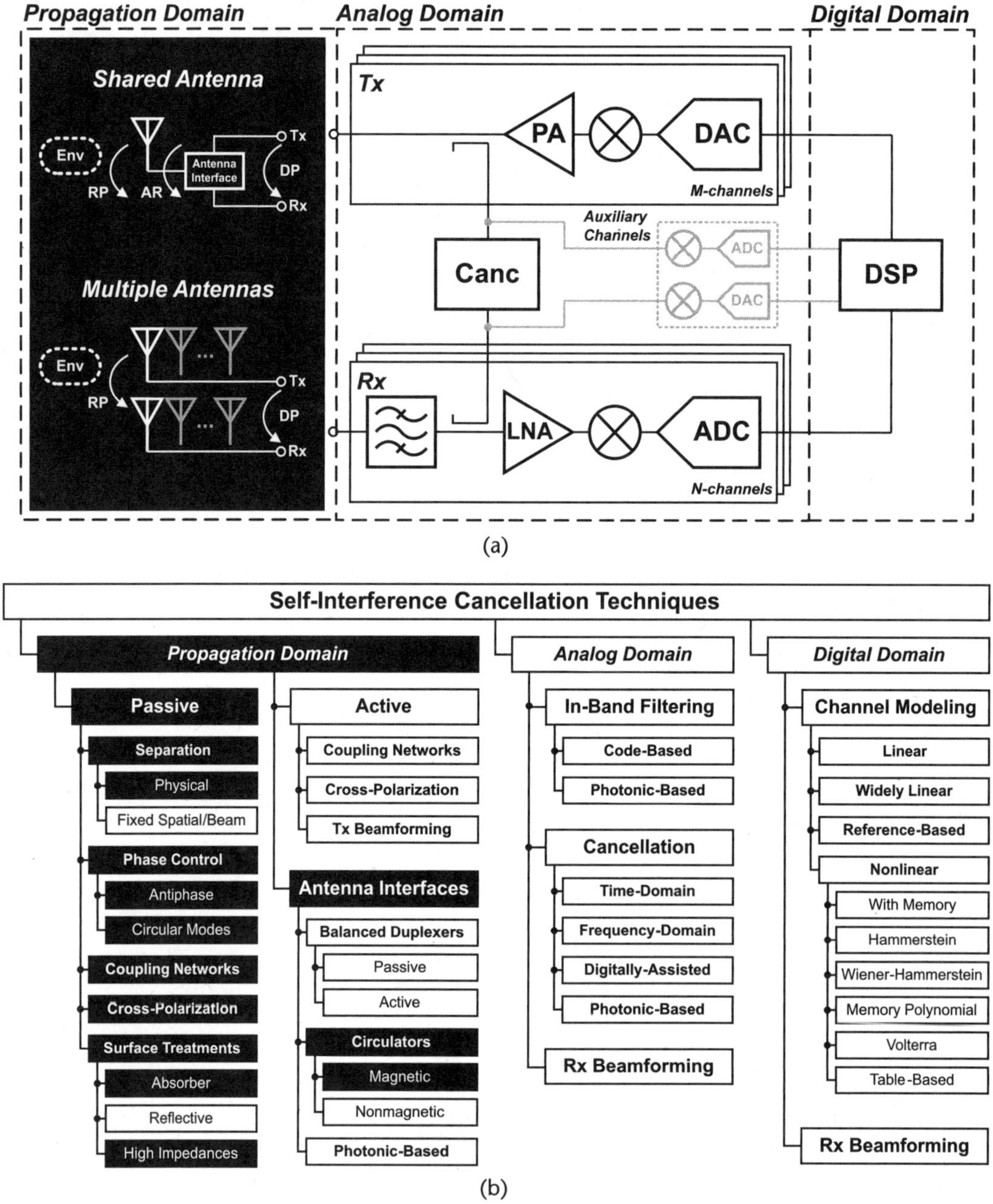

Figure 2.1 Antenna subsystems and specific chapter focus highlighted within (a) a generic IBFD transceiver and (b) a SIC techniques tree diagram.

from scatterers and nearby obstacles. The issue is the overall subsystem size, which for many space-constrained platforms is strictly prohibitive. The quasi-monostatic configuration, a unique blend of the two preceding techniques, mitigates the space constraints of bistatic while minimizing sensitivity to imbalances and asymmetries of a monostatic subsystem. Wide bandwidth, dual-IBFD-polarized operation and high Tx power can be achieved. The drawback is the difference in antenna far-field patterns for Tx and Rx. For extreme bandwidths, the above-discussed techniques will straggle. Hence, a hybrid architecture relying on a clever amalgamation of monostatic and bistatic subsystems is viable with two decades of bandwidth.

2.2 Bistatic Antenna Subsystems

2.2.1 Introduction

The most commonly considered implementation of IBFD apertures is based on two physically separated antennas for transmit and receive. Simple and intuitive, this embodiment is often referred to as a bistatic STAR antenna system. A high level of RF isolation is achieved at the expense of a larger physical area needed to host both apertures. Typically, the antennas are mounted on a common platform or housed inside a shared enclosure and separated by a few wavelengths (approximately $<10\lambda$) [1–5]. For practical separations, the SI on a propagation level is still too high, making a single antenna system generally insufficient to achieve the overall IBFD system needs. Minimizing this SI is a critical objective in IBFD system design.

In a bistatic configuration, there are three main ways through which power is coupled between the Tx and Rx antennas, specifically: (1) space or radiation inclusive of the presence of nearby scatterers; (2) surface currents; and (3) feed networks. Good design addresses all these paths while ensuring antennas maintain required near-field and far-field features over the entire bandwidth. Based on antenna configurations, type, available space, method of feeding, and bandwidth of operation, various cancellation techniques can be applied. This section reviews a few commonly deployed decoupling techniques with a focus on wideband operational needs.

2.2.2 Decoupling Techniques

2.2.2.1 Spatial Separation and Spatial Orientation

In an unbounded space, the power transmitted by an antenna decreases as $1/r^2$, where r is the distance from the antenna. For aperture antennas mounted on a conducting ground plane, the power carried along the surface decays as $\propto \frac{1}{r^2}$ in the E-plane, and as $\propto \frac{1}{r^4}$ in the H-plane [2, 6]. This also holds when the ground plane is of finite size [7]. The simulated power decay in the E- and H-planes of an open-ended, WR90, waveguide demonstrates the above arguments (Figure 2.2). Hence, increasing the separation distance is a simple approach to improve isolation [8]. Orienting the antennas along respective H-planes provides higher isolation for a given separation distance. Combining these two approaches therefore has the potential to maximize the achievable isolation.

2.2.2.2 Metallic Corrugations, Bed of Nails, and Metallic Fence

Metallic corrugations with quarter-wave ($\lambda/4$) depth at the antiresonant frequency are often used to reduce the currents on the ground plane. These are singly periodic structures of $\lambda/4$ deep-shorted chokes, where the periodicity is much less than the wavelength [2, 8–12]. The impedance seen at the surface is a function of the

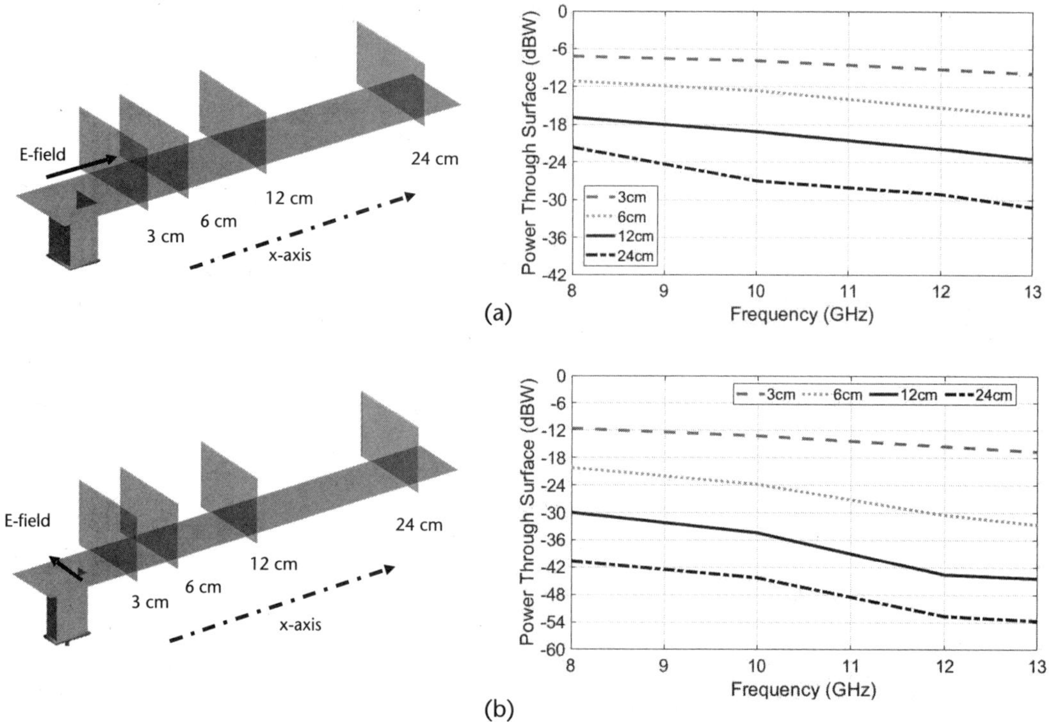

Figure 2.2 Computed power decay as a function of distance from the open-ended waveguide in the (a) E-plane and (b) H-plane.

corrugation depth; hence, $\lambda/4$ chokes remain capacitive for the TM_0 wave. Since the wave impedance of TM_0 is capacitive, corrugations break the transverse resonance conditions, resulting in radiation of surface currents and increased Tx to Rx isolation [1, 2, 5, 13].

The operational bandwidth of conventional corrugations is limited to an octave. In [13], the theory outlined in [14] is extended to demonstrate that varying corrugation depth (linear or exponential tapers) can overcome this bandwidth limitation and aid in achieving higher isolation (see Figure 2.3) over wider frequency range.

Analogously, a double periodic structure of metallic nails can be used to replicate corrugations (see Figure 2.4(a)). These are known as bed of nails (BON) [2, 15]. Here too, capacitive impedance for TM_0 can be obtained by $\lambda/4$ nails. In [2, 8], isolation >50 dB is achieved over 6–15 GHz using BON (Figure 2.4(b)).

Alternatively, a metal fence can be implemented (see Figure 2.5 [16]) in systems and at frequencies where there is no space for a high-impedance surface (HIS). This low-cost solution reduces the weight significantly and has a small impact on antenna near-field and far-field parameters. Refer to Chapter 12 for a discussion of an IBFD system built around this antenna.

2.2.2.3 Printed Reactive Surface

Printed reactive surfaces (PRS) are also double periodic structures consisting of metallic patches and vias, as depicted in Figure 2.6(a). These vias are essential for the

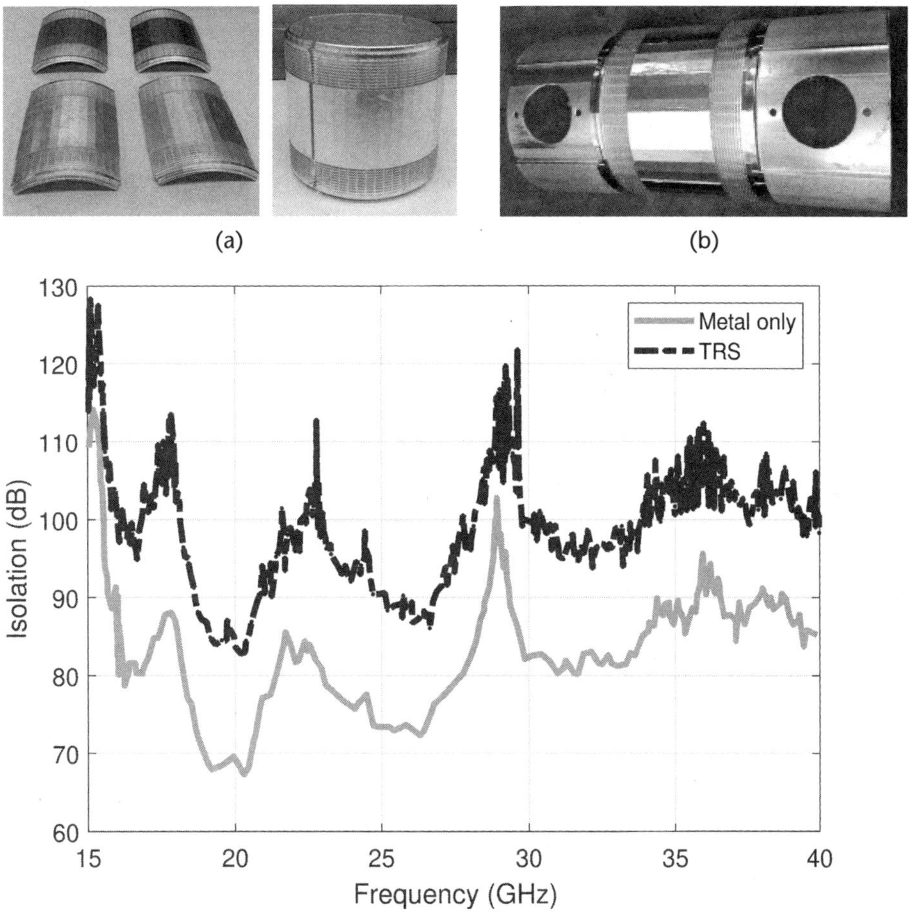

(c)

Figure 2.3 (a, b) Tapered ribbed surface (TRS) over metallic cylinder between two quad-ridge horns, and (c) measured isolation with and without TRS. (Adapted from [13].)

surface to be effective for TM_0 surface waves. Unlike bandgap structures, these are evaluated based on the surface impedance for TM_0 waves [2, 17]. These surfaces exhibit capacitive impedance over 3:1 bandwidth. Hence, PRSs are beneficial for achieving high isolation in wideband bistatic IBFD systems. In [2, 8], 60-dB isolation is achieved between two quad-ridge horn antennas using a printed HIS, as demonstrated in Figure 2.6(b). Additionally, these structures are lightweight, low-profile, and cost-effective when compared to metallic corrugations. Implementing a PRS at millimeter-wave frequencies is relatively simple when compared to quarter-wave chokes.

2.2.2.4 Absorbers

Carbon-loaded free-space plane wave and lossy magnetic absorbers can affect the radiation patterns and reduce the gain of an antenna. Therefore, the Rx antenna of a bistatic system can be recessed inside an absorber cavity to increase the isolation. This reduction in coupling is achieved at the expense of antenna radiation

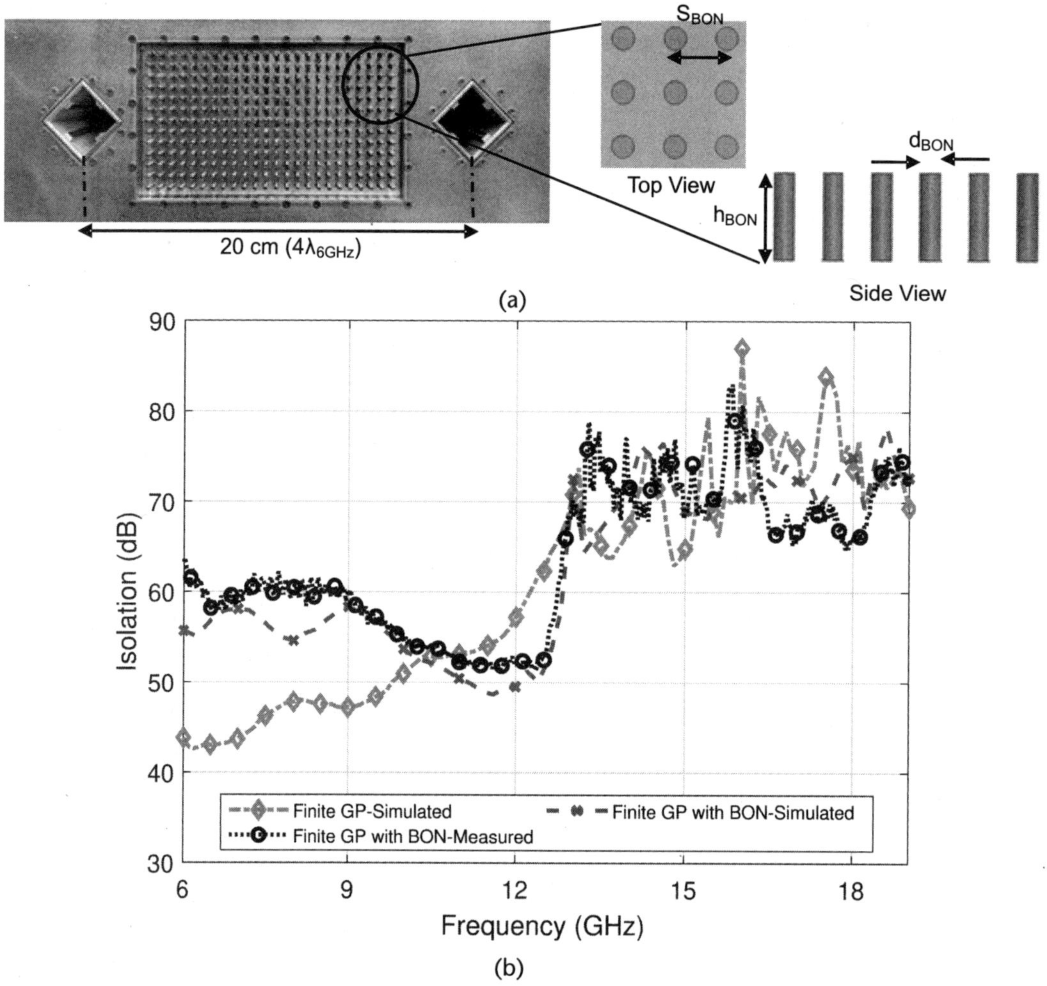

Figure 2.4 (a) Picture of computer-aided design (CAD) and fabricated model of BON, and (b) measured and simulated isolation between two quad-ridge antennas with and without BON. (© 2019. Reprinted, with permission, from [8].)

efficiency. Tapering the absorber cavity may increase the gain at boresight while maintaining the reduced radiated power at the horizon and thus increase both antenna efficiency and isolation. Similarly, thin sheet absorbers can be placed on the metal ground between the IBFD's Tx and Rx antennas to reduce the coupling not only from the near fields but also from the spurious feed network radiation. In [18], measured isolation >60 dB is achieved using LS26 absorber between the linear arrays of patch antennas.

2.2.3 Merits and Challenges

A bistatic IBFD antenna subsystem realization is generally simple. Tx and Rx antennas can be of different aperture sizes, types, and feed techniques. Bistatic arrays can scan over wide fields of view while maintaining high isolation, since isolation is not derived from signal cancellation.

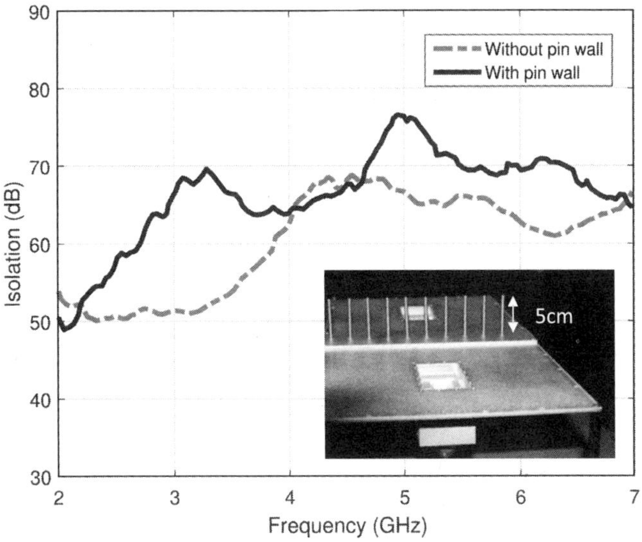

Figure 2.5 Measured isolation between two single polarized cavity-backed TEM horns with and without the metallic fence shown in the inset.

The amount of desired isolation influences overall system size. This problem amplifies at low frequencies because the reduction in wavelength corresponds to an increase in the physical separation to achieve the same isolation. The number of antenna elements, cost, and complexity in BFNs (for arrays) are double that of a monostatic system. The separate feeding mechanism required for Tx and Rx can pose further system design challenges.

2.3 Monostatic Antenna Subsystems

2.3.1 Introduction

In contrast to bistatic configurations, monostatic IBFD antenna subsystems share the aperture for both Tx and Rx, making them attractive for space-constrained platforms. The conventional approach to achieve monostatic IBFD is with a single circulator, as shown in Figure 2.7(a). Caveats to this configuration are related to the circulator's port-to-port leakage, which consists of a direct path (DP) between the Tx and Rx; antenna reflection (AR); and the environmental reflection paths (RPs), all of which are routed towards the Rx port, as discussed in Chapter 1. Instead, by taking advantage of geometrical and electrical symmetries of the antenna and signal routing within BFNs, it is theoretically possible to obtain infinite isolation. This section discusses several methods to achieve copolarized IBFD using a shared aperture. Any well-designed dual-polarized antenna could use polarization diversity for SIC, but this is suboptimal. The advantages of using circular polarization (CP) to mitigate environmental RPs [19] are wasted if the Rx antenna is oppositely polarized from its Tx counterpart. This section later discusses copolarized SIC techniques.

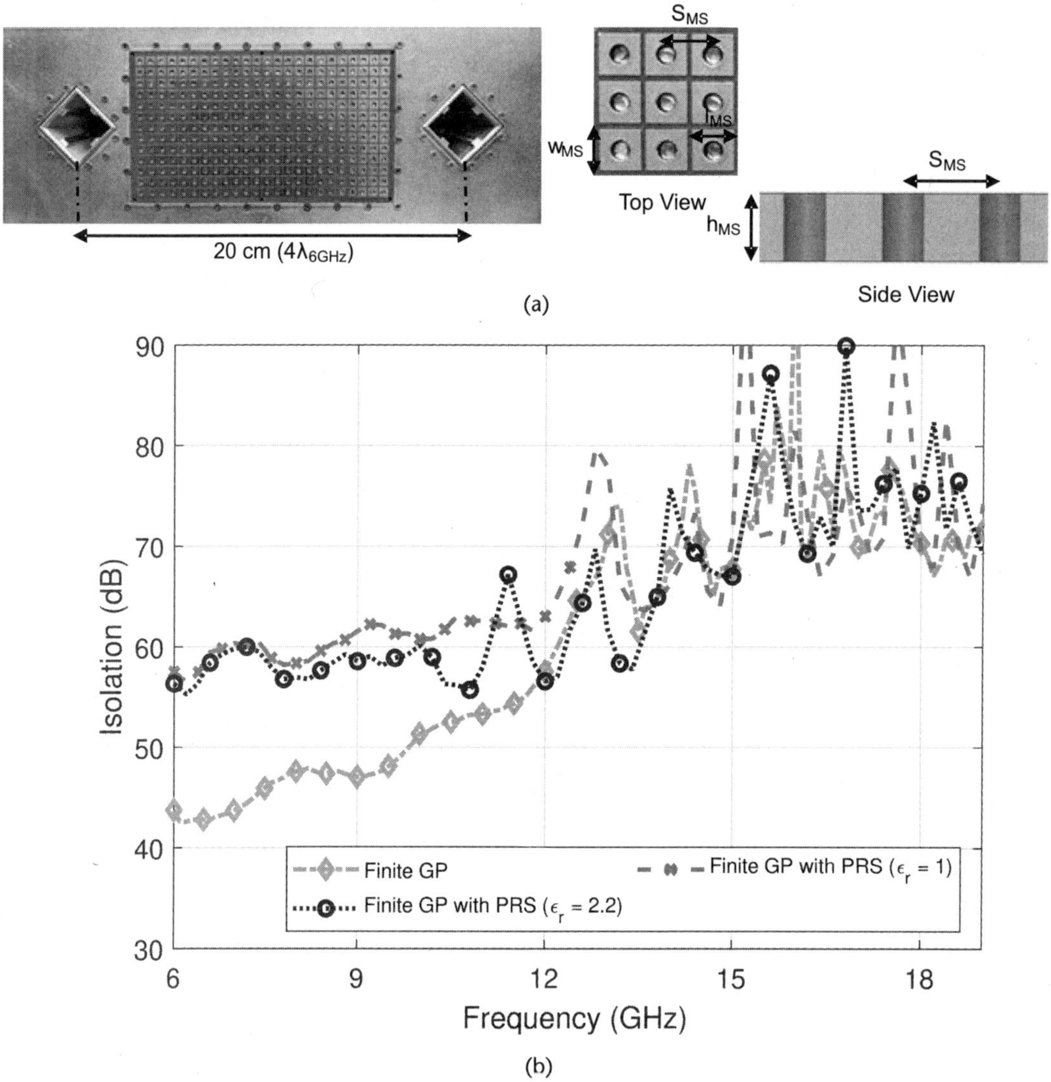

Figure 2.6 (a) Picture of CAD and fabricated model of a PRS, and (b) measured and simulated isolation between two quad-ridge antennas with and without PRS. (© 2019. Reprinted, with permission, from [8].)

2.3.2 Antiphase Spiral

The spiral is a well-known frequency-independent antenna and is often used for wide bandwidth applications. In [20], a four-arm spiral is reconfigured into a monostatic IBFD configuration by pairing the two sets of opposite arms into respective Tx and Rx channels, as shown in Figure 2.8(a). The symmetry of the spiral and the 180° phase difference between both arms makes the signals cancel at the ports to the spiral arms, resulting in theoretically infinite SIC between the Tx and Rx channels. The same principle is demonstrated with an 8-arm spiral, where multiple modes and different operational scenarios are also shown to be possible (see Figure 2.8(b)) [21]. The drawback with the antiphase spirals is reduced efficiency and radiation from the parasitic arms that distorts the polarization quality. As shown in [21], lens loading can be used to improve both the axial ratio and antenna efficiency. In [20],

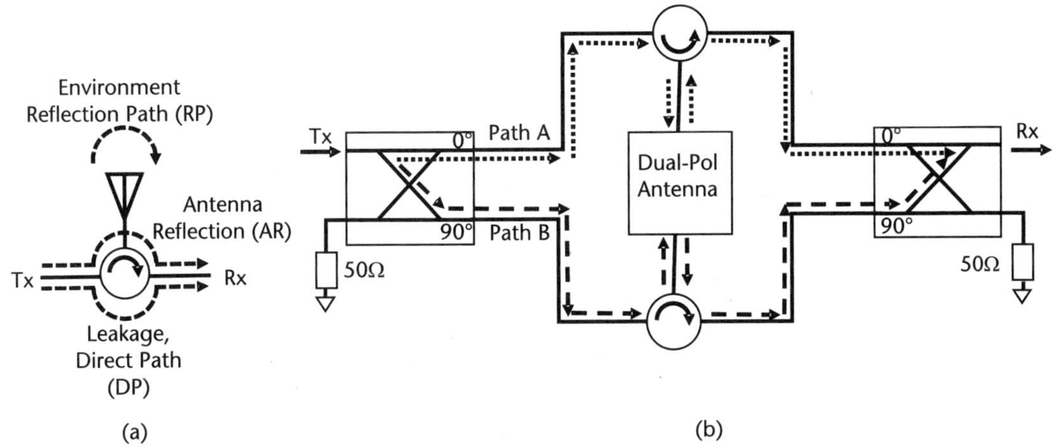

Figure 2.7 (a) Monostatic IBFD using a single circulator; and (b) network schematic for the balanced feed network.

a helix is used to connect the spiral with the ground plane through resistors, reducing the currents that reflect back and radiate in the opposite handedness. Isolation levels of >50 dB are measured over 6:1 bandwidth with antiphase spirals configured for monostatic IBFD operation. These antennas can also be used in arrays [22], but certain trade-offs in scan, bandwidth, isolation, and efficiency must be considered. The cancellation mechanism in antiphase spirals is independent of frequency and the imperfect SIC can be attributed to assembly or fabrication asymmetries.

2.3.3 Sequentially Rotated Array Antennas

A sequentially rotated 2 × 2 array comprising of linearly polarized (LP) elements excited with the same magnitude and (0, ±90°, 180°, ±270°) phase progression will generate CP radiation [24, 25]. Connecting each antenna port to a different BFN as seen in Figure 2.9(a) yields an antenna that takes the place of the circulator in Figure 2.7(a). Without imperfections, the complete SIC is obtained. The coupling paths in Figure 2.9(b) excited with the delays and phasing of the ports result in complete cancellation. To achieve dual-IBFD-polarized performance, Butler matrices are deployed in [24]. The compromise with this configuration is the reduced aperture efficiency, which for practical topologies is theoretically limited to 50%. This can be overcome by using CP-radiating components as in [25], where 2-arm spirals were used for demonstration. Furthermore, [26] rotates the elements as in Figure 2.10, avoiding the second BFN. This reduces system complexity at the expense of only singly copolarized operation.

2.3.4 Circular Arrays

Circular arrays with linearly polarized elements can be configured to achieve omnidirectional (i.e., dipole-like) IBFD operation without polarization multiplexing. To do so, they are excited with the modal phase progression through a Butler matrix-like feed where modal orthogonality and symmetries are utilized to achieve theoretically infinite isolation [26, 27].

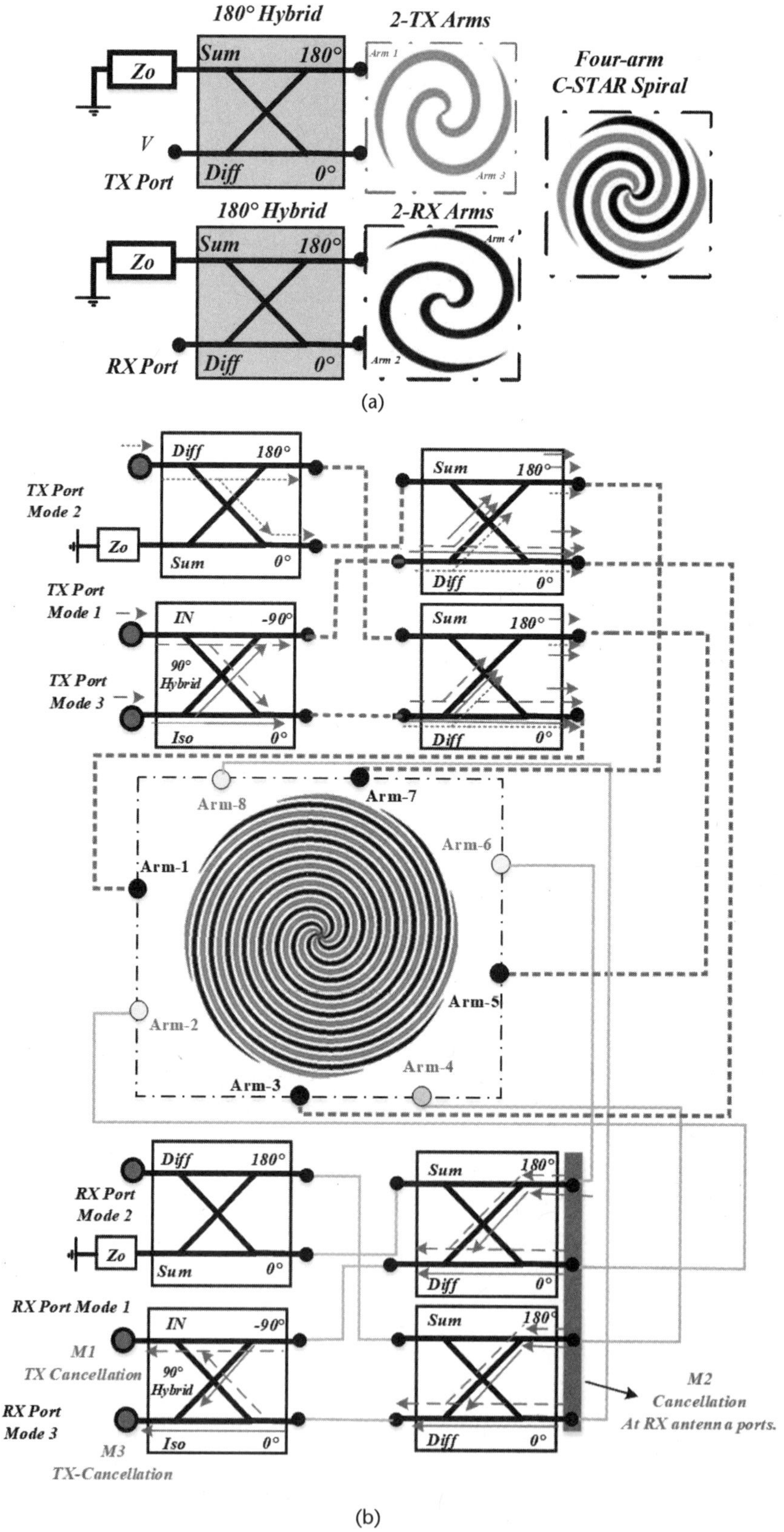

Figure 2.8 Schematics of (a) 4-arm and (b) 8-arm antiphase spiral. (Adapted from [23].)

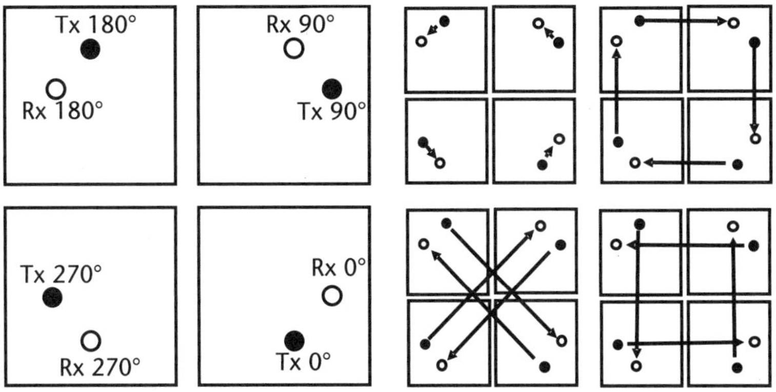

(a) (b)

Figure 2.9 (a) Excitations and (b) coupling paths for the sequentially rotated patches.

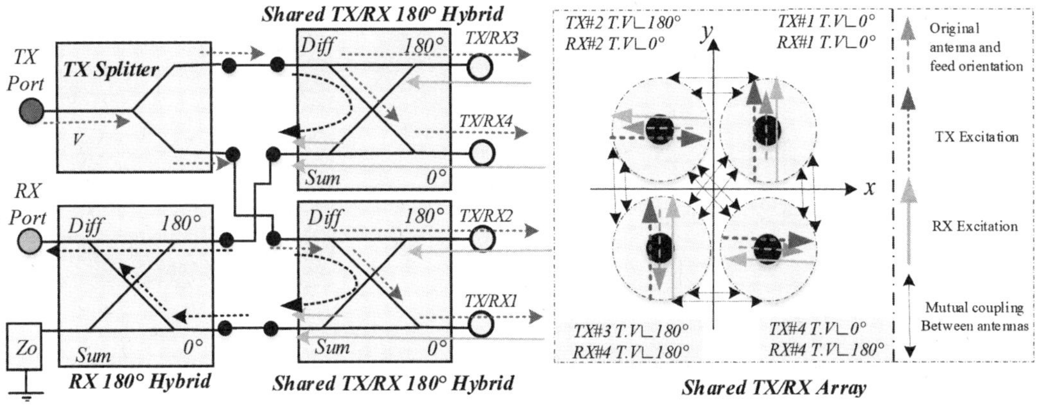

Figure 2.10 BFN for sequentially rotated spirals. (© 2018. Reprinted, with permission, from [23].)

The configuration in Figure 2.11(a) consists of a circular array with mode 1 (M1) phase progression and a center element. The symmetry and phase progression create a null at the center as the signals from pairs of array elements cancel out. This null does not exist for mode 0 (M0). An example of this technique using monopoles is seen in [26]. The resulting radiation characteristics are somewhat different due to the different Tx and Rx elements, but are both omnidirectional at the horizon. The configuration in Figure 2.11(b) works without a center element, instead relying on the orthogonality of circular modes. Two topologies are shown between Figures 2.12 and 2.13 [27]. The single-layer configuration in Figure 2.12 achieves theoretically infinite SIC using M1 for Tx and a superposition of modes 0 and 2 (M0 and M2) for Rx. The mixed-mode for Rx is necessitated by the leakage of real hybrids, which adds constructively at the Rx port with the BFN that gives M0 and M1 only. The configuration in Figure 2.13 allows using M0 for Tx and M1 for Rx while maintaining theoretically infinite SIC by combining the leakages through the difference port of another 180° hybrid. The single-mode approach (M0 Tx/M1 Rx) provides more similar Tx and Rx patterns when compared to the

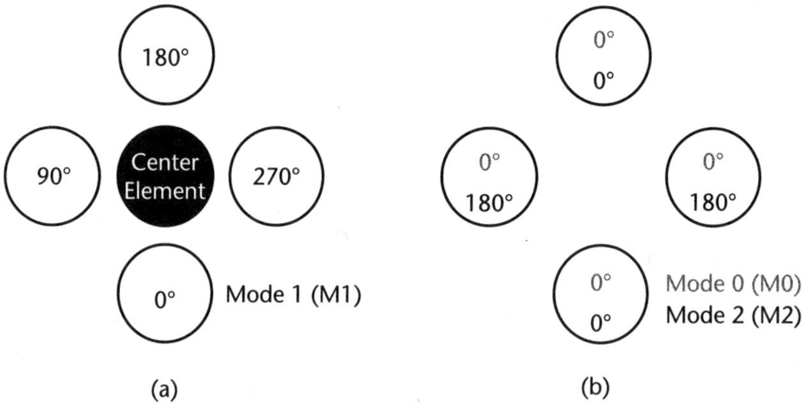

Figure 2.11 Circular arrays (a) with and (b) without a center element.

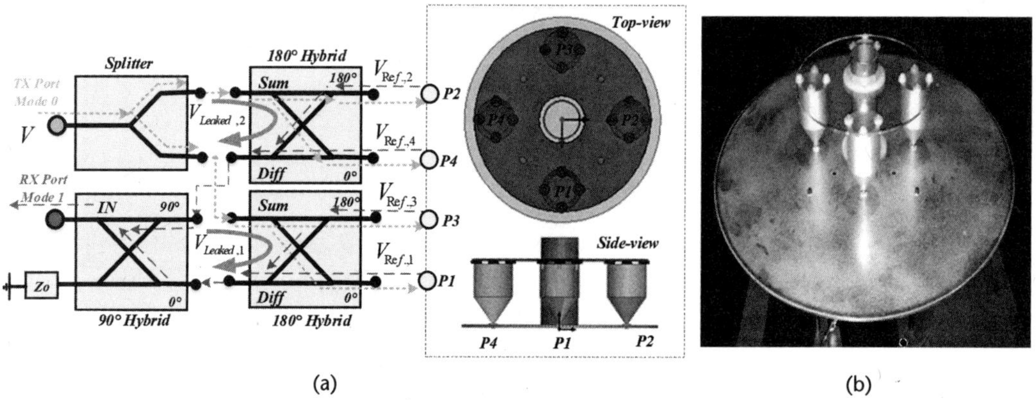

Figure 2.12 (a) Mixed mode BFN for (b) single-layer circular array. (Adapted from [23].)

mixed-mode approach (M1 Tx/M0+M2 Rx) for which the mixed-mode has nulls in the radiation pattern.

2.3.5 Balanced Feed Networks

Recent advances in conventional and unconventional circulators, including initiatives to develop them in monolithic microwave integrated circuits (as discussed in Chapter 4), have led to increased interest in topologies that utilize circulators while enabling even greater isolation enhancement than the isolation inherent to the device. One of those is a balanced-feed network approach that performs the cancellation in the antenna interface layer, as shown in Figure 2.7(b). This network consists of two hybrids and two circulators feeding a dual-polarized antenna. The SI due to AR can be obtained by tracing the signal paths in Figure 2.7(b) and expressed as (2.1), where $|B|$ is the magnitude of the circulator's leakage and Γ is the reflection coefficient at the antenna [19]

$$SI = PathA + PathB = \frac{1}{2}\Gamma\left(1-|B|^2\right) - \frac{1}{2}\Gamma\left(1-|B|^2\right) = 0\cdot \tag{2.1}$$

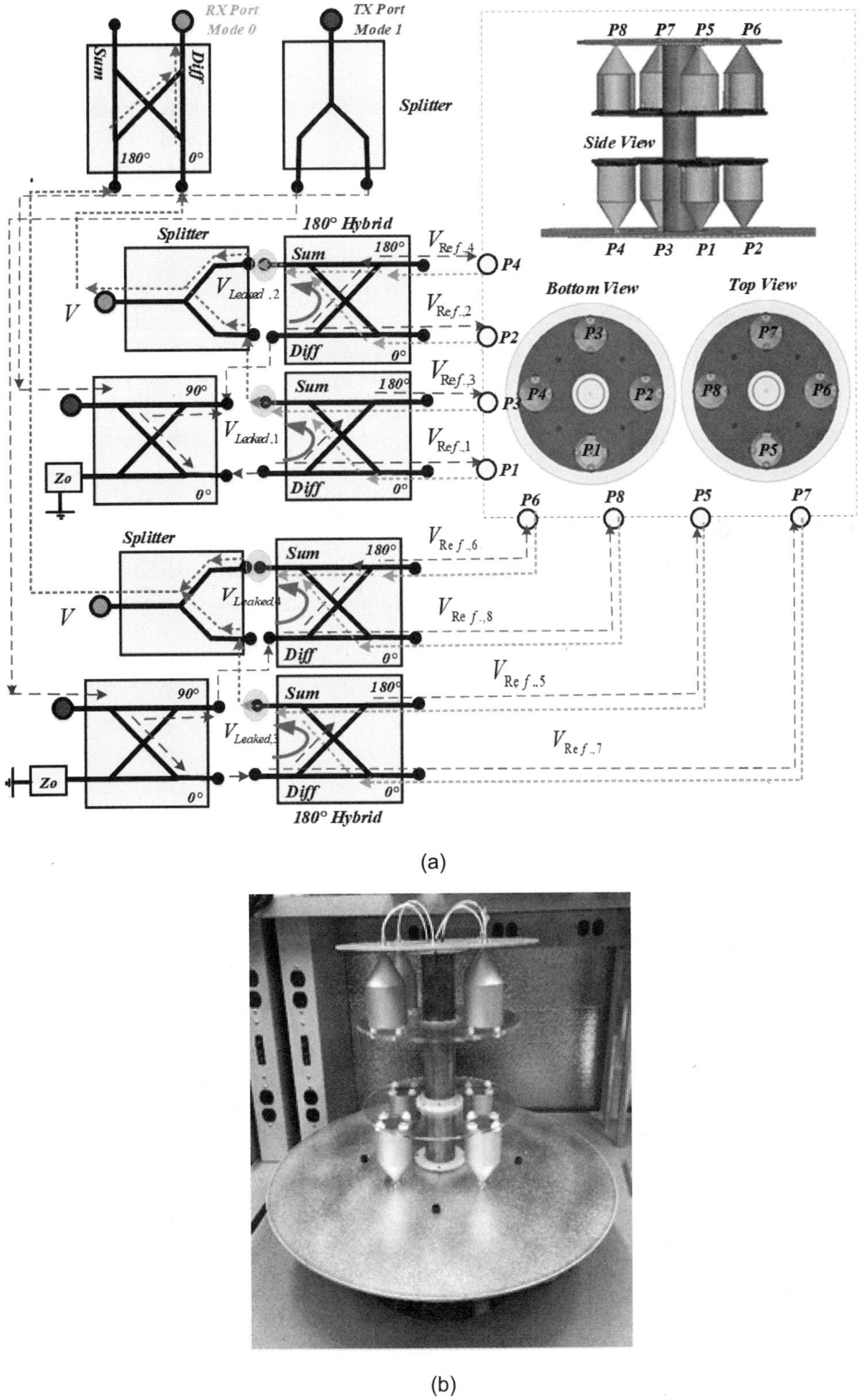

Figure 2.13 (a) BFN for (b) dual-layer monocone array. (Adapted from [23].)

The reflections and, similarly, the circulator leakage add destructively at the Rx port and constructively at the loaded port. The antenna used for this system must support two orthogonal linear polarizations that will be excited with a 90° phase difference. The reflection coefficients of the antenna ports should be identical and the ports isolated from each other; otherwise, the port-to-port coupling would show up as SI at the Rx port [28]. Dual-polarized circular patches [19] and a dual-polarized reflector antennas [29] are two among many possible implementations.

In [30], a different variation of the balanced-feed network is proposed, where the 90° hybrids are replaced with modified Butler matrices. This enables excitation of different spiral modes for the 4-arm spiral, as seen in Figure 2.14. The balanced-feed network provides theoretically infinite isolation, while the Butler matrix provides mode-to-mode isolation, allowing for simultaneous use of each spiral mode as different IBFD channels. The patterns are spatially diverse for different modes, as the spiral mode 1 transmits at broadside while modes 2 and 3 have conical spatial distributions.

2.3.6 Performance Comparison

As this book focuses on experimentally verified systems, Table 2.1 summarizes several articles on monostatic IBFD topologies and compares their measured performances. No system is a one-size-fits-all solution and the reader should dive

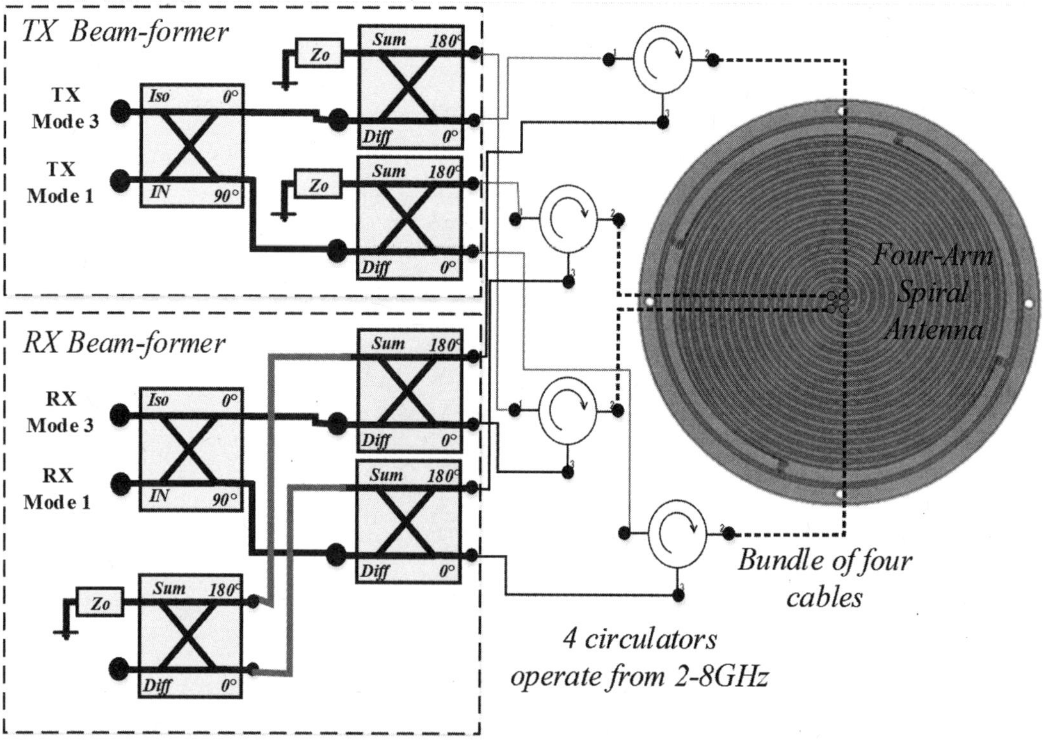

Figure 2.14 A 4-arm spiral with balanced feed network. (© 2018. Reprinted, with permission, from [23].)

Table 2.1 Comparison of SIC Achieved Through Various Monostatic Methods

Reference	*Method*	*Measured SIC*	*Frequency Band*	*Antenna Type*
[8]	Polarization diversity	>30 dB	6 to 19 GHz	OMT Fed Quad Ridge Horn
[20]	Antiphase spiral	>37 dB	0.5 to 2.5 GHz	4-Arm Archimedean Spiral
[21]	Antiphase spiral	>38 dB	0.5 to 2.5 GHz	8-Arm Archimedean Spiral
[22]	Antiphase spiral	>27 dB	0.65 to 2.75 GHz	4-Arm Spirals in Hexagonal 7 Element Array
[24]	Sequentially rotated array	>47 dB	2.4 to 2.5 GHz	2 × 2 Dual Polarized Patch Array
[25]	Sequentially rotated array	>38 dB	1.25 to 2.5 GHz	2 × 2 Archimedean Spiral Array
[26]	Circular array with center Element	>55 dB	2.4 to 2.5 GHz	Circular Monopole Array with Center Monopole
[27]	Orthogonal mode circular array	>40 dB (M1/M0+M2), >28 dB (M0/M1)	0.8 to 1.7 GHz	Circular Monocone Array
[19]	Balanced feed	>40 dB	902 to 928 MHz	Circular Patch
[30]	Balanced feed	>27 dB (Mode 1), >24 dB (Mode 3)	2 to 8 GHz	4-Arm Archimedean Spiral
[29]	Balanced feed	>30 dB	4 to 8 GHz	Coaxial Cavity Fed Reflector

deeper into the references to consider design beyond SIC when selecting an antenna subsystem.

2.4 Quasi-Monostatic Antenna Subsystems

2.4.1 Introduction

The preceding sections described bistatic and monostatic IBFD antenna subsystems in detail, illustrating their high isolation in theory and >30 dB in practice. Contrarily, bistatic systems are robust to assembly anomalies and can achieve much higher isolation at the expense of size. To partially mitigate issues of both topologies, a quasi-monostatic IBFD antenna subsystem is proposed [8, 31]. In this approach, two separate antennas are used for Tx and Rx, similarly to bistatic; however, the overall system is encapsulated inside the area occupied by one of the antennas (Tx, mostly). Importantly, isolation is obtained by combining signal cancellation techniques and one or more decoupling schemes of a bistatic subsystem. A wideband dual-polarized (linear or circular) performance can be obtained at the expense of reduced far-field correlation between the Tx and Rx patterns.

2.4.2 Single Reflector

Reflector antennas are preferred for applications requiring high directivity. When a bistatic configuration is implemented using reflectors, the system size increases considerably, making the approach less viable. Hence, the quasi-monostatic technique is a useful alternative to consider. The prototype, architecture and schematic of a single reflector IBFD antenna are shown in Figure 2.15 [8]. Here, the Rx antenna is mounted behind the feed of the Tx reflector. To achieve high isolation and the

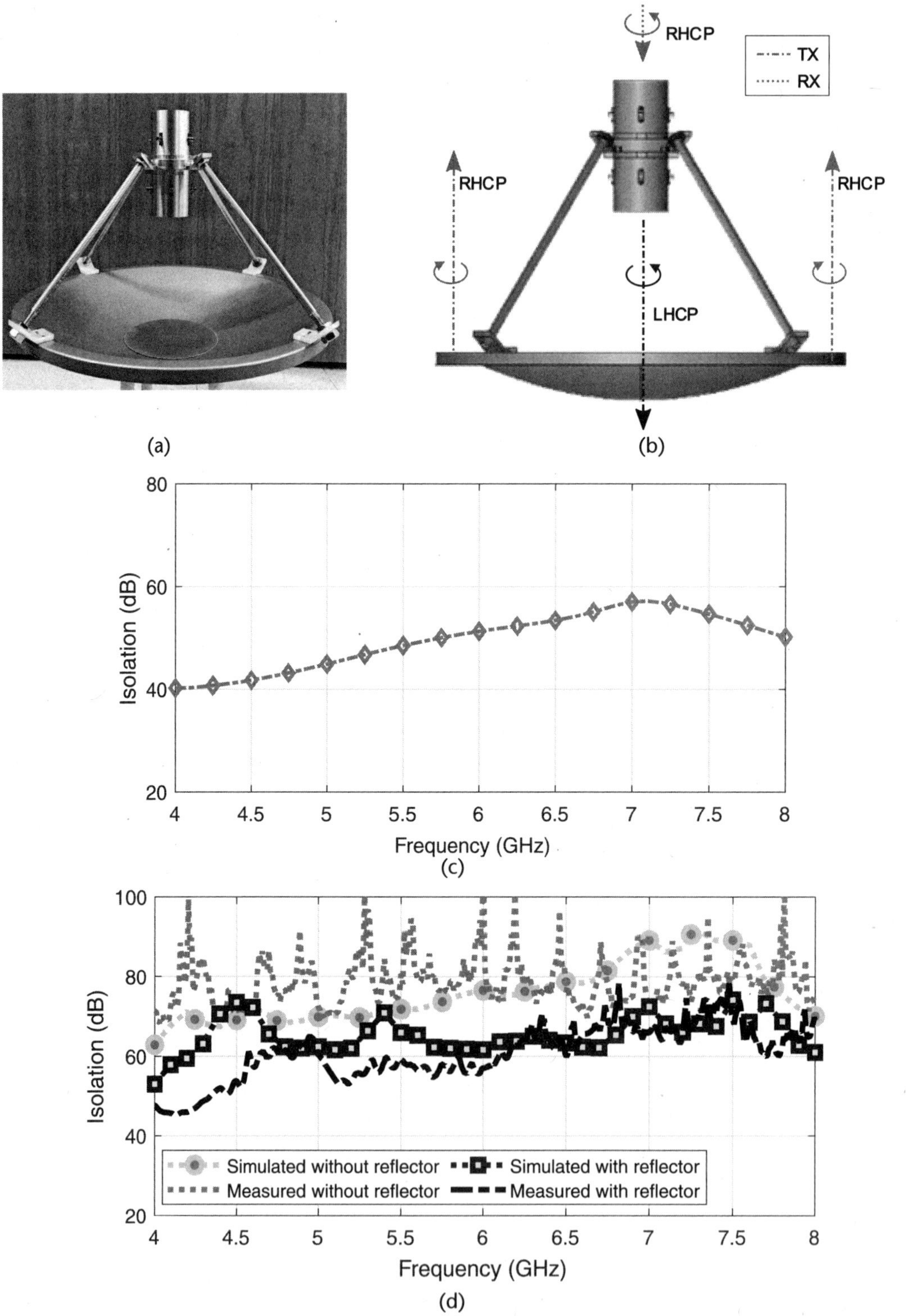

Figure 2.15 (a) Picture of fabricated reflector (Tx) and coaxial cavity antenna (Rx); (b) system's operational concept; (c) simulated isolation between the Tx and Rx, when the antennas are LP; and (d) measured and simulated isolation for CP, with and without reflector. (© 2019. Reprinted, with permission, from [8].)

same polarization operation, the Rx antenna and reflector feed are first designed for a high front-to-back ratio (F/B). Next, the Rx antenna and Tx (reflector) feed are operated in opposite CPs, meaning that the Tx feed is, for example, left-handed circular polarization (LHCP) and the Rx antenna is right-handed circular polarization (RHCP). The LHCP fields from the feed undergo polarization reversal after bouncing off the reflector, resulting in the same polarization for Tx and Rx. The prototyped system has 61-dB average measured isolation (see Figure 2.15) using commercial off-the-shelf (COTS) 90° and 180° hybrids with amplitude and phase imbalances of ±0.5 dB and ±3° and ±0.6 dB and ±10°, respectively. The isolation is 30 dB higher than that achieved in the monostatic system [29], which employed the same components, and a circulator. Additionally, the system has 40-dB isolation when operated in LP without polarization multiplexing. The high-level inherent isolation is the primary cause of the system's resilience to the asymmetry and imbalances. The drawback is the difference in radiation patterns (directivities) between the Tx and Rx antennas.

2.4.3 Dual Reflector

Because of the increased path loss, millimeter-wave systems require high gain for both the Tx, and Rx antennas. The system in [31] has a 15-dB difference between the Tx and Rx gains. Hence, an architecture with a reduced margin is attractive and often preferred. Moreover, the reflector feed is located farther away from the Tx front end, which contributes to additional losses and difficulties in waveguide routing. In [8], a quasi-monostatic dual reflector is designed to address these millimeter-wave challenges while maintaining high isolation. The operational concept, feeding network and geometry are illustrated in Figure 2.16. Here, the LHCP fields radiated from the Tx feed undergo a polarization flip after reflecting from the subreflector. These RHCP fields illuminate the main reflector, resulting in the final LHCP Tx fields. The Rx antenna mounted behind the sub-reflector is LHCP. Thus, both the Tx and the Rx have the same polarization. The shielding from the sub-reflector and the null in Rx radiation patterns, because of CP, lead to higher system isolation. Given the asymmetries and tolerances in COTS components, the fabricated prototype had 63-dB isolation for CP, as seen in Figure 2.16. Analogous to [31], the system achieves >40-dB isolation for LP. Furthermore, the lens-loaded Rx antenna has directivity >16 dBic, which minimizes the difference in directivity to 9.69 dB (max). Importantly, this Cassegrain reflector quasi-monostatic antenna system further enforces the ruggedness of the approach in attaining high isolation.

2.5 Hybrid Antenna Subsystems

To achieve an extremely wideband IBFD antenna subsystem, standalone monostatic, bistatic, or quasi-monostatic are typically insufficient. Therefore, two or more of these need to be combined into a single, multiport antenna topology, referred herein as a hybrid IBFD antenna. The individual subsystems must be designed to take advantage of synergistic SIC effects while avoiding antagonistic interactions between the different parts.

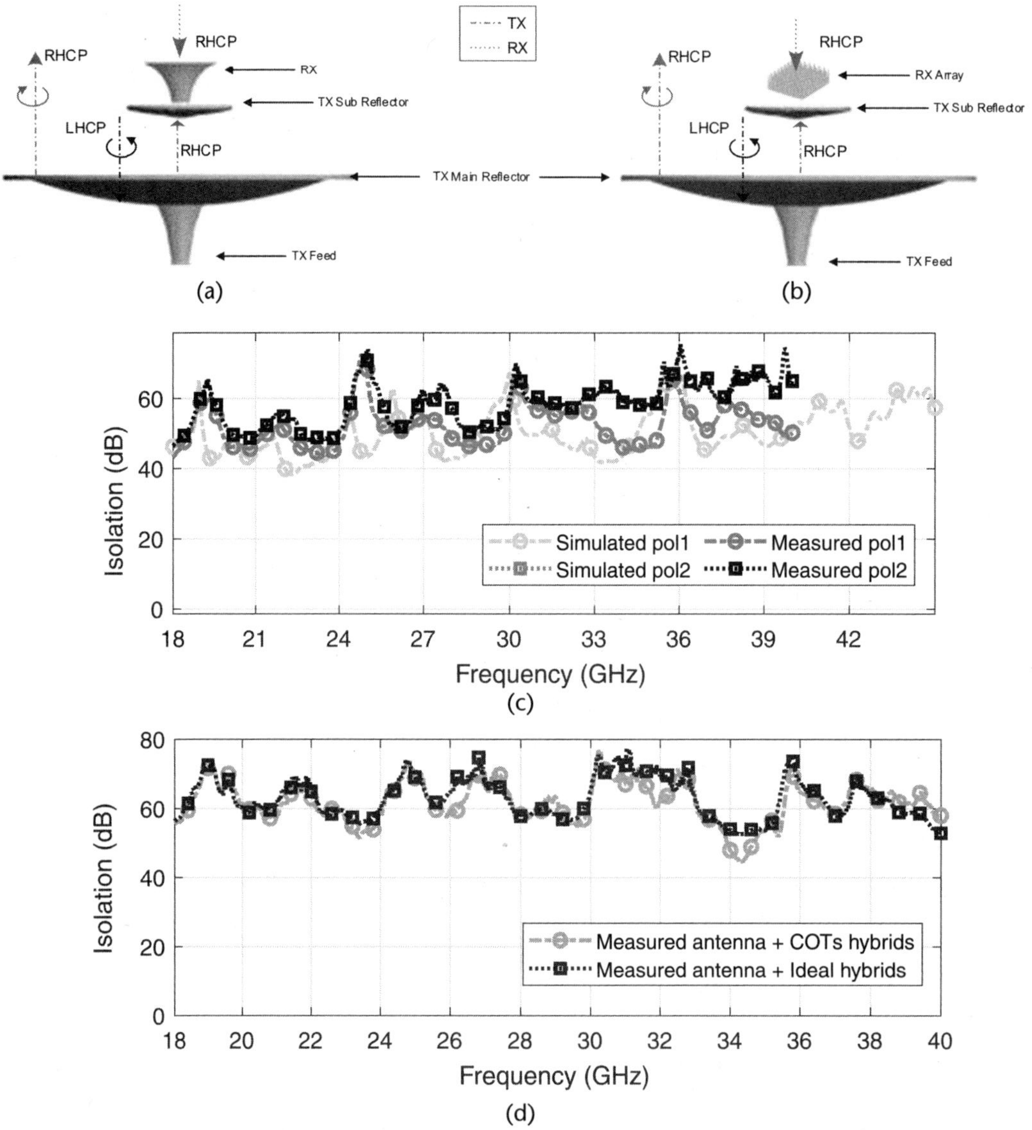

Figure 2.16 CAD models of quasi-monostatic dual reflector antenna systems illustrating the operational principle (a) Rx – quad-ridge horn; (b) Rx – tightly coupled phased array, measured isolation between the Tx and Rx; when the antennas are (c) LP and (d) CP with ideal and COTS components. (© 2019. Reprinted, with permission, from [8].)

A hybrid IBFD antenna configuration with different subsystems, each operating over a distinct band, is designed to work from 0.5 to 45 GHz in [2, 16] (see Figure 2.17). The highest frequencies, 18 to 45 GHz, are served with a bistatic IBFD antenna pair of dual-polarized quad-ridge horn antennas. The electrical distance between the Tx and Rx horns is sufficient to enable high SIC. For lower frequencies, the electrical distance available on the shared ground plane is reduced, so BONs were utilized to aid SIC. As seen, partially recessed BONs are introduced between the systems covering 6 to 19 and 2 to 7 GHz. The addition of these reactive-impedance surfaces (RISs) improved isolation by up to 15 dB at some frequencies while marginally affecting the radiation patterns of all antennas. The lowest band,

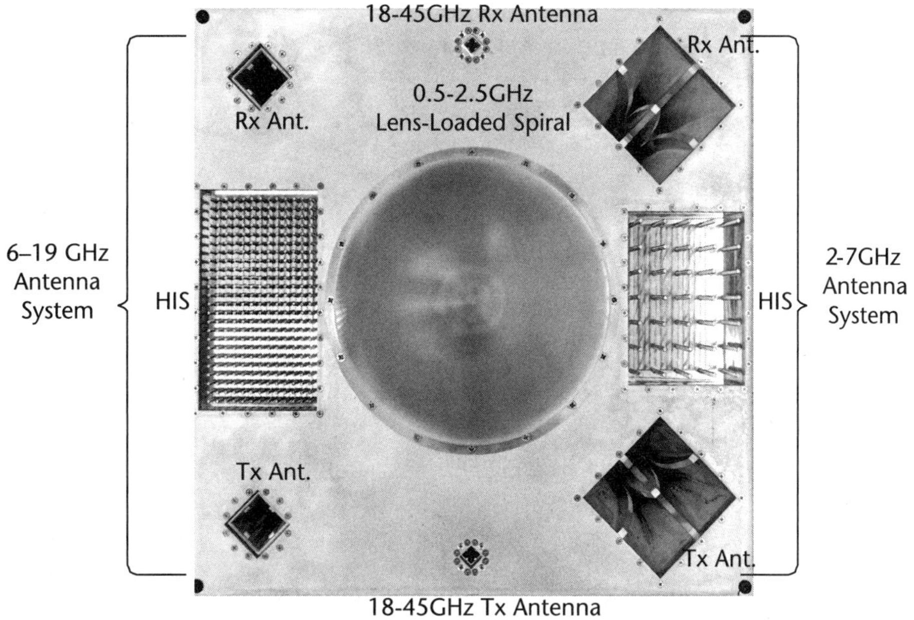

Figure 2.17 Hybrid IBFD. The overall size of the ground plane is 18 × 18 inches.

0.5 to 2.5 GHz, is serviced by a monostatic antiphase lens-loaded spiral, such as the ones discussed in Section 2.4.2. The carefully engineered monostatic spiral achieves over 50 dB measured SIC throughout the bandwidth while operating as a right-handed elliptically polarized antenna in both Tx and Rx modes. Its size determines the overall footprint of the hybrid aperture and is chosen to allow for acceptable efficiency over the lowest octave. The lens enables increased efficiency and higher directivity over the spiral's upper band. However, the lens also supports a strong surface wave between the bistatic Tx and Rx antennas if placed on the diagonal. Therefore, as shown in Figure 2.17, the bistatic antennas for 2 to 7-GHz and 6 to 19-GHz bands are along the same sides of the square ground plane.

A variation of a hybrid IBFD aperture with a smaller ground plane (only 12 × 12 inches), but the singly (co)polarized Tx and Rx operation from 0.5 to 110 GHz is shown in Figure 2.18 [16]. The 0.5 to 2.5-GHz band is covered by the same spiral from Figure 2.17; however, the aperture is raised above the ground to augment SIC between the bistatic systems. Four metallic pillars are included to diffract currents propagating on the ground plane. Ridge-horn antennas are used for bistatic subsystems over other bands, and their location on the ground and polarization are chosen to minimize coupling (no lens on the spiral) and overall size of the hybrid aperture. SIC better than 50 dB is obtained for the antiphase spiral and >60 dB for most of the band for the bistatic subsystems. The radiation pattern changes include some asymmetry, ripple, beamsquint, and increased cross-polarization; however, all these effects were found to be acceptable for the considered use.

A hybrid IBFD antenna composed of a monocone and two circular arrays is shown in Figure 2.19 [32]. Herein, the monocone (Tx) and Vivaldi (Rx) array are vertically polarized and rely on the monocone ground plane, symmetry, and physical distance to reduce mutual coupling (20-dB worst-case isolation). A horizontally polarized, tightly coupled dipole array (TCDA) alongside the Vivaldi array form a

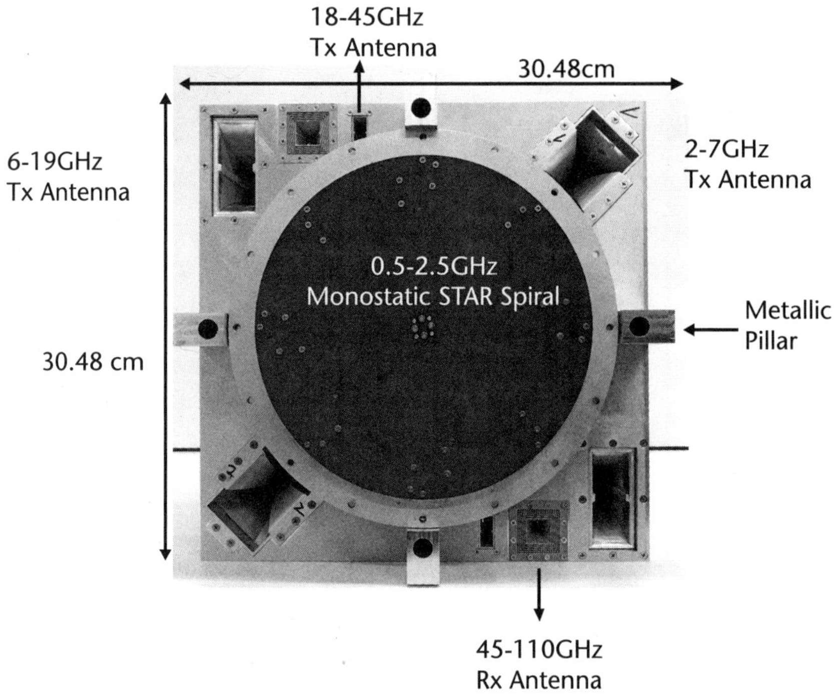

Figure 2.18 Singly polarized system. The overall size of the ground plane is 12 × 12 inches.

(a)

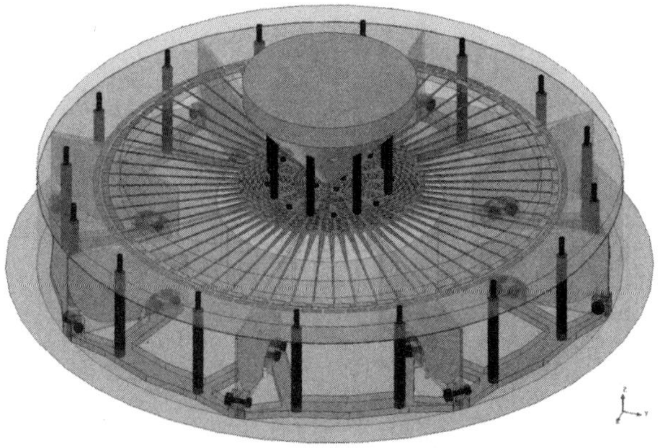

(b)

Figure 2.19 (a) Fabricated system and (b) CAD model to show the tightly coupled dipole array (TCDA) and tapered slot antenna (TSA), which are obscured by the structure. (Adapted from [32].)

direction of arrival sensor with polarimetric capability over a 2-octave bandwidth. The worst-case isolation between the monocone and TCDA is 26.5 dB. The SIC between the monocone and the Vivaldi array can be improved by 20 dB or more by alternating the phases of the elements in the circular array.

2.6 Practical Considerations

Antenna systems with IBFD capability remain a tremendous challenge to design, fabricate, and deploy. The great level of structural detail and symmetry enforcement needed to ensure maximum levels of SIC complicate their electromagnetic modeling, which, when coupled with electro-thermal effects and induced mechanical stresses, produces a compound effect in the initial development stage. Geometrical tolerances, particularly when IBFD systems are considered for millimeter-wave applications, are critically important and must be accounted for during the design for manufacturing. Statistical analysis of amplitude and phase imbalances within the feeding network and their impact on isolation need to be included to plan for the worst case. Radome design and integration with the aperture must be appraised from the start to ensure that their later addition (as is often done) does not detrimentally affect SIC. The presence of random objects nearby or broken symmetry in the aperture near field can also affect the level of isolation attained with the antenna system and some consideration of those must be included. Treatment of joints is important when antennas are built from multiple parts. Size, weight, power, and cost constraints are conventional considerations that often limit the choice of antenna systems and must be discussed from the start.

2.7 Conclusion

2.7.1 Summary

In this chapter, we focused on antenna subsystems and techniques used to enhance the SIC from the Tx to Rx channel (arrays of which are discussed in Chapters 13 and 14). Considered are four different classes of configurations with brief descriptions of various methods within each, along with their practical constraints and pros and cons. We showed that the polarization, space, and beam-multiplexing techniques can result in high levels of isolation at the expense of suboptimal use of available RF resources. Singly copolarized, monostatic STAR antennas can achieve high levels of SIC over wide bandwidths, at the expense of radiation efficiency and pattern quality. Sequentially rotated antennas and quasi-monostatic approaches can be used for dual-copolarized IBFD application over a range of bandwidths and carrier frequencies, as well as Tx powers. Extremely wide bandwidths are best-served with hybrid configurations, where different IBFD antenna subsystems cover different bands while sharing the same space.

2.7.2 Future Directions

Future research and development of IBFD antenna subsystems will be driven by specific applications, as a universal solution may never be available. The traditional goals that antenna engineers have been facing over many decades, such as small size, wide bandwidth, low cost, and high power, will need to be combined with the overall system-level objectives and included in the respective trade-off space. Antenna integration, monolithic or hybrid, with integrated IBFD analog and digital circuits, as well as packaging of all subsystems into a functional RF system, will be a great focus in the near future. The majority of IBFD antenna subsystems demonstrated in literature can still be improved to serve the ever-changing needs of real applications. The release of new frequency bands and increasing requirements for greater digital speeds, human/machine to human/machine communications, national defense, and climate change monitoring, just to name a few, will continue to drive the developments in this area.

References

[1] Prasannakumar, P. V., M. A. Elmansouri, and D. S. Filipovic, "Wideband Decoupling Techniques for Dual-Polarized Bi-Static Simultaneous Transmit and Receive Antenna Subsystem," *IEEE Transactions on Antennas and Propagation*, Vol. 65, No. 10, October 2017, pp. 4991–5001.

[2] Elmansouri, M.., et al., "Single and Dual-Polarized Wideband Simultaneous Transmit and Receive Antenna System," *2017 IEEE Int. Symp. Antennas Propag. (APSURSI)*, San Diego, CA, 2017, pp. 1105–1106.

[3] Tianang, E. G., and D. S. Filipovic, "A Dipole Antenna System for Simultaneous Transmit and Receive," *2015 IEEE Int. Symp. Antennas Propag & USNC/URSI Nat. Radio Sci. Meet.*, Vancouver, BC, 2015, pp. 428–429.

[4] Krishnaswamy, H., et al., "Full-Duplex in a Hand-Held Device—From Fundamental Physics to Complex Integrated Circuits Systems and Networks: An Overview of the Columbia FlexICoN Project," *Proc. 50th Asilomar Conf. Signals Syst. Comput.*, November 2016, pp. 1563–1567.

[5] Manafi, S., M. A. Al-Tarifi, and D. S. Filipovic, "Isolation Improvement Techniques for Wideband Millimeter-Wave Repeaters," *IEEE Antennas Wireless Propag. Lett.*, Vol. 17, No. 2, February 2018, pp. 355–358.

[6] Lyon, J. A. M., et al., *Electromagnetic Coupling Reduction Techniques*, University of Michigan, Ann Arbor, MI, Tech Rep. AFAL-TR-68-132, 1968, http://hdl.handle.net/2027.42/6402.

[7] Hannan, P., and S. Litt, "Capacitive Ground Plane for a Phased Array Antenna," *Proc. Antennas Propag. Soc. Int. Symp.,* 1968, pp. 115–123.

[8] Prasannakumar, P. V., "Wideband Bi-Static and Monostatic STAR Antenna Systems," Ph.D. dissertation, Dept. Elect. Comput. Energy Eng., CU Boulder, Boulder, 2019, https://scholar.colorado.edu/concern/graduate_thesis_or_dissertations/3f462569j

[9] Cutler, C. C., *Electromagnetic Waves Guided by Corrugated Conducting Surfaces*, Bell Laboratories, Report No. MM-44-160-218, October 25,1944.

[10] Watkins, D. A., *Topics in Electromagnetic Theory*, New York: John Wiley & Sons, 1958.

[11] Kildal, P. -S., "Artificially Soft and Hard Surfaces in Electromagnetics," *IEEE Transactions on Antennas and Propagation*, Vol. 38, No. 10, October 1990, pp. 1537–1544.

[12] Sievenpiper, D., "High-Impedance Electromagnetic Surfaces," Ph.D. dissertation, Dept. Elect. Eng., University of California, Los Angeles, CA, 1999.

[13] Allen, B. F., "Isolation Enhancement for Cylindrical Structure Millimeter-Wave Repeaters," M.S. Thesis, Dept. Elect. Comput. Energy Eng., CU Boulder, Boulder, 2018, https://scholar.colorado.edu/concern/graduate_thesis_or_dissertations/pv63g0609

[14] Tereshin, O., "Decoupling of Two Slot Antennas by Means of a Ribbed Structure in the Plane of the Slots," *Radio Eng. and Electron.*, Vol. 5, No. 12, 1960, pp. 99–108.

[15] King, R. J., and K. S. Park, "Synthesis of Surface Reactances Using Grounded Pin Bed Structure," *Electron. Lett.*, Vol. 17, No. 1, January 1981, pp. 52–53.

[16] Filipovic, D. S., et al., *Signal Processing Electronic Attack RFIC (SPEAR)*, Final Project Report to Naval Research Laboratory, N00173-15-C-2021, 2018.

[17] Scarborough, C. P., et al., "Demonstration of an Octave-Bandwidth Negligible- Loss Metamaterial Horn Antenna for Satellite Applications," *IEEE Transactions on Antennas and Propagation*, Vol. 61, No. 3, March 2013, pp. 1081–1088.

[18] Samaiyar, A., et al., "Phased Array Antenna for Bi-Static Simultaneous Transmit and Receive (STAR) Systems," *2019 IEEE Int. Symp. Phased Array Syst. Tech. (PAST)*, Waltham, MA, 2019, pp. 1–5.

[19] Knox, M. E., "Single Antenna Full Duplex Communications Using a Common Carrier," *WAMICON 2012 IEEE Wireless Microw. Tech. Conf.*, Cocoa Beach, FL, 2012, pp. 1–6.

[20] Etellisi, E. A., M. A. Elmansouri, and D. S. Filipovic, "Wideband Monostatic Simultaneous Transmit and Receive (STAR) Antenna," *IEEE Transactions on Antennas and Propagation*, Vol. 64, No. 1, January 2016, pp. 6–15.

[21] Etellisi, E. A., M. A. Elmansouri, and D. S. Filipovic, "In-Band Full-Duplex Multimode Lens-Loaded Eight-Arm Spiral Antenna," *IEEE Transactions on Antennas and Propagation*, Vol. 66, No. 4, April 2018, pp. 2084–2089.

[22] Elmansouri, M. A., A. J. Kee, and D. S. Filipovic, "Wideband Antenna Array for Simultaneous Transmit and Receive (STAR) Applications," *IEEE Antennas Wireless Propag. Lett.*, Vol. 16, 2017, pp. 1277–1280.

[23] Etellisi, E., "Wideband Monostatic Co-Channel Simultaneous Transmit and Receive (C-STAR) Antenna and Array Systems," Ph.D. dissertation, Dept. Elect. Comput. Energy Eng., CU Boulder, Boulder, 2018.

[24] Ha, J., et al., "Monostatic Co-Polarized Full-Duplex Antenna with Left- or Right-Hand Circular Polarization," *IEEE Transactions on Antennas and Propagation*, Vol. 65, No. 10, October 2017, pp. 5103–5111.

[25] Etellisi, E. A., M. A. Elmansouri, and D. S. Filipovic, "Wideband Monostatic Co-Polarized Co-Channel Simultaneous Transmit and Receive Broadside Circular Array Antenna," *IEEE Transactions on Antennas and Propagation*, Vol. 67, No. 2, February 2019, pp. 843–852.

[26] Kolodziej, K. E., et al., "Ring Array Antenna with Optimized Beamformer for Simultaneous Transmit and Receive," *Proc. 2012 IEEE Int. Symp. Antennas Propag*, Chicago, IL, 2012, pp. 1–2.

[27] Etellisi, E. A., M. A. Elmansouri, and D. Filipovic, "Broadband Full-Duplex Monostatic Circular-Antenna Arrays: Circular Arrays Reaching Simultaneous Transmit and Receive Operation," *IEEE Antennas and Propagation Magazine*, Vol. 60, No. 5, October 2018, pp. 62–77.

[28] Mulero Hernandez, C. A., M. Elmansouri, and D. S. Filipovic, "High-Directivity Beam-Steerable Lens Antenna for Simultaneous Transmit and Receive," *2019 IEEE Int. Symp. Phased Array Syst. Technol. (PAST)*, Waltham, MA, 2019, pp. 1–5.

[29] Prasannakumar, P. V., M. A. Elmansouri, and D. S. Filipovic, "Broadband Reflector Antenna with High Isolation Feed for Full-Duplex Applications," *IEEE Transactions on Antennas and Propagation*, Vol. 66, No. 5, May 2018, pp. 2281–2290.

[30] Etellisi, E. A., M. A. Elmansouri, and D. S. Filipovic, "Wideband Multimode Monostatic Spiral Antenna STAR Subsystem," *IEEE Transactions on Antennas and Propagation*, Vol. 65, No. 4, April 2017, pp. 1845–1854.

[31] Prasannakumar, P. V., et al., "Wideband High-Directivity Quasi-Monostatic Simultaneous Transmit And Receive Antenna," *IEEE Transactions on Antennas and Propagation*, Vol. 68, No. 4, April 2020, pp. 2630–2637.

[32] Pack, R., "Wideband Dual-Polarized Digital Direction of Arrival Sensors," Ph.D. dissertation, Dept. Elect. Comput. Energy Eng., CU Boulder, Boulder, 2019, https://scholar.colorado.edu/concern/graduate_thesis_or_dissertations/sx61dm45j.

CHAPTER 3

Electrical Balance Duplexers

Leo Laughlin

3.1 Introduction

As discussed in Chapter 1, IBFD transceivers typically achieve high levels of transmit-to-receive (Tx-Rx) isolation by combining multiple stages of SI avoidance and/or cancellation in the propagation, analog circuit, and digital domains. Moreover, due to the limited dynamic range of practical receivers, obtaining isolation in the propagation and/or analog domain is an indispensable requirement, as the IBFD front end must prevent the receiver from being saturated by SI. However, propagation and analog-based IBFD front-end designs rely on specialized hardware such as multiple antennas and/or multitap RF cancellers, as discussed in Chapters 2 and 8, respectively. This additional hardware can result in complex, large, and costly systems, and while this may be acceptable in some use cases (e.g., infrastructure and defense applications), others, such as consumer electronics, call for much smaller and simpler designs.

One such design is the subject of this chapter: the electrical balance duplexer (EBD). This circuit connects a transmitter and receiver to a shared antenna, but exploits the electrical balancing of signals within a hybrid junction to provide high Tx-Rx isolation. It can be implemented in an integrated circuit and does not rely on multiple antennas for duplexing, making it an attractive choice for low-cost, small-form factor devices. EBDs have been demonstrated at various frequencies, from <1 GHz to millimeter wave. An EBD is also tunable over wide frequency ranges, making it suitable for multiband front-end designs, as well as the dynamic spectrum access systems envisaged for the future. Fundamentally, the EBD separates signals based on direction of travel at the antenna port and is inseparable from the antenna. It can therefore be considered a propagation domain technique, as highlighted within the IBFD transceiver diagram of Figure 3.1(a) and the SIC techniques tree of Figure 3.1(b).

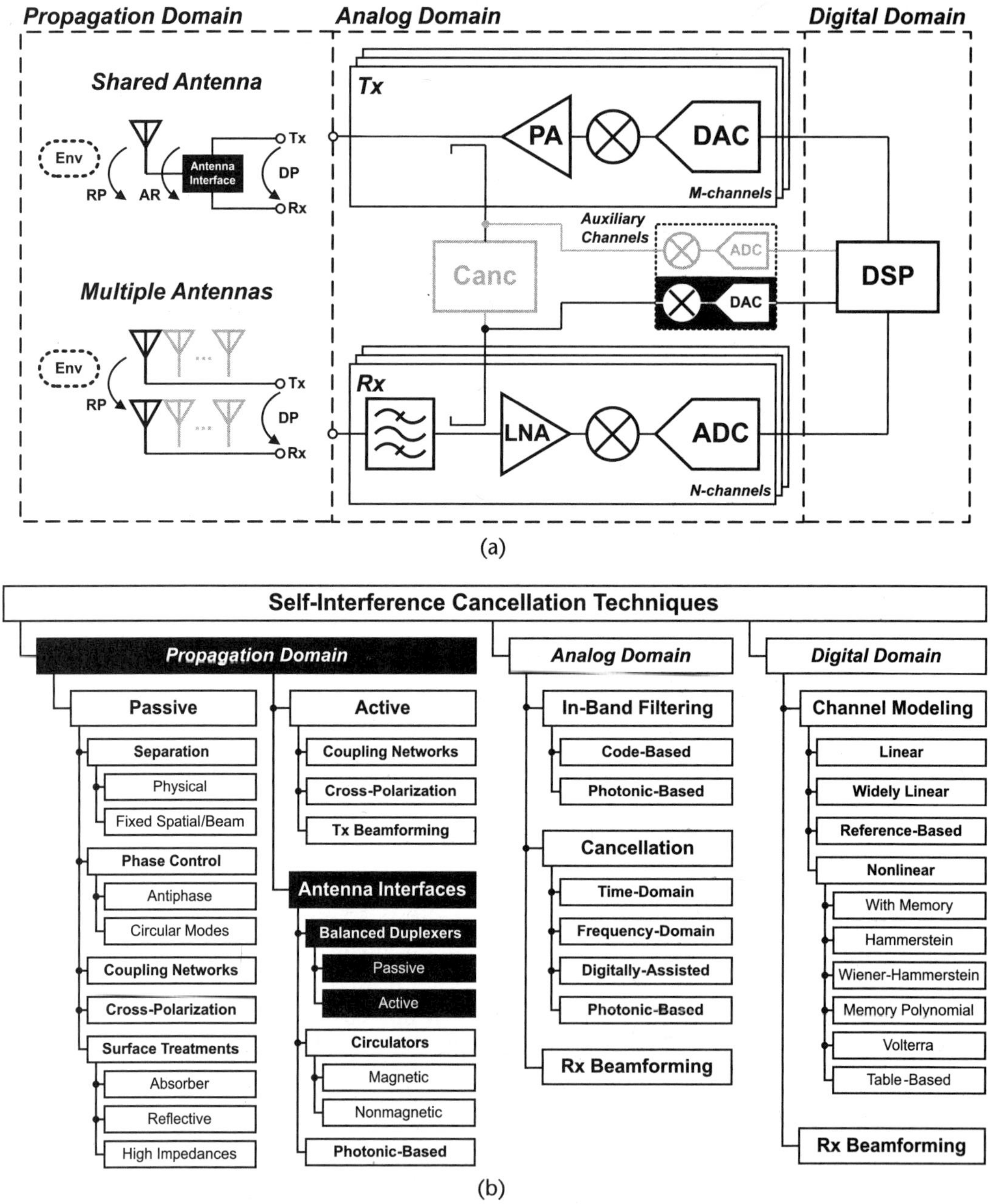

Figure 3.1 EBD and specific chapter focus highlighted within (a) a generic IBFD transceiver and (b) a SIC techniques tree diagram.

3.2 Principles of Operation

3.2.1 Pre-Electronic Telephone

The EBD is based on the balancing of signals in a hybrid junction. The hybrid junction is a widely used component in electronic and microwave engineering, with the defining property being that, when the ports are correctly impedance-matched, a signal entering any port is routed to the two adjacent ports, but not to the remaining (opposite) port. This property of coupled and isolated ports can be exploited for the purposes of duplexing signals. Indeed, the pre-electronic telephone of the early twentieth century used a hybrid transformer (Figure 3.2); signals entering the center tap of the primary winding are split, flowing in opposite directions through each half of the coil, appearing at each end of the coil, but not in the secondary winding. The currents in the primary winding create equal but opposite magnetic fluxes, meaning no net current is induced in the secondary winding (Figure 3.2(a)). The tapping of the primary winding need not be symmetrical, but may be asymmetric with tapping ratio r. This alters the power split in the hybrid (later discussed).

The telephone duplexer (Figure 3.2(b)) used a hybrid transformer to couple the microphone and earpiece to a common telephone line, but isolate them from each other, such that a user's own speech would not be heard in the earpiece. The fourth port of a telephone duplexer, the balance port, is terminated such that the currents

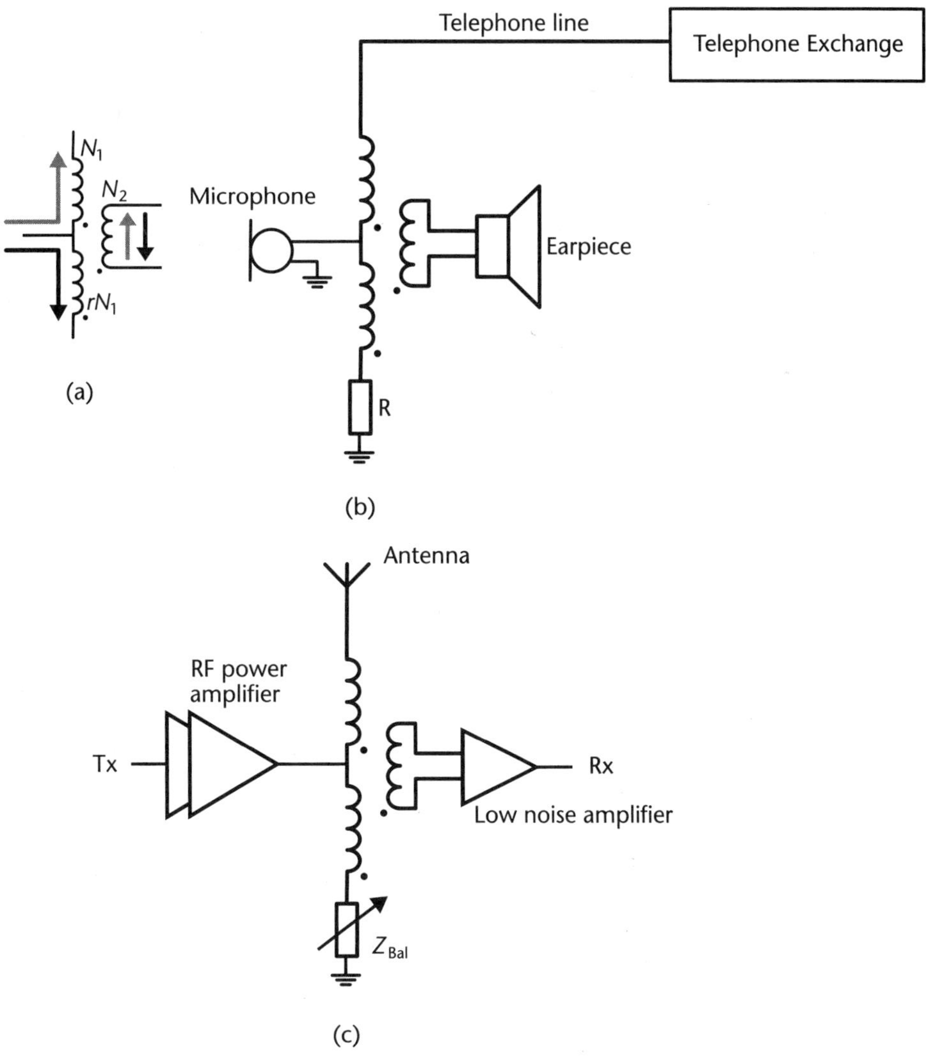

Figure 3.2 (a) The hybrid transformer and arrows showing direction of current flow components, leading to cancellation in the secondary winding; (b) the telephone duplexer; and (c) the wireless EBD [1].

in the primary winding are correctly balanced, thus providing isolation. The wireless EBD (Figure 3.2(c)) follows the same principle as the telephone duplexer, but substitutes the microphone, earpiece, telephone line, and resistor with a transmitter, receiver, antenna, and balancing impedance, respectively.

3.2.2 Tx-Rx Isolation

The Tx-Rx isolation provided by the EBD arises due to the specific signal coupling behavior of the hybrid junction, and depends on the impedances of the antenna and balancing network, as shown in the following theoretical analysis (originally derived in [2]).

The S-matrix for a 180° hybrid junction is given as

$$\begin{bmatrix} b_{Tx} \\ b_{Rx} \\ b_{Ant} \\ b_{Bal} \end{bmatrix} = \begin{bmatrix} 0 & 0 & c & d \\ 0 & 0 & d & -c \\ c & d & 0 & 0 \\ d & -c & 0 & 0 \end{bmatrix} \begin{bmatrix} a_{Tx} \\ a_{Rx} \\ a_{Ant} \\ a_{Bal} \end{bmatrix}, \tag{3.1}$$

where a_{Tx}, a_{Rx}, a_{Ant}, and a_{Bal} are the transmit, receive, antenna, and balancing ports, respectively, with b_{Tx}, b_{Rx}, b_{Ant}, and b_{Bal} being the corresponding scattered signals. The coupling coefficient, c, determines the ratio by which power is split between adjacent ports, and $d = \sqrt{1-c^2}$ for a lossless hybrid. In the case of a hybrid transformer-based implementation, as depicted in Figure 3.2, the coupling coefficient depends on the transformer tapping ratio according to $c = \frac{\sqrt{r}}{\sqrt{1+r}}$.

Expanding (3.1) gives:

$$b_{Tx} = ca_{Ant} + da_{Bal} \tag{3.2}$$

$$b_{Rx} = da_{Ant} - ca_{Bal} \tag{3.3}$$

$$b_{Ant} = ca_{Tx} + da_{Rx} \tag{3.4}$$

$$b_{Bal} = da_{Tx} - ca_{Rx} \tag{3.5}$$

Assuming the low-noise amplifier (LNA) input impedance is matched (and hence no energy is reflected at the receiver port), and noting that a receiver input does not transmit a signal, there is no signal incident at the hybrid junction receiver port; thus,

$$a_{Rx} = 0. \tag{3.6}$$

Ignoring any wireless signals received by the antenna, we may also define the incident signals at the antenna and balancing ports according to the reflection coefficients at those ports, such that

$$a_{Ant} = b_{Ant}\Gamma_{Ant} \tag{3.7}$$

$$a_{Bal} = b_{Bal}\Gamma_{Bal}, \tag{3.8}$$

where Γ_{Ant} and Γ_{Bal} are the antenna and balancing reflection coefficients, respectively. Substituting (3.6) through (3.8) into (3.2) through (3.5) gives

$$b_{Tx} = cb_{Ant}\Gamma_{Ant} + db_{Bal}\Gamma_{Bal} \tag{3.9}$$

$$b_{Rx} = db_{Ant}\Gamma_{Ant} - cb_{Bal}\Gamma_{Bal} \tag{3.10}$$

$$b_{Ant} = ca_{Tx} \tag{3.11}$$

$$b_{Bal} = da_{Tx}. \tag{3.12}$$

Substituting (3.11) and (3.12) into (3.10) yields an expression for the scattered signal at the receiver port

$$b_{Rx} = dca_{Tx}\Gamma_{Ant} - cda_{Tx}\Gamma_{Bal}, \tag{3.13}$$

which can be rearranged to form

$$\frac{b_{Rx}}{a_{Tx}} = dc(\Gamma_{Ant} - \Gamma_{Bal}). \tag{3.14}$$

Since b_{Rx} is the input signal to the LNA and a_{TX} is the output of the power amplifier (PA), the above expression gives the transmit-to-receive gain of the EBD, and we may rewrite this as

$$G_{Tx-Rx} = L(\Gamma_{Ant} - \Gamma_{Bal}). \tag{3.15}$$

The Tx-Rx gain (and hence the isolation) is therefore proportional to the difference between the antenna and balancing reflection coefficients, with a constant of proportionality, $L = dc$ that depends on the symmetry of the hybrid junction. For a symmetrical hybrid, $d = c = 1/\sqrt{2}$, and hence $L = 1/2$. Isolation is obtained by adjusting the balancing reflection coefficient to equal that of the antenna. This is evident in (3.15): when $\Gamma_{Bal} = \Gamma_{Ant}$, the gain expression sums to zero, and the duplexer is balanced, giving theoretically infinite isolation. The EBD can be thought of as a form of feed-forward RF cancellation, implemented within the antenna feed network; there are essentially two paths between the transmitter and receiver: one that depends on the antenna reflection (AR) coefficient, and the other that depends on the balancing reflection coefficient, and the balancing network is adjusted such that these two paths cancel out.

The analysis above is for a 180° hybrid. In the case of a quadrature hybrid, isolation is achieved when the balancing reflection coefficient is the inverse of the AR coefficient, due to the different phase relationships in the hybrid.

3.2.3 Tx and Rx Insertion Losses

A hybrid splits input power on any one port between two output ports. Due to this, the EBD suffers from unavoidable losses in the Tx and Rx paths, which for a symmetrical hybrid is 3 dB in both; however, in asymmetric hybrids, these losses are traded off against each other. The relationship of the Tx and Rx insertion losses to the coupling coefficient, *c*, in an ideal EBD, is shown in Figure 3.3. Since the ideal hybrid junction is lossless, this relationship is simply governed by the conservation of energy.

These unavoidable Tx and Rx insertion losses are one of the most significant drawbacks of the EBD, due to the impact this has on Tx efficiency and Rx sensitivity, respectively. It is arguable that these losses make the basic EBD design impractical. However, these losses can be mitigated by codesigning the EBD and the LNA using noise matching rather than impedance matching, such that the signal-to-noise ratio (SNR) at the LNA is improved. Details of such a design are shown in [3], which skewed the hybrid to favor the Tx, and used noise matching to mitigate Rx noise, achieving a Tx insertion loss of 2.5 dB and an antenna-referred noise figure (NF) of 5 dB. While losses are still incurred in this architecture, this may be comparable to some other RF cancellation architectures that add additional hardware in the Tx/Rx paths (e.g., splitters, couplers, and circulators), as discussed in Chapters 8 through 10.

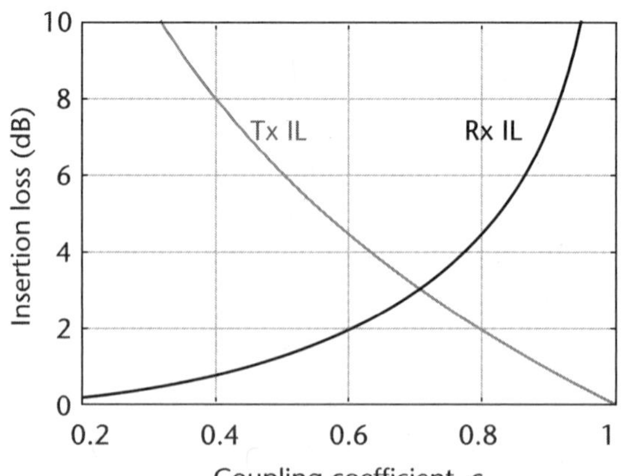

Figure 3.3 The relationship between the Tx and Rx insertion losses and the hybrid junction coupling coefficient, *c*.

3.3 Balancing Networks

3.3.1 Wideband Balancing

In practice, antennas are inherently resonant components, and the AR coefficient is a complex function of frequency. The same is true for the balancing reflection coefficient. Any practical hybrid implementation will also have a limited bandwidth and add further variations to the frequency response of the Tx-Rx gain. To take these factors into account, we may rewrite (3.15) with all quantities as functions of frequency

$$G_{Tx-Rx}(\omega) = L(\omega)\left(\Gamma_{Ant}(\omega) - \Gamma_{Bal}(\omega)\right). \tag{3.16}$$

From this expression, it is clear that in order to perfectly balance the EBD over a given frequency range, $\omega_l \leq \omega \leq \omega_h$, requiring that the antenna and balancing reflection coefficients are equal over this frequency range,

$$\Gamma_{Ant}(\omega) = \Gamma_{Bal}(\omega) \quad \text{for} \quad \omega_l \leq \omega \leq \omega_h. \tag{3.17}$$

Thus, in order to provide perfect isolation over a given bandwidth, the balancing impedance network must exactly mimic the antenna impedance over that bandwidth. For practical balancing networks, this limits the isolation bandwidth, as described in the following section.

3.3.2 Passive Balancing Networks

A passive balancing network has several requirements:

- It must be tunable over the full range of expected antenna impedances and operating frequencies.
- The tuning must have fine granularity. Very fine tuning is required to achieve high levels of isolation.
- The network must be capable of withstanding large voltage swings and dissipating substantial power, as a substantial portion of the Tx power (50% for a symmetrical EBD) is sunk into this network.
- It must have adequate linearity in order to limit distortion. This is especially important for IBFD applications, where further stages of cancellation are applied; nonlinearity in the balancing network will result in additional nonlinear SI needing to be subsequently cancelled [4].

Substantial advances in the power handling and linearity of integrated EBD implementations have been demonstrated, achieving +30-dBm maximum Tx power handling and +70-dBm input third-order-intercept (IIP3) performance using partially depleted RF silicon-on-insulator (SOI) complementary metal-oxide semiconductor (CMOS) [5]. Due to the complexity of the above requirements and despite advances in integrated implementations, networks with many tunable elements remain unattractive due to their size and cost. Balancing impedance networks that balance the EBD at a single frequency point have been used in a number of prototype EBD implementations for their simplicity [3, 6]. For example, a parallel RC balancing network provided a 50-dB isolation bandwidth (the bandwidth over which >50-dB isolation is achieved) of over 200 MHz in [6]. Some early works on wireless EBDs [3, 6], while demonstrating deep submicron CMOS implementations with the required power handling for cellular handset applications, did not include antennas, but instead terminated the antenna port with a 50Ω (or near 50Ω) load. As a result, the resulting Tx-Rx isolation is overoptimistic, as it does not include the detrimental impact of the frequency domain antenna impedance variation. This is shown in Figure 3.4, which plots measurements from a discrete EBD prototype [7]. The Smith chart plot shows measured antenna and balancing reflection coefficients, and illustrates the problem of balancing over a wide bandwidth; the antenna and balancing reflection coefficients are well-balanced at the center frequency, but moving away from this point, the reflection coefficients diverge on the Smith chart, leading to the notch-like Tx-Rx isolation frequency response. As can be seen, rather than achieving 50-dB isolation bandwidths in the hundreds of megahertz, the practical antenna impedance severely limits the isolation bandwidth; for the system in [7] (Figure 3.4), the 50-dB isolation bandwidth was just a few megahertz.

One way of increasing the isolation bandwidth is to include more tunable elements in the balancing network [5, 8]. For example, the prototype reported in [8] achieved a mean isolation of 50 dB over an 80-MHz bandwidth, but entailed substantial complexity, using 7 tunable elements in the balancing network. This system also included an antenna impedance tuner with an additional 3 tunable elements, and relied on transmission-line impedance inverters to emulate tunable inductors using tunable capacitors; the size and complexity of this would be impractical in some applications. A further drawback of increasing the number of tunable elements is that balancing becomes more complicated, adding dimensions to the optimization problem that must be solved.

3.3.3 Active Balancing Networks

Due to the aforementioned difficulties with passive balancing, active circuits have also been investigated as a method of balancing the EBD. An active balancing

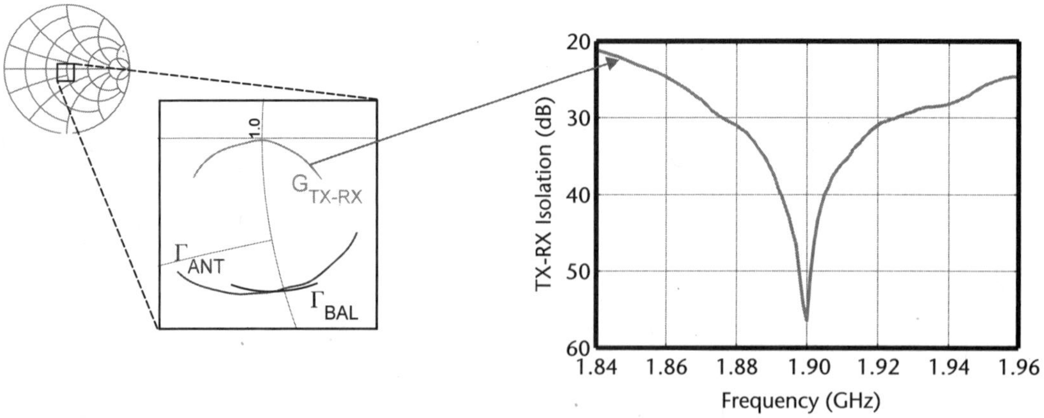

Figure 3.4 Measurements from a discrete EBD prototype, showing the divergent antenna and balancing reflection coefficients on the Smith chart, and the limited isolation bandwidth that results [7].

network is essentially an implementation of active SI cancellation, using a separate transmitter chain to generate a cancellation signal (similar to the RF cancellation approach discussed in Chapter 10). This has the benefit that high-order digital filtering can be applied to generate the cancellation signal, giving excellent SI modeling accuracy. This is in contrast to passive analog feed-forward cancellation techniques, where the high cost and size of the analog RF hardware (amplitude/phase shifters) limit the number of cancellation taps that can be practically applied, thereby limiting the modeling accuracy, as discussed in Chapters 8 and 9. Instead, active cancellation is limited by noise and distortion introduced by the RF hardware (mainly the Tx chains), which does not cancel out and remains as SI. Active cancellation also has the drawback that it requires an additional Tx chain, increasing complexity and power consumption.

Figure 3.5 depicts (among others) the passive and active embodiments of the EBD. As described earlier, the passive EBD implements a form of passive feed-forward SI cancellation, with the tunable elements in the balancing network being analogous to the taps of a conventional time-domain canceller (discussed in Chapter 8). In contrast, for the active EBD, the effective reflection coefficient at the balancing port of the hybrid is determined by the signal being fed into that port. Thus, the isolation is not limited by the number of tunable elements (there are not any in an active EBD), but by the accuracy of this balancing signal; and as is the case for active cancellation in general, this is primarily determined by the hardware imperfections of the transmit chains. Thus, active balancing has the potential to provide much wider isolation bandwidths compared to passive balancing; it moves the signal processing complexity from the RF domain to the digital baseband domain, where high-order filtering is readily applied. However, it cannot cancel Tx noise.

Passive and active techniques can also be combined in the EBD. Figure 3.5 also depicts a passive EBD, combined with a further stage of active cancellation. This would be considered the conventional method of combining cancellation techniques, with the cancellation processes occurring in separate, sequential stages. This architecture is an effective combination of techniques; for example, in [7], a passive EBD providing 45-dB isolation over a 20-MHz bandwidth was combined

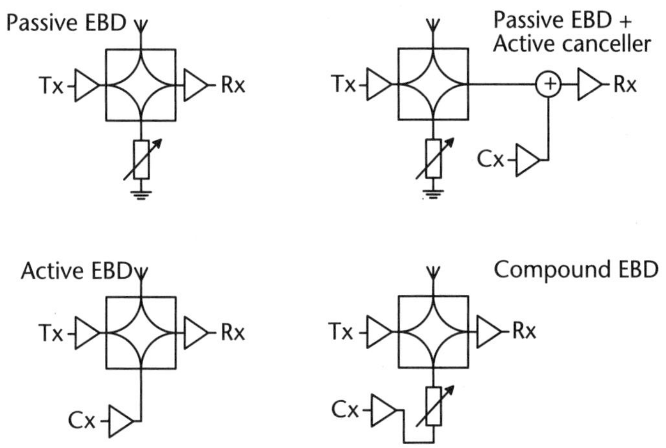

Figure 3.5 Four different EBD-based front-end architectures applying and/or combining passive and active cancellation.

with an active canceller to increase isolation to 83 dB of combined passive and active cancellation.

Passive and active EBDs can also be combined to create the compound EBD (Figure 3.5) [9]. A passive impedance matching network is tuned to provide a first level of passive isolation, as limited by the capability of that passive network to mimic the antenna impedance. At the same time, a balancing signal is also actively fed into the passive balancing network, to actively cancel the SI that remains after passive balancing, thereby increasing the overall isolation.

3.3.4 Prototype and Measurement Results

A hardware prototype implementing both passive, active and compound EBD architectures was reported in [9]. A block diagram of this system is shown in Figure 3.6, using a National Instruments vector signal transceiver platform to implement the RF transceiver hardware and baseband processing. The orthogonal frequency-division multiplexing (OFDM) physical layer allows cancellation signal generation to be readily performed using frequency-domain equalization (FDE) to apply a complex multiplication to each subcarrier, thereby achieving wideband cancellation. FDE coefficients can be calculated based on measurements of the coupling channels between both of the transmitters and the receiver [9]. The Tx power for this system was 10 dBm and the operating band was 80 MHz wide centered at 1.9 GHz.

Measurement results from this system are shown in Figure 3.7. For passive EBD measurement, the balancing impedance was tuned to maximize passive isolation, giving 58-dB peak isolation at the band center and 35.8-dB average isolation over an 80-MHz bandwidth. For the active measurement, the balancing impedance was set to 50Ω, giving only 18 dB of (untuned) passive isolation, with the active cancellation increasing this to a total of 44.6 dB. In the compound EBD, the balancing impedance was tuned to maximize passive isolation (as above), and the active canceller was then also used to further increase isolation, achieving a substantially higher total isolation of 81.5 dB. Notably, the residual SI in the active and

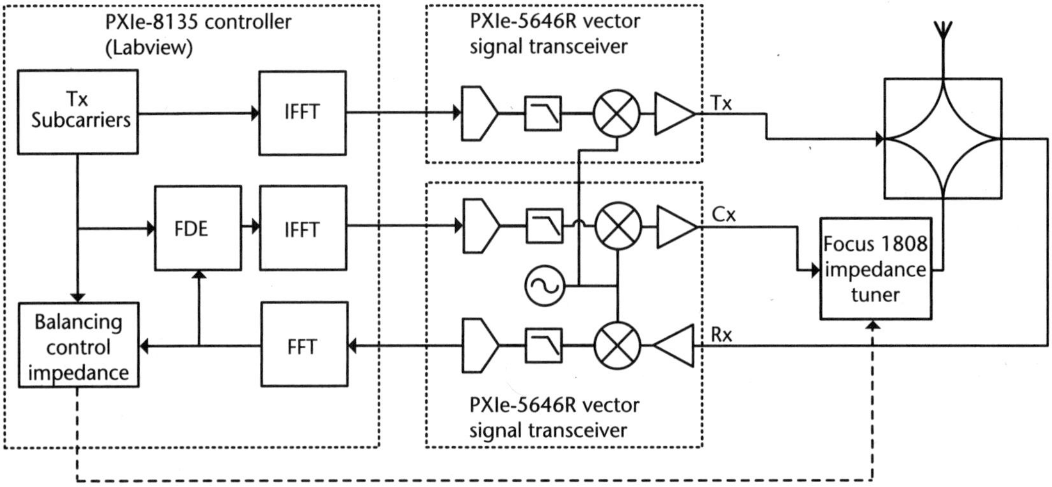

Figure 3.6 A block diagram of an EBD combining passive and active cancellation techniques, as reported in [9].

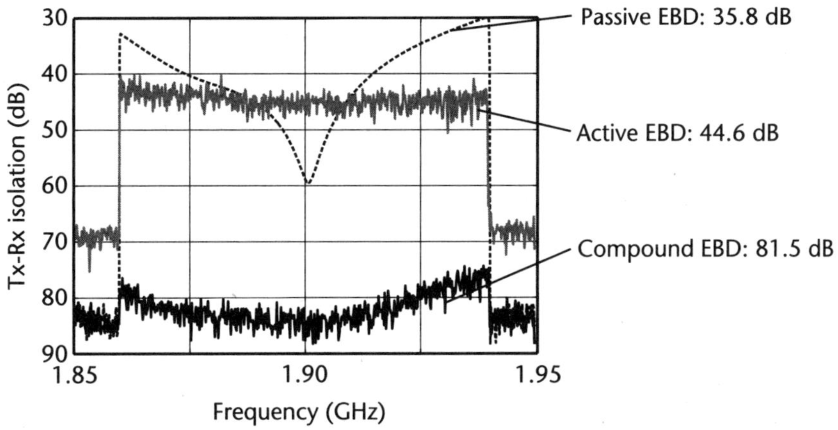

Figure 3.7 Measured Tx-Rx isolation for the passive, active and compound EBDs [9].

compound EBDs has a relatively flat spectrum compared to the passive EBD, demonstrating the effectiveness of the frequency selective cancellation applied. Table 3.1 provides a performance comparison of several EBDs using different balancing networks as reported in the literature.

3.4 Dynamic Adaptation

3.4.1 Environmental Interaction

It is well-known that objects in the local environment can have an impact on the impedance of an antenna. Moreover, the antenna impedance and associated AR SI may vary with time due to motion of the device or objects in the environment (e.g., the user). This requires the balancing impedance to adapt to changes in the antenna impedance in order to maintain balance, and the Tx-Rx isolation could potentially be reduced if the adaptation process cannot adequately track the antenna impedance in the time domain.

Figure 3.8 shows a passive EBD with adaptive balancing. The impact of the environment on EBD balancing is best understood by considering how this impacts

Table 3.1 Performance Comparison of Several EBD Prototypes Using Different Balancing Networks

Reference	*Type*	*Tunable Dimensions*	*Mean Isolation/ Bandwidth*	*Comments*
[7]	Passive	2	45 dB/20 MHz	—
[10]	Passive	4	>50 dB/20 MHz	Antenna and balance network codesigned for improved isolation
[8]	Passive	10	50 dB/80 MHz	Includes antenna impedance tuning
[9]	Active	>10,000 (FDE)	45 dB/80 MHz	Tx noise/distortion remains as residual SI
[9]	Compound	2 (passive) + >10,000 (FDE)	81 dB/80 MHz	Tx noise/distortion remains as residual SI

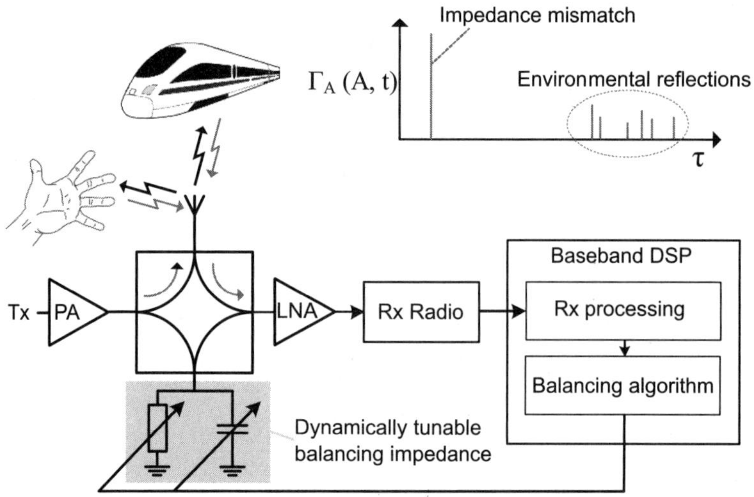

Figure 3.8 An EBD receiving reflections from environmental objects and utilizing dynamic impedance adaptation. Inset graph shows the impact of environmental reflections on the AR coefficient power-delay profile.

the AR coefficient power-delay profile: transmitted energy reflected from objects in the environment and received back at the antenna results in additional time-delayed components in the AR coefficient (see the inset plot in Figure 3.8). This can reduce the mean isolation across the band of interest, as some of this reflected energy may not be cancelled. Furthermore, in a dynamic environment, the amplitude and delay of these reflections will vary, causing variation in the overall AR coefficient frequency response, requiring dynamic adaptation of the balancing impedance to maintain optimum Tx-Rx isolation.

The process by which the correct balancing impedance settings are attained and maintained is known as the balancing algorithm. Various algorithms have been reported, ranging from gradient-based iterative optimization algorithms [11, 12], to semi-deterministic algorithms and deterministic algorithms, which include an element of system characterization [9, 11]. To specify and design such algorithms, however, the requirements for EBD adaptation in dynamic environments must be established.

3.4.2 Dynamic Simulations

The impact of dynamic environments on EBD Tx-Rx isolation can be investigated by combining measurements of AR coefficients with basic EBD circuit simulations. As described above, the environment affects EBD isolation because transmitted energy is reflected and received back at the antenna, changing the effective reflection coefficient at the antenna port. Thus, when measuring the reflection coefficient (e.g., using a vector network analyzer (VNA)), the measurement captures not only the reflection due to any mismatch with the antenna itself, but also these environmental reflections. Therefore, by taking successive measurements of the AR coefficient in a dynamic environment, the dynamics of that environment, as relevant to EBD isolation, are also captured.

This approach was used in [1, 13, 14]. In these works, a VNA was configured to sample the AR coefficient frequency response repeatedly, thereby obtaining time-frequency AR coefficient data. From this, the Tx-Rx frequency response of the EBD isolation can be estimated as follows. To model a dynamic EBD, we can extend (3.16) to make the antenna and balancing network reflection coefficients both functions of time as well as frequency; thus,

$$G_{Tx-Rx}(\omega,t) = L(\omega)\left(\Gamma_{Ant}(\omega,t) - \Gamma_{Bal}(\omega,t)\right). \quad (3.18)$$

This provides the basis for a basic EBD circuit simulation [13], where values for $\Gamma_{Ant}(\omega, t)$ are taken from the measured time-frequency antenna data. Values for $\Gamma_{Bal}(\omega, t)$ are calculated using basic circuit theory (e.g., assuming a single pole RC circuit for balancing), and dynamic balancing behavior can be simulated by controlling the simulated balancing network parameters. Three types of dynamic adaptation were simulated: ideal balancing, where the balancing network parameters are always optimal; limited rate adaptation (LRA), where $\Gamma_{Bal}(\omega, t)$ is updated to the optimum at a given interval, but is fixed in the intervening time, thereby simulating imperfect tracking behavior; and static balancing, where $\Gamma_{Bal}(\omega, t)$ is not dynamically updated at all.

3.4.3 Real-Time Adaptive Implementation

An adaptive EBD implementation was also reported in [14]. This was based on the same hardware platform as described earlier, but replaces the mechanical impedance tuner with a microelectromechanical system (MEMS)-based tunable impedance implementation, as shown in Figure 3.9. Dynamic balancing control is implemented in

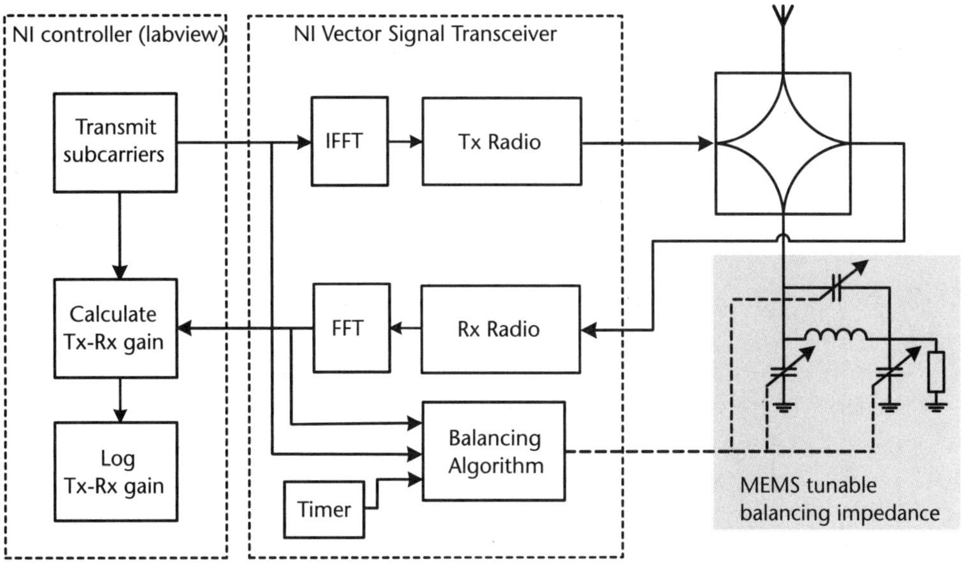

Figure 3.9 Block diagram of the real-time adaptive EBD implementation used for testing in dynamic environments.

the onboard field-programmable gate array (FPGA) using the balancing algorithm described in [9]. The balancing algorithm runs at a given interval, and in the intervening time, the Tx-Rx isolation is measured and logged. This therefore implements the same balancing functionality as the simulated EBD with limited rate adaptation.

3.4.4 Measurement Environments

Simulation analysis with antenna measurement data and hardware EBD measurements were performed in two environments: a train and a user interaction scenario.

In the train scenario [14], the antenna/EBD hardware was positioned close to the train window, representing the case where a user in a window seat is using a wireless device. It was noted that trains passing on the adjacent track were a particular worst case of interest in this environment. In this situation, a large metallic structure (the other train) is within 1m of the antenna and traveling at a very high relative speed, potentially having a substantial impact on dynamic EBD isolation. To investigate this, antenna and hardware EBD measurements were performed at times with and without a train passing on the adjacent track.

In the user interaction scenario [15], the proximity of the user's hand is known to cause substantial variations in antenna impedance. To investigate the impact on EBD isolation, the antenna was encased in a plastic mobile phone housing, and the effect of a person using the device was emulated by someone holding the mock device and making browsing or texting gestures with his or her fingers.

3.4.5 Results

Figure 3.10 shows results for the train scenario, plotted as cumulative distribution functions (CDFs) of the dynamic Tx-Rx isolation, with said isolation being the average isolation over a 20-MHz bandwidth centered at 1,900 MHz. The simulation results show that the passing train does have an impact on Tx-Rx isolation, but rebalancing the EBD at intervals of 5 ms is sufficient to maintain >50-dB isolation with 95% probability. Even in the static balancing case, the impact of the passing

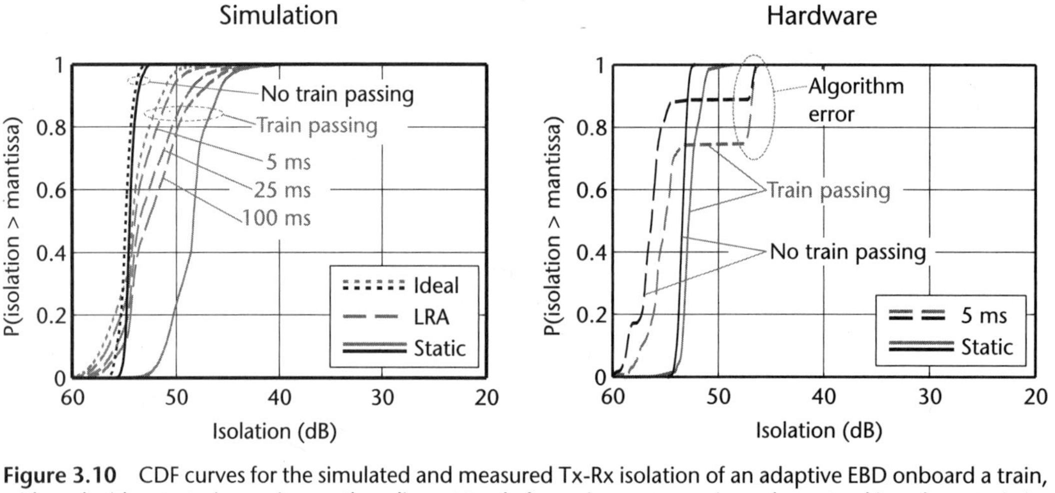

Figure 3.10 CDF curves for the simulated and measured Tx-Rx isolation of an adaptive EBD onboard a train, with and without a train passing on the adjacent track, for various antenna impedance tracking characteristics [14].

train was not catastrophic, and >45-dB isolation was maintained with 95% probability. This shows that the reflections from the passing train have a limited impact on EBD isolation (due to the attenuation suffered in the reflection path), and, thus, very high-speed impedance tracking is not necessary to achieve this level of isolation in this environment. That said, these reflections will be highly significant for subsequent stages of cancellation in an IBFD transceiver. The hardware results also presented in Figure 3.11 show agreement with the simulation results, validating the results and simulation methodology. A notable difference in the hardware results is a step change in the CDF curve. This is caused by errors in the balancing algorithm; in some instances, the algorithm provided a nonoptimal balancing impedance setting, reducing isolation.

Figure 3.11 shows simulated and measured results for the user-interaction scenario, likewise as a CDF of mean isolation over a 20-MHz bandwidth centered at 1,900 MHz. Results show that, in this scenario, dynamic adaptation is critical. In the case of static balancing, the Tx-Rx isolation was observed to fall below 30 dB for both the simulation and hardware measurements. The simulation suggests that a 5-ms rebalancing interval is again sufficient to maintain isolation close to its maximum value. In general, the hardware results also show good agreement with the simulation, although in the case of 5-ms limited rate adaptation, the isolation provided by the hardware EBD was not quite as good as the simulation predicted. This can be attributed to the practical limitations of the real-time adaptive balancing implementation.

3.5 Practical Considerations

One of the reasons that EBD has been a topic of research in recent years is due to the practical feasibility of integrated implementations with its required power handling, linearity, and tuning capabilities, as demonstrated in [3, 5, 6]. Furthermore, as shown here, the requirements for dynamic tuning in real environments are also

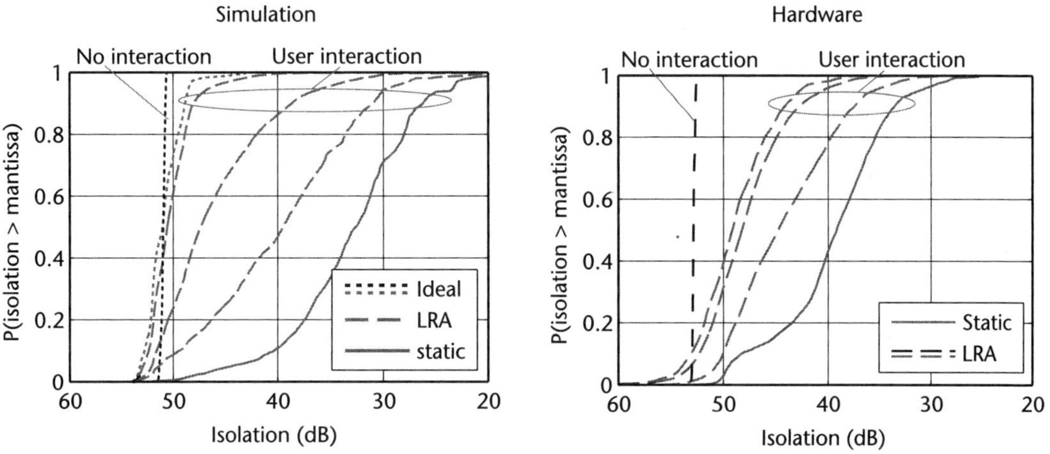

Figure 3.11 CDF curves for the simulated and measured Tx-Rx isolation of an adaptive EBD in a user interaction scenario, with and without user interaction, for various antenna impedance tracking characteristics [15].

practical: a 5-ms rebalancing interval, which would correspond to a tracking loop bandwidth of ~200 Hz, is achievable with basic electronics and digital signal processing (DSP).

Fully understanding how much cancellation can be achieved by a particular cancellation technique or IBFD transceiver design requires systems to be built and tested in the real world; purely theoretical analysis is of limited value when it comes to predicting practical performance. There are two distinct mechanisms that govern practical performance, however: the implementation and the environment.

The capability of a particular implementation of SIC is in part determined by the practical limitations of the specific hardware employed, and in this respect, there is no substitute for prototyping. As shown in this chapter, it is possible to decouple the impact of the environment from a particular implementation. Measurements that characterize the SI channel and, within this, the environment and its dynamics, can be usefully combined with very basic modeling (i.e., without considering hardware imperfections). As shown above, this enables analysis of the impact of the environment itself and can inform the specification and design of SIC hardware and DSP, which will be effective for those environments.

3.6 Conclusion

3.6.1 Summary

This chapter has provided an overview of the EBD, covering the basic principle of signal coupling in hybrid junctions, various methods for balancing EBDs and their limitations, and practically achievable performance in real-world dynamic environments. The EBD can be implemented on-chip and utilizes a shared antenna for Tx and Rx, making it suited to low-cost, small-form factor implementation. The EBD suffers from relatively high loss in the transmit and receive paths, although this can be mitigated through design enhancements. Dynamic adaptation of the balancing impedance is critical in handheld device applications, due to the effect of the user on the antenna impedance, but this can be effectively mitigated with real-time antenna impedance tracking. The breadth of work on EBDs that has been published thus far has demonstrated that this circuit, among others, is a practical and feasible implementation of SIC, which offers a particular trade-off of size, cost, bandwidth, isolation, and loss. Challenges remain, as discussed in the following section.

3.6.2 Future Directions

Achieving wideband isolation with practical balancing networks is one of the longstanding challenges in EBD design, which presents a trade-off of complexity and isolation bandwidth. While active balancing was shown to be effective, passive balancing is preferable for IBFD applications, as this provides cancellation of Tx noise and nonlinearities. There is therefore a scope for passive balancing network design improvements to increase isolation bandwidth while mitigating complexity.

Almost all radio systems require some form of RF filtering in the Tx and Rx path to limit out-of-band emissions and mitigate out-of-band interference. Thus, in a practical deployment, it may be necessary to add filters to an EBD-based front end (or, indeed, any other IBFD design). The losses inherent in the EBD leave little budget for adding insertion loss from additional RF components. Methods of reducing or mitigating insertion loss are an important future direction for EBD design. One way of mitigating loss would be to design the hybrid junction itself to be narrowband, and perhaps even tunable, to provide some level of RF filtering, eliminating separate filters to reduce overall loss.

In addition to the fundamental performance parameters mentioned above, there is potential to expand the EBD concept to a multiple-input multiple-output (MIMO) implementation. This would enable it to be used for MIMO communication, or potentially even be combined with MIMO-based SIC schemes, as later discussed in Chapter 14.

Although further work is required to mitigate the various drawbacks of the EBD, loss, bandwidth, and complexity to and implement additional wireless features, these problems do not currently appear insurmountable, and work goes on towards the goal of a fully integrated duplexer.

References

[1] Laughlin, L., et al., "Electrical Balance Duplexing for Small Form Factor Realization of In-Band Full Duplex," *IEEE Communications Magazine*, Vol. 53, No. 5, May 2015, pp. 102–110.

[2] Carlin, H. J., and A. B. Giordano, *Network Theory: An Introduction to Reciprocal and Non-Reciprocal Circuits*, Upper Saddle River, NJ: Prentice-Hall, 1964.

[3] Mikhemar, M., H. Darabi, and A. A. Abidi, "A Multiband RF Antenna Duplexer on CMOS: Design and Performance," *IEEE J. Solid-State Circuits*, Vol. 48, No. 9, 2013, pp. 2067–2077.

[4] van Liempd, B., et al., "An Electrical-Balance Duplexer for In-Band Full-Duplex with -85dBm In-Band Distortion at 10dBm TX-Power," *ESSCIRC Conference 2015 - 41st European Solid-State Circuits Conference (ESSCIRC)*, September 2015, pp. 176–179.

[5] van Liempd, B., et al., "A +70-dBm IIP3 Electrical-Balance Duplexer for Highly Integrated Tunable Front-Ends," *IEEE Transactions on Microwave Theory and Techniques*, 2016, pp. 1–13.

[6] Abdelhalem, S. H., P. S. Gudem, and L. E. Larson, "Hybrid Transformer-Based Tunable Differential Duplexer in a 90-nm CMOS Process," *IEEE Transactions on Microwave Theory and Techniques*, Vol. 61, No. 3, 2013, pp. 1316–1326.

[7] Laughlin, L., et al., "A Widely Tunable Full Duplex Transceiver Combining Electrical Balance Isolation and Active Analog Cancellation," *2015 IEEE 81st Vehicular Technology Conference (VTC Spring)*, May 2015, pp. 1–5.

[8] Manuzzato, E., et al., "Digitally-Controlled Electrical Balance Duplexer for Transmitter-Receiver Isolation in Full-Duplex Radio," *European Wireless Conference*, 2016.

[9] Laughlin, L., et al., "Passive and Active Electrical Balance Duplexers," *IEEE Transactions on Circuits Syst. II Express Briefs*, Vol. 63, No. 1, January 2016, pp. 94–98.

[10] van Liempd, B., et al., "A Dual-Notch +27dBm Tx-Power Electrical-Balance Duplexer," *ESSCIRC 2014 - 40th European Solid State Circuits Conference (ESSCIRC)*, September 2014, pp. 463–466.

[11] Castellano, G., et al., "A Low Power Control System for Real-Time Tuning of a Hybrid Transformer-Based Receiver," *2016 IEEE International Conference on Electronics, Circuits and Systems (ICECS)*, December 2016, pp. 328–331.

[12] Vermeulen, T., et al., "Real-Time RF Self-Interference Cancellation for In-Band Full Duplex," *2015 IEEE International Symposium on Dynamic Spectrum Access Networks (DySPAN)*, September 2015, pp. 275–276.

[13] Laughlin, L., et al., "Dynamic Performance of Electrical Balance Duplexing in a Vehicular Scenario," *IEEE Antennas Wireless Propag. Lett.*, Vol. 16, 2017, pp. 844–847.

[14] Laughlin, L., et al., "Electrical Balance Duplexer Field Trials in High-Speed Rail Scenarios," *IEEE Transactions on Antennas and Propagation*, Vol. 65, No. 11, 2017.

[15] Laughlin, L., et al., "The Effect of User Interaction on Electrical Balance Duplexer Isolation," Submitt. Press, 2020, pp. 1–4.

CHAPTER 4

Nonmagnetic Circulators

Negar Reiskarimian

4.1 Introduction

Various propagation domain techniques and antenna interfaces for IBFD wireless systems are discussed in Chapters 2 and 3. Here, another candidate for the antenna interface of full-duplex systems, namely circulators, is introduced, as highlighted within the IBFD transceiver diagram of Figure 4.1(a) and SIC techniques tree of Figure 4.1(b). N-port circulators are a class of nonreciprocal components in which the signal can only travel in one direction (from port 1 to 2, 2 to 3, ..., $N-1$ to N, N to 1). A three-port circulator can be used in the antenna interface of full-duplex systems to connect the Tx and the Rx to a single antenna while isolating the two from each other, as shown in Figure 4.2. This chapter focuses on high-performance nonmagnetic circulators that can be implemented within a compact footprint and describes various approaches and performance metrics that are relevant to IBFD applications.

The IBFD antenna interfaces introduced in Chapters 2 and 3, namely antenna pairs and electrical balance duplexers (EBDs), respectively, are reciprocal in nature. Lorentz reciprocity is a fundamental physical precept of electronic structures that are linear time-invariant (LTI), passive, and based on conventional materials with symmetric permittivity and permeability tensors. In a reciprocal component, the signal transmission profile between any two ports is identical in opposite directions.

Breaking reciprocity enables the realization of nonreciprocal components such as circulators. To achieve nonreciprocal wave propagation, one of the necessary conditions of the Lorentz reciprocity must be violated. These conditions include using materials with symmetric constituent tensors, linearity, or time invariance. Violation of each of these conditions results in a family of nonreciprocal components with unique features and capabilities.

Ferrite materials exhibiting nonreciprocity via the Faraday effect while in the presence of an external magnetic field have historically been used to produce nonreciprocal circulators [1]. These ferrite circulators are bulky, expensive, and incompatible with CMOS fabrication processes. Moreover, active-biased transistors and material/component nonlinearity have been exploited to create nonmagnetic

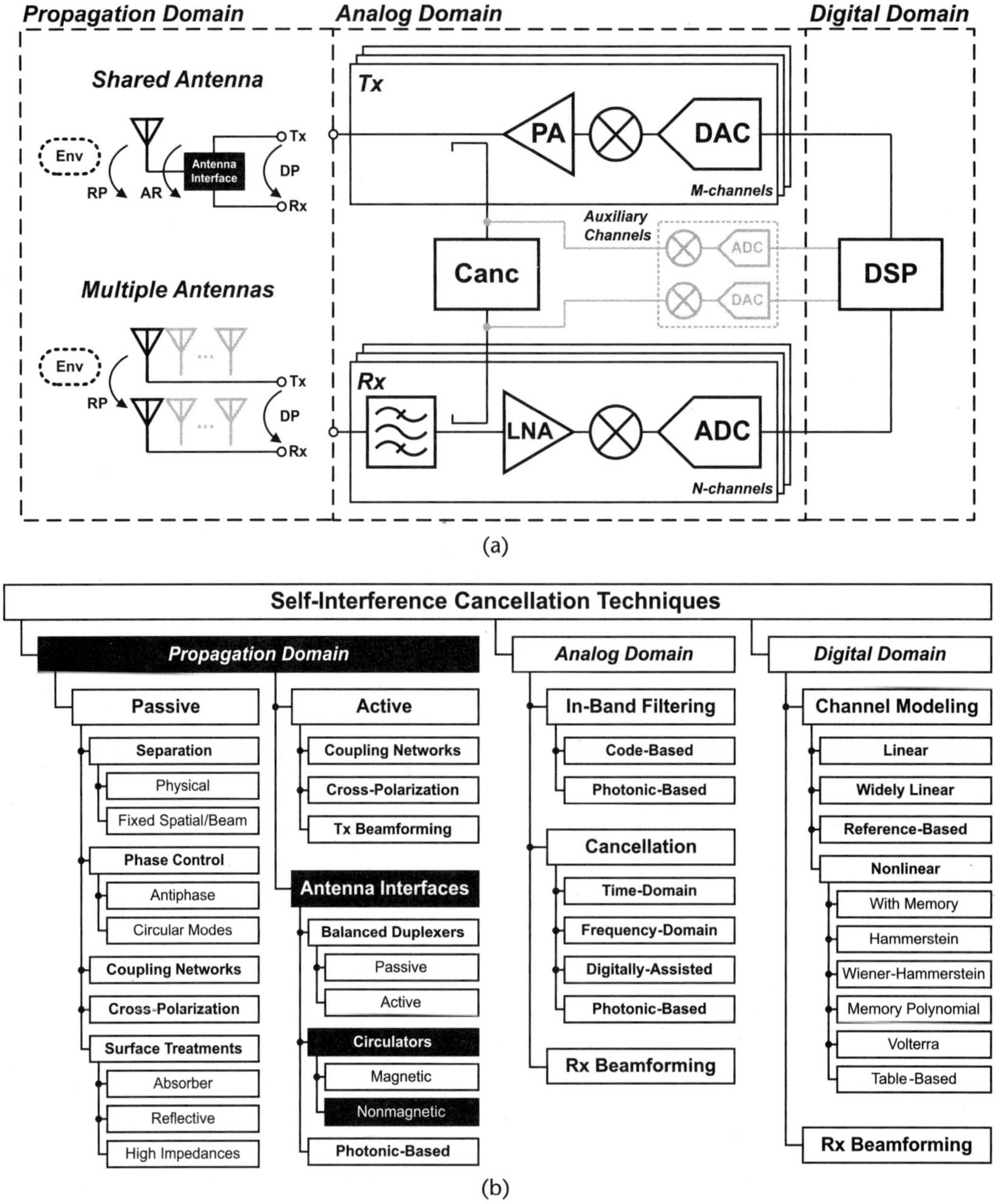

Figure 4.1 Nonmagnetic circulator and specific chapter focus highlighted within (a) a generic IBFD transceiver and (b) a SIC techniques tree diagram.

circulators. Active approaches are limited by the noise and nonlinearity introduced by the active devices, while nonlinear approaches are signal-power-dependent, aside from their inherent nonlinear behavior. As a result, these techniques do not find utility at the front end of transceivers for IBFD applications, where Tx power handling and Rx noise performance are paramount. For additional information about these techniques, please refer to [1, 2] and the references therein.

The main focus of this chapter is nonmagnetic circulators based on spatio-temporal modulation of material properties such as permittivity (using modulated

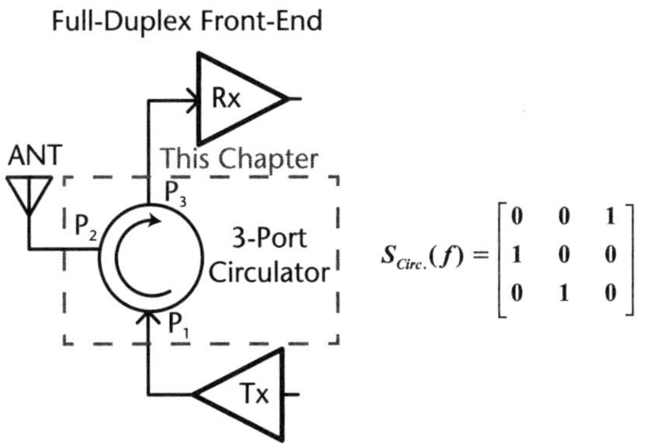

Figure 4.2 A three-port nonreciprocal circulator used in the antenna interface of a full-duplex system. Signals can only propagate from Port 1 (Tx) to Port 2 (ANT), and from Port 2 (ANT) to Port 3 (Rx). Tx and Rx are isolated from each other. The S-parameters of an ideal circulator are shown on the right.

capacitors) and conductivity (using modulated resistors). Nonreciprocal components based on the spatio-temporal modulation approach are compatible with CMOS integration, are theoretically loss/noise-free, and can be linear to the desired signal, making them exciting candidates for the antenna interface of IBFD systems.

Many of the early IBFD demonstrations relied on bulky/off-the-shelf antenna interfaces (such as antenna pairs and magnetic circulators), which are not suitable for form-factor-constrained mobile applications, especially in sub-6-GHz frequency bands. Compact single-antenna IBFD antenna interfaces such as EBDs and nonmagnetic circulators are also more compatible with MIMO and diversity applications. Furthermore, single-antenna interfaces ensure wireless channel reciprocity, which is useful at the higher layers[1] [3].

The Tx-to-Rx isolation in any three-port IBFD single-antenna interface is limited to the quality of matching at the antenna port. In a practical IBFD system, antenna matching depends heavily on environmental reflections, as mentioned in Chapter 1. As a result, antenna balancing techniques are necessary to maintain Tx-to-Rx isolation across ANT impedance variations.

Furthermore, basic network theory states that a passive-matched three-port reciprocal device cannot be lossless, and hence EBDs feature a theoretical 3-dB loss[2] (from Tx to antenna and antenna to Rx) associated with their operation, as discussed in Chapter 3. In EBDs, the losses can be made asymmetric to favor either Tx-ANT or Rx noise performance [4]. However, three-port circulators can be concurrently matched and lossless, and many commercial magnetic circulators report sub-1-dB losses.

In the following sections, nonmagnetic circulators based on the spatio-temporal modulation of permittivity and conductivity are discussed in more detail.

1. Wireless channel reciprocity means that the wireless channel between two nodes does not change, regardless of which node is transmitting or receiving. It should not be confused with the nonreciprocity that is required to implement circulators.
2. Typically higher when parasitic losses are factored in.

4.2 Nonmagnetic Nonreciprocal Circulators Based on Permittivity Modulation

As mentioned previously, violating time invariance can lead to nonreciprocity, the earliest examples of which include nonreciprocal parametric amplifiers [5], an example of which is shown in Figure 4.3. Nonreciprocal parametric amplifiers were based on two central concepts: (1) the availability of variable capacitors (varactors) that can be temporally modulated using a pump signal; and (2) the introduction of a dynamic pump in time and space (note the two varactors in Figure 4.3). Consider a nonlinear voltage-controlled varactor whose charge-voltage relationship is expressed by the following Taylor series:

$$q(t) = C_0 v(t) + a_2 v^2(t) + a_3 v^3(t) + \dots, \tag{4.1}$$

where C_0 is the static capacitance and the a_i's are the higher-order nonlinear coefficients. Ignoring all of the higher-order terms except the first-order and second-order ones, a modulation voltage of $v_m\cos(\omega_m t + \phi_m)$ across the capacitor results in a voltage-dependent variable capacitance of $C(v, t) = C_0 + a_2 v_m \cos(\omega_m t + \phi_m)$. For a single-tone input voltage at frequency ω_{in}, it can be shown that the current that passes through the capacitor contains tones at ω_{in} as well as $\omega_m \pm \omega_{in}$.

While the basic operating principle of a parametric amplifier has been studied with such a nonlinear device, the nonlinearity is not central to achieving a parametric response. Assuming that we can build a voltage-independent varactor where the modulation is caused by another force, for example, a mechanical force changing the distance between the plates of a parallel-plate capacitor as in a MEMS varactor, similar frequency translation properties are attainable.

Many recent approaches have focused on permittivity as the modulated parameter [6–10]. In the RF domain, permittivity modulation occurs using varactors that exhibit a limited modulation index (about 2 to 4, e.g., $\frac{C_{max}}{C_{min}} \sim 1.5$ in the implementation of [6]). In general, permittivity modulation is quite weak and associated with loss, particularly as the frequency increases.

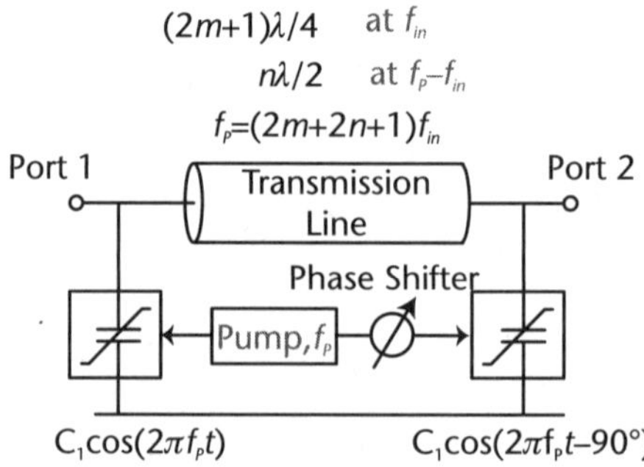

Figure 4.3 The nonreciprocal parametric amplifier proposed in [5]. Varactors are modulated through a sinusoidal pump signal with a frequency f_p. The transmission line is designed to provide the desired electrical lengths as described in the figure at f_{in} and $f_p - f_{in}$.

4.2.1 Circulators Based on Parametric Modulation

A more recent example of a parametric design was demonstrated in [6], where the parametric modulation is performed along a quasi-distributed transmission line. The transmission line is loaded with varactors modulated by a traveling single-tone carrier such that $C(z, t) = C_0(1 + \xi_\varepsilon \cos(\omega_m t - \beta_m Z + \phi_m))$, as shown in Figure 4.4, in which C_0 is the average capacitance, $\xi_\varepsilon = \frac{C_m}{C_0}$ is the modulation index, C_m is the amplitude of the modulated capacitance, and β_m and ϕ_m are the wave vector and phase constant of the modulating carrier, respectively.

A simplified analysis of such a structure was carried out in [6], which shows that input signals launched in the same direction as the modulation carrier gradually convert to the frequency-shifted side bands as z increases. At $z = \frac{\lambda_{in}}{\xi_\varepsilon \sqrt{2}}$, where λ_{in} is the wavelength of the input signal, all of the signal power is converted to the mixing terms. The smaller the modulation index ξ_ε, the longer the length required for a full harmonic conversion, increasing the dimensions of the device. If the modulation frequency is higher than the signal frequency, the signal can also have a gain proportional to the frequency ratio of the upconverted and input signals.

For signals propagating in the opposite direction, it has been shown that no harmonic conversion takes place. A trade-off exists between the choice of f_m, the power consumption in the modulation path, and the gain of the frequency-translated signal. Additionally, the noise performance of such approaches depends on the loaded quality factor of the transmission line, the quality factor of the individual varactors, and the permittivity modulation index. Although not discussed in [6], phase noise in the modulation signal can also impact noise performance. Finally, the effect of noise folding from the sidebands should be considered and can be minimized by a lowpass filter at the input.

As mentioned previously, at RF, varactors are able to achieve low modulation ratios, resulting in structures that are of the order of a wavelength. Another disadvantage of approaches based on permittivity modulation in parametric structures is that the mode/frequency conversion necessitates the use of additional filters or diplexers.

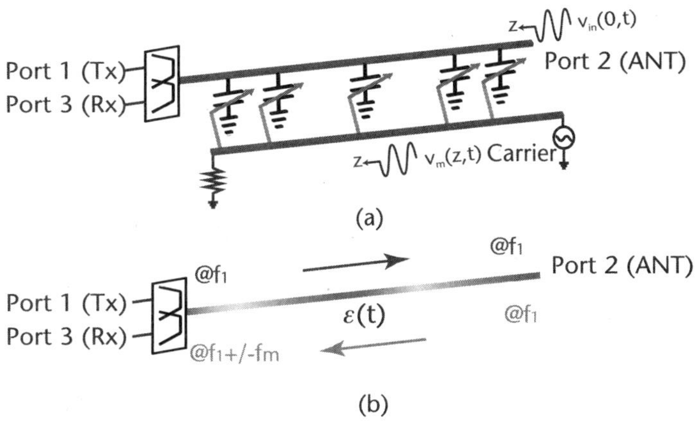

Figure 4.4 (a) The concept and (b) schematic of a parametric, time-varying transmission line and its frequency conversions as demonstrated in [6]. It should be noted that the Tx and Rx should be separated by using additional filters or diplexers.

The circulator in [6] is implemented on a printed circuit board (PCB), is driven with f_m = 4.23 GHz, and has an area of 128 × 10 mm ($\frac{\lambda}{2.3} \times \frac{\lambda}{30}$ for a center frequency of 1 GHz). It achieves a maximum of about 2-dB loss in the Tx-ANT path and a receiving gain (ANT to Rx) of 1 dB to –4 dB across 0.45 to 1.8 GHz.

4.2.2 Circulators Based on Angular Momentum Biasing

Inspired by how magnetic circulators operate and the Faraday rotation concept [1], a nonreciprocal response can be achieved using angular-momentum biasing of a resonant ring [7–10]. An effective electronic spin is enabled by the spatiotemporal permittivity modulation with a traveling wave along the ring, as shown in Figure 4.5. Additionally, resonators can be used to miniaturize the size of the ring significantly while boosting the weak permittivity modulation effect because of the limited modulation ratio, resulting in a stronger nonreciprocal response at the subwavelength scale. As shown in Figure 4.5, the resonance frequency of each resonator is modulated in a rotational fashion by a phase-shifted version of the modulation signal. The resonators (bandpass or bandstop) can be connected together in various forms, such as delta or wye topologies, and can be either configured single-ended or differentially [7–9]. The resonators must exhibit high loaded Q to achieve significant impedance change, and this limits the operation's bandwidth. Additionally, the inductors and the varactors must have even higher unloaded Q to achieve low loss, which is fundamentally challenging, particularly on integrated semiconductor substrates.

It is shown in [9] that the bandwidth of any pseudo-LTI, magnetic-free circulator is smaller than that of twice the modulating bandwidth (f_m). This implies that, for larger bandwidths, a higher modulation frequency is required. By increasing f_m, however, the dynamic power consumption is increased. Additionally, varactors, and permittivity modulation in general, also exhibit a trade-off between modulation index and loss, especially as the operating frequency grows. Similar to parametric approaches, the noise performance in angular-momentum-based devices depends on the resonators' loaded Q and the unloaded Q of the varactors. Phase noise in the modulation signal can also impact the noise figure by creating random variation in the harmonic S-parameters. Proper design of the modulation circuitry

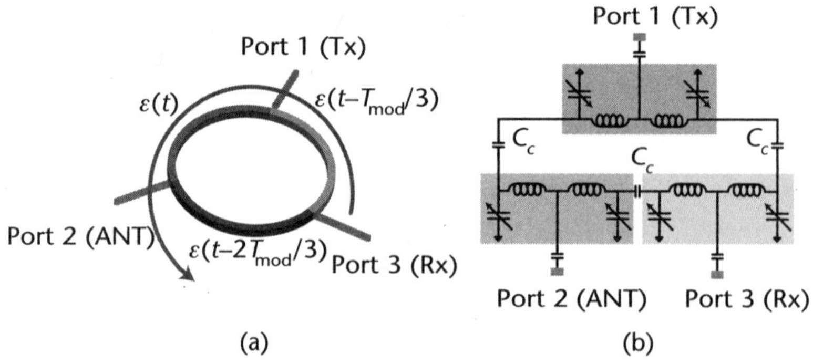

Figure 4.5 (a) Spatio-temporal permittivity modulation in a ring resonator; and (b) the implementation of a ring resonator by three separate inductor-capacitor (LC) tanks, as shown in [7].

can lower such undesired effects, and a noise figure close to the insertion loss can be achieved [9].

Another key challenge with spatio-temporal parametric modulation approaches in general is that the property that enables modulation (varactors) often represents a nonlinearity to the signal itself, particularly as the modulation ratio becomes higher, resulting in nonlinear distortion at higher signal levels. To further improve the linearity, the DC bias of the varactors should be increased or advanced varactor configurations should be used.

Various implementations of the angular momentum biasing technique have been reported in the literature using inductor-capacitor (LC) resonators [7–10]. These circulators have been implemented on PCBs [7–9] and CMOS platforms [10]. The state-of-the-art angular-momentum-based circulator achieves <2-dB insertion loss, 20-dB isolation bandwidth of 2.3%, 2.5-dB noise figure, and +31-dBm input third-order intercept point (IIP3) operating at 1 GHz with a modulation frequency of 100 MHz and a size of 143 mm^2 $\left(\frac{\lambda}{23}\times\frac{\lambda}{27}\right)$.

Furthermore, the angular momentum biasing technique has been explored by using other types of resonators such as MEMS thin-film bulk acoustic resonators (FBAR) [11]. Because MEMS resonators have much higher Qs and better integrability, they can extend the range of operation frequencies and reduce the modulation frequency and therefore circulator's power consumption, as well as additional potential benefits in terms of insertion loss, isolation and form factor.

In [11], an FBAR circulator based on the concept of angular momentum biasing was demonstrated. FBARs work at higher frequencies compared to other acoustic resonators such as surface acoustic wave (SAW) resonators. The 2.5-GHz prototype is built using three FBARs along with parallel inductors to modulate each resonator at a much lower $\frac{f_{in}}{f_m}$ ratio compared to their LC counterparts. The implemented circulator achieves 4-dB loss, 15-dB isolation, 35-MHz bandwidth, and +40-dBm IIP3 for a $\frac{f_{in}}{f_m}$ ratio of 62.5 and a form factor of 225 mm^2 $\left(\frac{\lambda}{8}\times\frac{\lambda}{8}\right)$.

4.3 Nonmagnetic Nonreciprocal Circulators Based on Conductivity Modulation

Conductivity modulation has been introduced as an alternative approach to achieve nonreciprocity in time-varying systems [12]. In contrast to the low-permittivity modulation index of varactors, a passive transistor switch in a modern CMOS process can provide orders of magnitude higher conductivity modulation index (in the order of 10^3–10^5). In this section, two conductivity modulation-based approaches will be discussed in greater detail.

4.3.1 Switched-Capacitor-Based Structures

In the solid-states circuits community, temporally modulated systems are commonly known as linear periodically time-varying (LPTV) circuits. An example of an LPTV circuit is introduced in [13], called an N-path filter. N-path filters have reemerged as an innovative means of enabling the incorporation of tunable, high-quality-factor filters at RF in nanoscale CMOS integrated circuit technology. They are a class of LPTV networks where the signal is periodically commutated with passive transistor switches through a bank of capacitors, as illustrated in Figure 4.6(a). As a result of the commutation action with a modulation frequency of f_{LO}, a second-order bandpass filter profile is created through the translation of the lowpass filtering of the capacitor to a center frequency defined by the clock frequency.

Applying a relative phase shift to the nonoverlapping clocks responsible for driving the input and output switch sets of a two-port N-path filter has been shown to create a nonreciprocal phase shift to the signals traveling forward and reverse because they see a different ordering of the phase-shifted switches (lead-lag instead of lag-lead). The magnitude response remains unchanged and reciprocal, and there is also some low-loss tunable filtering, similar to traditional N-path filters, indicated in Figure 4.6(a).

One approach to create a three-port circulator is to place the N-path-filter with ±90° phase shift within a transmission-line loop with a length of $3\lambda/4$, as shown in Figure 4.6(b). Combining both the nonreciprocal phase shift of the N-path filter and the reciprocal phase shift of the transmission line creates a unidirectional wave. Ports can be placed anywhere along the loop, which creates a three-post circulator as long as they maintain a $\lambda/4$ circumferential distance between them. Maximum linearity with respect to the Tx port is achieved if the circulator is designed asymmetrically and the Rx port is placed adjacent to the N-path filter, because the inherent Tx-Rx isolation suppresses the voltage swing on both sides of the N-path filter, thus enhancing its linearity to Tx signals. Another approach to build a circulator by using an N-path-filter with nonreciprocal phase response in conjunction with a hybrid coupler is reported in [14].

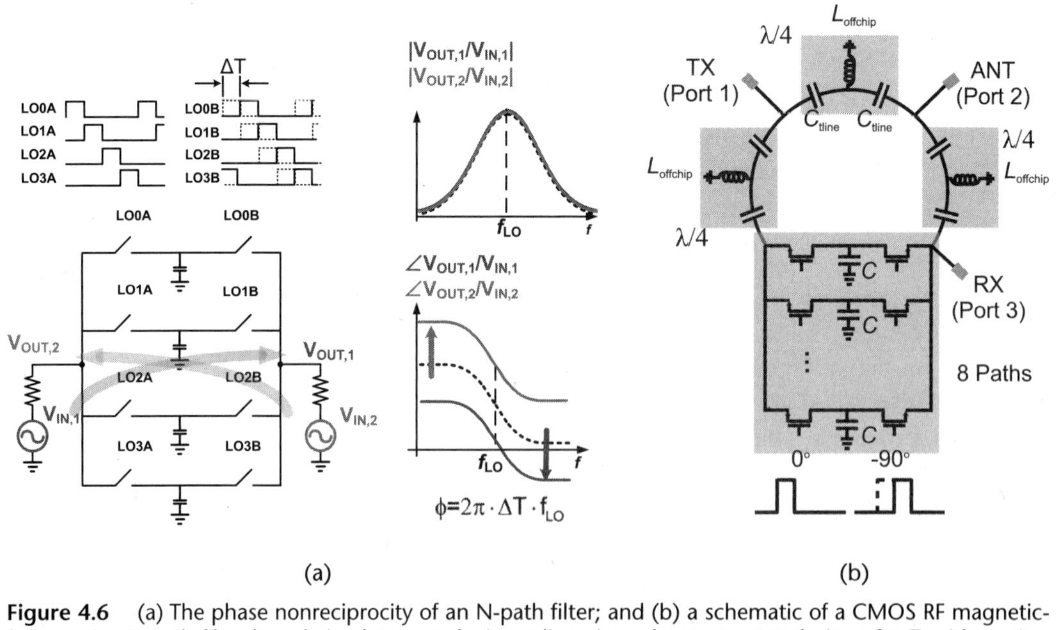

Figure 4.6 (a) The phase nonreciprocity of an N-path filter; and (b) a schematic of a CMOS RF magnetic-free passive N-path filter-based circulator employing a linearity-enhancement technique for Tx-side excitations [3, 12].

The first integrated, magnetic-free, conductivity-based passive circulator was demonstrated in a 65-nm CMOS technology in [3, 12]. It has been designed for an operation frequency of 750 MHz $\left(\frac{f_{in}}{f_{LO}} = 1\right)$ and achieves sub-2-dB loss between its Tx-ANT and ANT-Rx ports. It also takes advantage of the linearity enhancement technique for Tx-side excitations as described previously, improving the Tx-ANT IIP3 by roughly two orders of magnitude compared to ANT-Rx IIP3 based on the measured performance.

Various implementations of IBFD systems with integrated N-path filter-based nonreciprocal antenna interfaces have been shown [15–18]. In [16], the N-path filter in the circulator is repurposed as a downconverting mixer, thereby directly providing the baseband received signals on the N-path filter capacitors and lowering both power consumption and noise figure since the antenna interface and mixing functionalities occur within the same block. It additionally allows the antenna interface to be codesigned with the Rx and to embed Tx-Rx isolation-enhancing techniques to combat antenna impedance variations.

Furthermore, as discussed in Chapter 14, IBFD phased-array front ends based on N-path-filter-based circulator receivers developed in [16] enable baseband beamforming with minimal overhead, by virtue of the multiphase outputs of N-path-filter-based circulators [17]. An 8-element IBFD phased-array based on the nonmagnetic circulators has been showcased in [17] that achieves 100-dB overall array SIC, including digital SIC supporting +16.5-dBm Tx array power handling. Finally, a scalable 2-element MIMO IBFD receiver array with integrated N-path-filter-based circulators was implemented in [18] incorporating RF and baseband SIC. The prototyped MIMO system exhibits up to 45-dB average SIC across the 20-MHz bandwidth and more than 53-dB average crosstalk-interference cancellation (XIC) across the 20-MHz bandwidth through passive RF cancellation embedded in the nonmagnetic circulator and additional baseband cancellation.

4.3.2 Switched-Delay Structures

Inspired by the N-path filter-based RF circulator discussed previously, a generalized conductivity modulation concept is proposed using switched transmission lines [19]. This structure consists of two sets of differential mixer-quad switches on either side of a differential transmission-line delay (replacing the commutated capacitors), as shown in Figure 4.7(a). The switches are clocked at a modulation frequency f_m, and the delay of the line is equal to $T_m/4$ ($\frac{1}{4}f_m$). The switches are clocked using 50% duty-cycle square-wave clocks with a relative phase shift of $T_m/4$. Waves that travel from left to right therefore experience the transmission-line delay with no sign flips in either half of the clock period. Conversely, however, waves traveling from right to left experience the transmission-line delay alongside one sign flip.

In this structure, transmission in both directions is perfectly lossless, and there is an infinitely broadband 180° nonreciprocal phase difference. The critical element of this architecture is that the infinite bandwidth of the device implies that the signal frequency and modulation frequency are completely decoupled. An arbitrarily low-modulation frequency can be used, with the only restriction being an associated increase in the transmission line length and, hence, loss. This advantage has

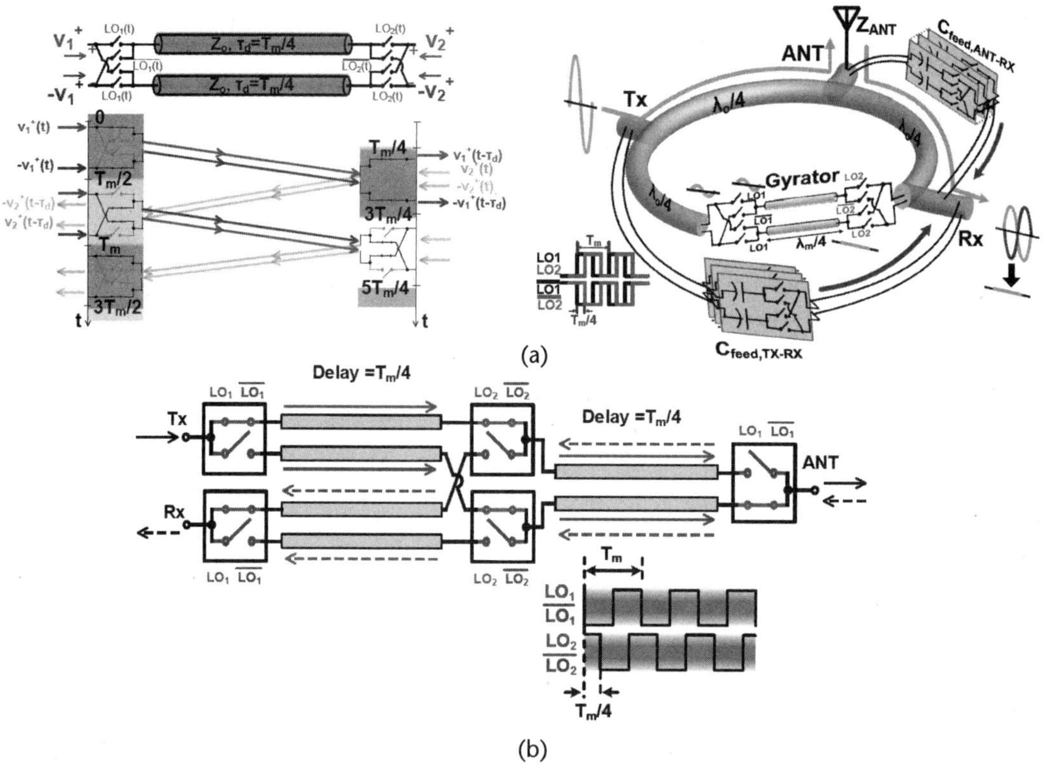

Figure 4.7 (a) Switched-transmission-line concept reported in [19] and a three-port RF circulator based on switch-delay structure [20]; and (b) wideband circulator based on sequentially switched delay lines with a modulation period of T_m [21].

been leveraged to implement circulators at millimeter wave while the modulation frequency is still at RF [19] and to implement RF circulators with high-power handling through the use of high-voltage switches [20, 22].

Another important feature of this infinitely broadband device is that, while it is internally time-varying, it appears externally LTI, with no mixing products between the signal and the modulation seen at the ports. As a result, there is no folding of noise from other frequencies to the signal frequency of interest and, hence, low noise figure.

In [20], a highly linear RF circulator based on the switched-transmission line structure was reported, demonstrating superior power handling and linearity at a center frequency of 1 GHz. The switches are modulated at 333 MHz ($\frac{f_{in}}{f_m} = 3$), which enables the use of thick-oxide devices in 180-nm silicon-on-insulator (SOI) CMOS technology to boost power handling close to +1W. The Tx-ANT IIP3 exceeds +50 dBm, enabling this technique to approach the performance required for commercial Wi-Fi and cellular applications.

Additionally, an ANT-balancing technique can be employed by using feed capacitors between the Tx-Rx and ANT-Rx ports to overcome the effect of limited isolation due to mismatches at the ANT port. Because there is a –90° inherent phase difference between the signals at the Tx and ANT ports and thanks to a differential implementation, orthogonal currents can be generated at the Rx port;

hence, a complete coverage of a voltage standing-wave ratio (VSWR) of up to 2.33 is achieved in [22] without any inductive or resistive elements.

Similar ideas related to switched transmission lines used in realizing broadband nonreciprocity have been explored in [21, 23], as shown in Figure 4.7(b). In [23], a wideband circulator based on sequentially switched delay lines is introduced and implemented on a 0.2-μm gallium nitride (GaN) high-electron-mobility transistor (HEMT) technology that operates from 0.1 to 1 GHz with a modulation frequency of 1.25 GHz. The reported insertion loss from Tx to ANT and ANT to Rx is from 3 dB at 100 MHz to 4.1 dB at 1 GHz. The output power compression point is +14 dBm and the power consumption is 4.1W.

To further increase the modulation frequency ratio ($\frac{f_{in}}{f_m}$) and lower the modulation path power consumption at the expense of operating bandwidth, recent work has focused on using electroacoustic components such as SAW devices, which provide larger group delays compared to on-chip and surface-mount device passives.

In [24], a 1.95-GHz circulator is designed and implemented based on the conductivity modulation technique, with a SAW filter used as the delay medium between two sets of quad switches. The modulation frequency is chosen to be 50 MHz, and the measurements have shown close to a 3-dB loss and more than 40-dB isolation between the Tx and Rx ports. In [25], a similar approach has been taken to create a four-port circulator that operates around 900 MHz with a modulation frequency of 12.6 MHz. Measurement results reveal 2.9-dB insertion loss, 2.6% fractional bandwidth, and +41-dBm IIP3.

The noise generated in conductivity-modulation-based approaches comes from three main sources: (1) the switch resistance of the transistors and limited Q of the transmission lines; (2) the noise folding from the harmonics; and (3) the effect of the phase noise of the modulating clock [9]. It has been shown that the noise figure of conductivity-modulation-based approaches can be as low as their insertion loss and, hence, mostly limited to the effect of switch resistance and the limited Q of the passives. Additionally, the performance of the nonreciprocal devices should be maintained, even under the application of a large-signal Tx or interferers.

4.4 Practical Considerations

An ideal antenna interface for IBFD should be compact, have no loss and noise, and provide extremely high-power handling without any additional power consumption. While no antenna interface can achieve all the metrics at once, the nonmagnetic circulators discussed in this chapter have the potential to become the mainstream antenna interface, especially for mobile IBFD devices. The state-of-the-art nonmagnetic circulators reported in this chapter can achieve sub-2-dB losses, multiwatt power handling, and >+50-dBm IIP3 with <100-mW power consumption. Additionally, the introduced techniques are compatible with CMOS integration and are compact compared to other alternatives such as antenna pairs and magnetic circulators.

To compare the various antenna interfaces introduced, a figure of merit (FOM) called antenna interface efficiency (η_{ANT}) is defined in [20] as:

$$\eta_{ANT} = \frac{P_{out,PA} \times (S_{TX-ANT} / NF)}{P_{dc,PA} + P_{dc,interface}} \times \frac{1}{\eta_{PA}} \times 100\%, \tag{4.2}$$

where η_{PA} is the efficiency of the PA, $P_{out,PA}$ is the PA output power and $P_{dc,PA}$ is the DC power consumption of the PA. Moreover, S_{TX-ANT} and NF are the Tx-ANT loss and ANT-Rx noise figure of the antenna interface and $P_{dc,interface}$ is its DC power consumption. For an ideal reciprocal interface (with a theoretical 3-dB loss and $P_{dc,interface}$ = 0 mW) such as an EBD, η_{ANT} reaches a maximum 25%. For real reciprocal interfaces, η_{ANT} can be significantly lower than 25%. The state-of-the-art nonmagnetic circulator reported in [22] has achieved a superior η_{ANT} of 27.2% when compared with EBDs, due to the low loss and high power-handling.

4.5 Conclusion

4.5.1 Summary

IBFD radios employing a pair of antennas, one for transmit and one for receive, experience a direct trade-off between form factor and transmit-receive isolation arising from the antenna spacing and design. Single-antenna IBFD systems are highly desired in form-factor-constrained IBFD applications and can be implemented by using antenna interfaces such as EBDs or nonreciprocal circulators. Among the various approaches that enable nonreciprocity, the spatiotemporal modulation of material properties (e.g., permittivity and conductivity) has gained the most attention in the recent past because of its potential for noise-free, highly linear operation. These magnetic-free approaches are also compatible with integrated circuit technology, have much smaller form factors, and enable high performance and reconfigurable implementations.

4.5.2 Future Directions

Research on nonmagnetic circulators have seen a surge over the last few years. The performance of the CMOS-based nonmagnetic circulators has improved significantly since its inception: the traditional gap between magnetic circulators and nonmagnetic circulators' performance, especially in terms of linearity and power handling, is closing, paving the way for a new era of single-antenna IBFD systems.

Additionally, many of the nonmagnetic circulators described in this chapter, specifically those based on conductivity modulation, have been implemented in IBFD systems showcasing the strength of the techniques discussed. There is currently ongoing research within various groups to further study and characterize a broader range of nonmagnetic circulators in IBFD systems. This includes extensive research on isolation enhancement and antenna balancing techniques to cover large antenna voltage standing-wave ratios, codesign and cooptimization of the nonmagnetic circulator, and the receiver front end at the same time and adaptive cancellation techniques for isolation tuning across frequency.

Additional research directions include heterogeneous integration of gallium arsenide (GaAs)/GaN technologies with CMOS to simultaneously take advantage of the highly linear GaAs/GaN transistors in addition to the unique capabilities of CMOS to generate the complex modulation schemes.

In the previous sections, various circulators based on electromechanical resonators and electro-acoustic delays were mentioned. While these implementations can provide much higher modulation frequency ratios and lower power consumption, their performance in terms of insertion loss and linearity is still far from the performances of the CMOS integrated, purely electronic-based approaches discussed in this chapter. Future developments to enhance the performance of nonreciprocal electromechanical/electro-acoustic circulators are required before they can be considered as a serious contender in IBFD applications.

References

[1] Reiskarimian, N., et al., "Non-Reciprocal Electronic Devices—A Hypothesis Turned into Reality," *IEEE Microwave Magazine*, Vol. 20, No. 4, April 2019, pp. 94–111.

[2] Nagulu, A., et al., "Non-Reciprocal Electronics Based on Temporal Modulation," *Nature Electronics*, Vol. 3, May 2020, pp. 241–250.

[3] Reiskarimian, N., et al., "A CMOS Passive LPTV Nonmagnetic Circulator and Its Application in a Full-Duplex Receiver," *IEEE Journal of Solid-State Circuits*, Vol. 52, No. 5, May 2017, pp. 1358–1372.

[4] van Liempd, B., et al., "Adaptive RF Front-Ends Using Electrical-Balance Duplexers and Tuned SAW Resonators," *IEEE Transactions on Microwave Theory and Techniques*, Vol. 65, No. 11, November 2017, pp. 4621–4628.

[5] Hamasaki, J., "A Theory of a Unilateral Parametric Amplifier Using Two Diodes," *Bell System Technical Journal*, Vol. 43, No. 3, 1964, pp. 1123–1147.

[6] Qin, S., et al., "Nonreciprocal Components with Distributedly Modulated Capacitors," *IEEE Transactions on Microwave Theory and Techniques*, Vol. 62, No. 10, October 2014, pp. 2260–2272.

[7] Kord, A., et al., "Magnet-Less Circulators Based on Spatiotemporal Modulation of Bandstop Filters in a Delta Topology," *IEEE Transactions on Microwave Theory and Techniques*, Vol. 66, No. 2, February 2018, pp. 911–926.

[8] Kord, A., et al., "Pseudo-Linear Time-Invariant Magnetless Circulators Based on Differential Spatiotemporal Modulation of Resonant Junctions," *IEEE Transactions on Microwave Theory and Techniques*, Vol. 66, No. 6, June 2018, pp. 2731–2745.

[9] Kord, A., et al, "Broadband Cyclic-Symmetric Magnetless Circulators and Theoretical Bounds on Their Bandwidth," *IEEE Transactions on Microwave Theory and Techniques*, Vol. 66, No. 12, December 2018, pp. 5472–5481.

[10] Kord, A., et al, "CMOS Integrated Magnetless Circulators Based on Spatiotemporal Modulation Angular-Momentum Biasing," *IEEE Transactions on Microwave Theory and Techniques*, Vol. 67, No. 7, July 2019, pp. 2649–2662.

[11] Yu, Y., et al., "Highly-Linear Magnet-Free Microelectromechanical Circulators," *IEEE Journal of Microelectromechanical Systems*, Vol. 28, No. 6, December 2019, pp. 933–940.

[12] Reiskarimian, N., and H. Krishnaswamy, "Magnetic-Free Non-Reciprocity Based on Staggered Commutation," *Nature Communications*, Vol. 7, No. 4, April 2016.

[13] Ghaffari, A., et al., "Tunable High-Q N-Path Band-Pass Filters: Modeling and Verification," *IEEE Journal of Solid-State Circuits*, Vol. 46, No. 5, May 2011, pp. 998–1010.

[14] Jain, S., et al., "A 0.55-to-0.9GHz 2.7dB NF Full-Duplex Hybrid-Coupler Circulator with 56MHz 40dB TX SI Suppression," *IEEE International Solid-State Circuits Conference*, February 2018, pp. 400–402.

[15] Reiskarimian, N., et al., "One-Way Ramp to a Two-Way Highway: Integrated Magnetic-Free Nonreciprocal Antenna Interfaces for Full-Duplex Wireless," *IEEE Microwave Magazine*, Vol. 20, No. 2, February 2019, pp. 56–75.

[16] Reiskarimian, N., et al., "Analysis and Design of Commutation-Based Circulator-Receivers for Integrated Full-Duplex Wireless," *IEEE Journal of Solid-State Circuits*, Vol. 53, No. 8, August 2018, pp. 2190–2201.

[17] Baraani Dastjerdi, M., et al., "Full Duplex Circulator-Receiver Phased-Array Employing Self-Interference Cancellation Via Beamforming," *IEEE Radio Frequency Integrated Circuits Symposium*, June 2018, pp. 108–111.

[18] Baraani Dastjerdi, M., et al., "Analysis and Design of Full-Duplex 2x2 MIMO Circulator-Receiver with High TX Power Handling Exploiting MIMO RF and Shared-Delay Baseband Self-Interference Cancellation," *IEEE Journal of Solid-State Circuits*, Vol. 54, No. 12, December 2019, pp. 3525–3540.

[19] Dinc, T., et al., "Synchronized Conductivity Modulation to Realize Broadband Lossless Magnetic-Free Nonreciprocity," *Nature Communications*, Vol. 8, No. 10, October 2017.

[20] Nagulu, A., and H. Krishnaswamy, "Non-Magnetic CMOS Switched-Transmission-Line Circulators with High Power Handling and Antenna Balancing: Theory and Implementation," *IEEE Journal of Solid-State Circuits*, Vol. 54, No. 5, May 2019, pp. 1288–1303.

[21] Biedka, M., et al., "Ultra-Wide Band Non-Reciprocity Through Sequentially-Switched Delay Lines," *Scientific Reports*, Vol. 7, 2017.

[22] Nagulu, A., et al., "Non-Magnetic 0.18μm SOI Circulator with Multi-Watt Power Handling Based on Switched-Capacitor Clock Boosting," *IEEE International Solid-State Circuits Conference*, February 2020, pp. 444–445.

[23] Biedka, M., et al., "100MHz to 1GHz On-Chip Circulator with Integrated Driver Amplifiers," *2019 IEEE MTT-S International Microwave Symposium*, June 2019, pp. 1488–1491.

[24] Bahamonde, J., et al., "1.95-GHz Circulator Based on a Time-Modulated Electro-Acoustic Gyrator," *Proc. IEEE Int. Symposium on Antennas and Propagation and USNC-URSI Radio Science Meeting*, 2018.

[25] Yu, Y., et al., "Radio Frequency Magnet-Free Circulators Based on Spatiotemporal Modulation of Surface Acoustic Wave Filters," *IEEE Transactions on Microwave Theory and Techniques*, Vol. 67, No. 12, December 2019, pp. 4773–4782.

CHAPTER 5

Photonic-Based Antenna Interfaces

Edward I. Ackerman and Charles H. Cox III

5.1 Introduction

While we have designed a number of complete STAR systems using either electronic or photonic components, the attention here is on several photonic architectures for implementing analog front-end cancellation via unique antenna interfaces. It will be shown that these photonic methods provide significantly more bandwidth capability compared to the electronic-based antenna interfaces discussed in Chapters 3 and 4. The focus of this chapter is highlighted within the IBFD transceiver diagram of Figure 5.1(a) and SIC techniques tree of Figure 5.1(b). For systems with such shared antennas for receiving and transmitting, a magnetic circulator is often suggested as a way to provide transmit cancellation right at the antenna interface. However, this option is typically ruled out due to its narrow bandwidth and relatively inadequate degree of cancellation.

When designing a system to enable STAR or IBFD, a reasonable question to ask is: why not just design the linear range of the receiver front end to handle both the high-power transmit signal and the lower-power receive signal? The answer is that doing so would require a dynamic range that is well beyond the present state of the art, and for a fairly fundamental reason: the transmit signal power typically exceeds the linear dynamic range of the first electronic component that it encounters. Consider the differential amplifier, a simplified schematic of which is shown in Figure 5.2. The output of such an amplifier is proportional to the difference between its two inputs. The subtraction occurs at the output, which is after the signals to be subtracted have encountered the nonlinear inputs of the transistors.

Therefore, to be effective, we need the analog front-end canceller to be right at the front of the front end (i.e., right at the output of the antenna), so that the initial cancellation occurs before the combined receive and transmit signal encounters any nonlinear stage. Over the years, we have developed three technologically distinct methods of realizing a front-end canceller, each of which can be implemented either electronically or photonically. In this chapter, we discuss the photonic implementations of two such methods.

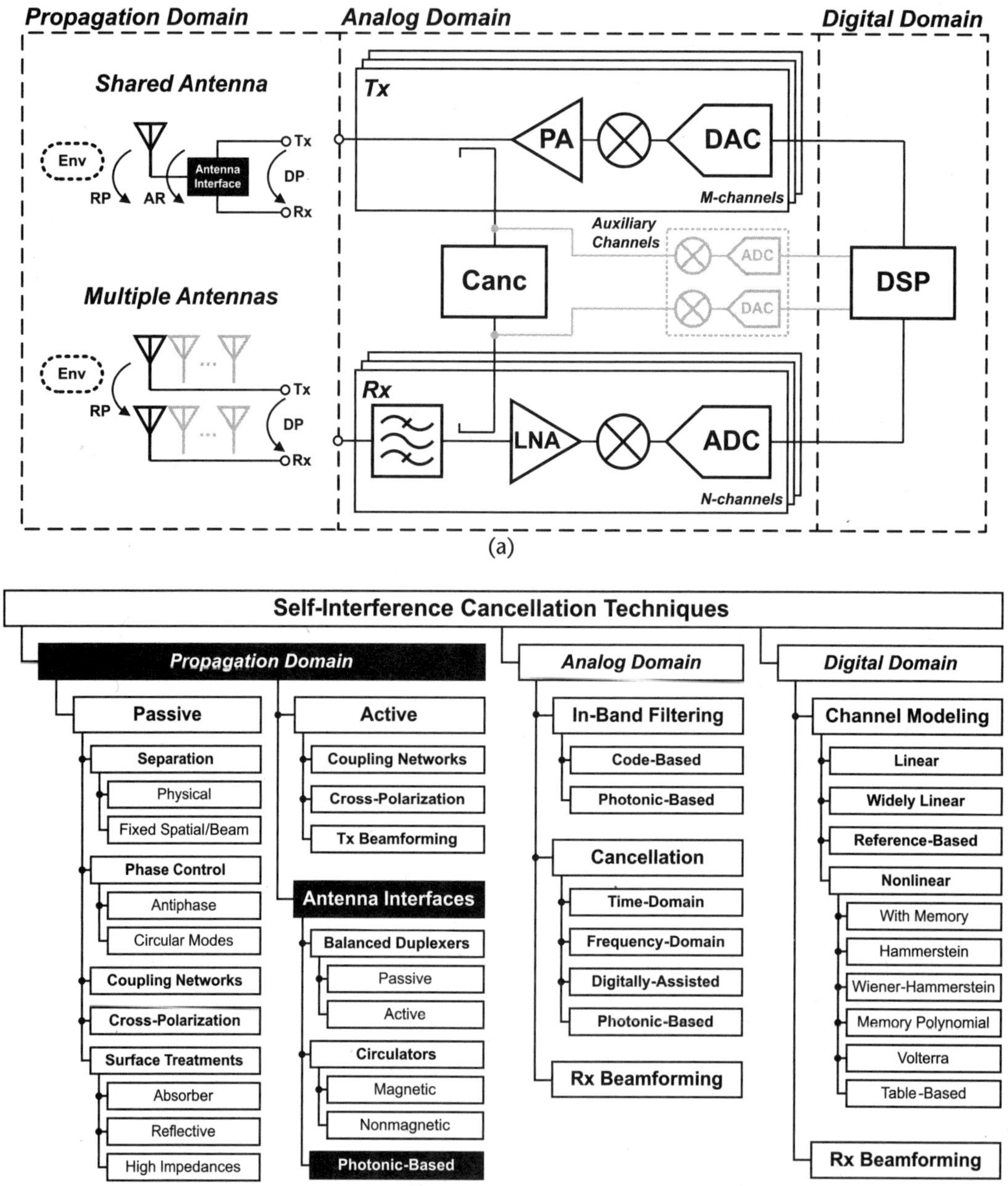

Figure 5.1 Photonic-based antenna interface and specific chapter focus highlighted within (a) a generic IBFD transceiver and (b) a SIC techniques tree diagram.

It is tempting to refer to this front-end canceller as a new form of circulator. These new circulators are generally active, as opposed to a conventional magnetic circulator, which is passive. The active aspect often, but not always, means that specific functions are assigned to specific ports (i.e., the receiver, transmit signal source, and antenna need to be connected to ports specifically designed for the receiver, transmit signal source, and antenna, respectively), whereas the ports of a magnetic circulator are general, so long as a specific order of assigning the functions is observed. These aspects are captured in the present Institute of Electrical

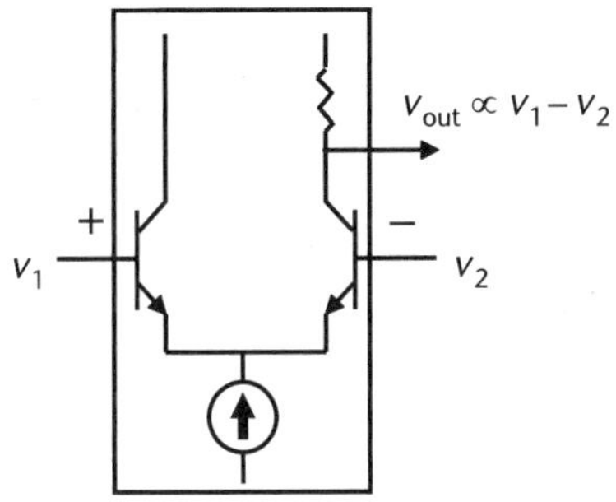

Figure 5.2 Simplified schematic diagram of a differential amplifier, which can function as a front end that performs cancellation of interference signals if they have sufficiently small amplitude.

and Electronics Engineers (IEEE) definition of a circulator. Hence, we have chosen to refer to the photonic implementations of our front-end cancellers not as circulators but rather as transmit-isolating photonic receivers (TIPRx). For the schematic symbol of a TIPRx, we replace the conventional circular symbol for a circulator with a hexagon containing a similar directional arrow, which conveys the functional similarity to, but makes a clear technical distinction from, a magnetic circulator [1, 2].

5.2 Analog Photonic Front-End Cancellation Techniques

5.2.1 Counterpropagating Cancellation

For more than 40 years, analog fiber-optic links, which, as shown in Figure 5.3(a), consist of an electrical-to-optical converter (EOC) at the input end connected via an optical fiber to an optical-to-electrical converter (OEC) on the output end, have been used for antenna remoting (i.e., conveying the signal detected by an antenna to a remotely located receiver). Early intrinsic links, links without any RF amplification, had significant RF loss and consequently high noise figure. Thus, it was common to precede the link with an LNA to compensate for the loss and reduce the noise figure of the LNA + link. This technique worked well, but limited the LNA + link to receive-only (i.e., nonduplexed) applications.

Applying incremental or small-signal modeling to analog links, Cox [3] first showed that RF loss and noise figure were a result of the choice of the performance and operating levels of the link components. Consequently, with an appropriate selection of component parameters and operating levels, one could reduce the link loss or even have positive available power gain [3]. As a result, the analog link could replace the LNA, as discussed in Chapter 7 [4].

Eliminating the need for an LNA between the antenna and input to the analog fiber-optic link introduced the possibility of redesigning the EOC to provide new capabilities, among them for the link to provide analog front-end cancellation before the large signal encountered any nonlinearity. Next we present three implementations of one general technique and two implementations of another in Section 5.2.2.

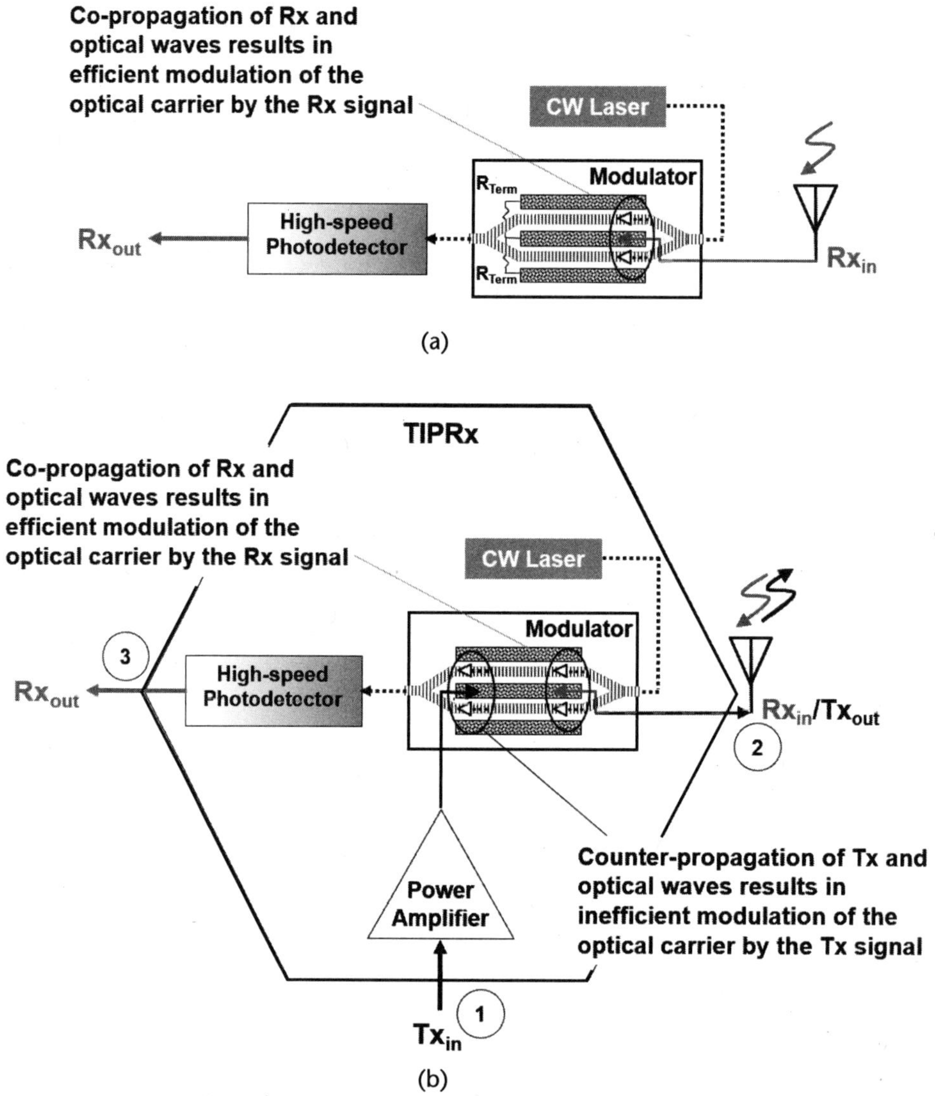

Figure 5.3 (a) Analog fiber-optic link in which the intensity of light from a laser is efficiently modulated by signals received by an antenna, which copropagate along the length of the traveling-wave electrodes of a Mach-Zehnder interferometric modulator; and (b) TIPRx, which conveys receive signals from an antenna at port 2 to a receiver at port 3 while this same antenna transmits signals fed to port 1, which propagate along the length of the modulator's traveling-wave electrodes in the direction opposite to that of the light.

5.2.1.1 Standard Electrodes

Figure 5.3(a) shows an analog fiber-optic link using the most common type of EOC, an integrated optical form of the two-beam-splitter interferometer first demonstrated by Mach [5] and Zehnder [6], and which is therefore called a Mach-Zehnder modulator (MZM). Each arm of the MZM is a phase modulator, which is a strictly linear modulation process between applied electrical voltage and the resulting optical phase shift. The outputs of these two phase modulators are combined at the output coupler, with the result that the phase modulation is converted to intensity modulation. It is at this phase-to-intensity modulation conversion juncture that the modulating signal encounters its first nonlinear process.

For low enough RF modulation frequencies that we can neglect the time it takes for light to travel the length of the electrodes, L, we can treat the electrodes as lumped elements. When the optical transit time past the electrodes cannot be neglected, typically for RF modulation above a few gigahertz, it is then important to treat the MZM electrodes as a traveling-wave structure. It can be shown [7] that the highest modulation efficiency, expressed as degrees of optical phase shift per volt of electrical modulation, occurs under the velocity-match condition (i.e., when the propagation velocity of the RF signal on the modulator electrodes matches the propagation velocity of the optical carrier in the optical waveguides). As is common with traveling-wave structures, the output end of the RF electrodes is terminated in the characteristic impedance of the electrodes.

Figure 5.3(b) shows a modification of the receive-only link that enables the link to simultaneously convey a transmit signal to the antenna while continuing to convey the signal received by the antenna to the receiver. This combined functionality is accomplished by replacing the passive termination of the modulator electrodes with the output impedance of the transmit PA.

We begin the presentation of this configuration with an analysis that establishes the fundamental viability of the topology. The addition of a transmit source means that, together with the receive source, we have two, independent sources. A common circuit technique for analyzing linear circuits with multiple, independent sources is to apply the superposition principle, which states that the total response of a network with multiple sources is equal to the sum of the responses of the network when excited by each source individually, with the remaining independent sources set to zero [8]. Figure 5.4(a) shows the connections of Figure 5.3(b) with the receive and transmit sources represented by their Thévenin equivalent circuits, which consist of ideal voltage sources and resistive impedances. To analyze the operation with the receive source active, we set the transmit source to zero; a voltage source with zero voltage across it is a short circuit; and the result is the connection shown in Figure 5.4(b). The response of the network is that the modulator electrodes are terminated by the output impedance of the transmit PA. Thus, the receive signal modulates the optical carrier with the same efficiency as it did when the electrodes were passively terminated, as shown in Figure 5.3(a). To analyze the operation with the transmit source active, we set the receive source to zero, which again, because it is a voltage source, gets replaced by a short circuit, as shown in Figure 5.4(c). In this case, the power from the transmit PA is conveyed to the impedance of the receive device, which in the case where the receive device is an antenna, means that the transmit power is conveyed to the antenna and radiated, as desired. The superposition principle states that the total response of the connection shown in Figure 5.4(a) is the sum of the individual responses obtained in Figure 5.4(b, c). The photonics-based antenna interface shown in Figure 5.3(b) is therefore indeed capable of simultaneously transmitting and receiving.

The efficiency with which the transmit signal modulates the optical carrier is much lower than the efficiency with which the receive signal modulates the optical carrier because, as also shown in Figure 5.3(b), the transmit signal on the modulator electrodes is counterpropagating with respect to the optical carrier, as opposed to copropagating with it as the receive signal does. Thus, instead of being well velocity-matched as the receive signal is, the transmit signal is poorly velocity-matched. Figure 5.5 is a plot of transmit-signal cancellation versus frequency. The

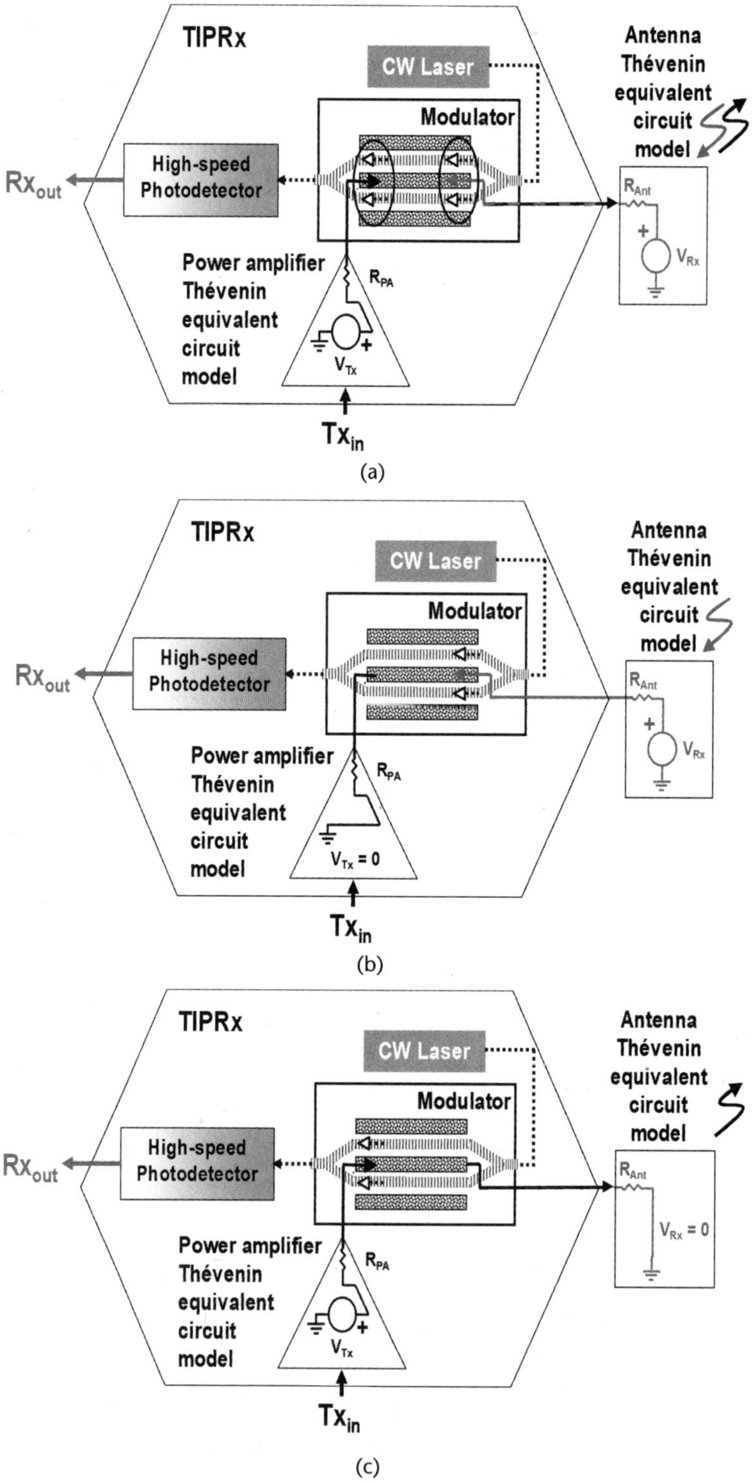

Figure 5.4 Illustration of the TIPRx operation using the principle of superposition. (a) TIPRx diagram with signal sources and their impedances shown explicitly; (b) TIPRx diagram with the receive signal source present but the transmit signal source shorted out; and (c) TIPRx diagram with the transmit signal source present but the receive signal source shorted out.

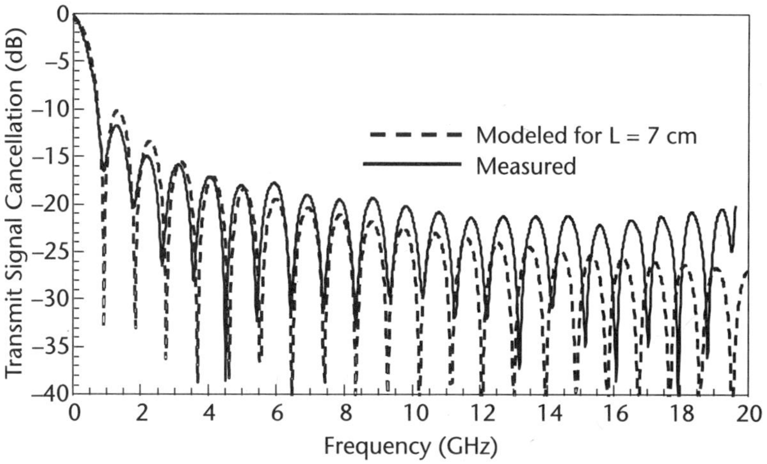

Figure 5.5 Modeled (dashed curve) and measured (solid curve) transmit signal cancellation versus frequency for a TIPRx using a MZM with electro-optic (E-O) interaction over a length of L = 7 cm.

measured cancellation (solid curve) is in good agreement with the modeled cancellation (dashed curve), which for low-loss electrodes behaves approximately according to the following expression (derived in [9]):

$$\textit{Transmit Signal Cancellation} = \left[\frac{\sin(x)}{x}\right]^2, \tag{5.1}$$

where

$$x \equiv \frac{2\pi f n L}{c}, \tag{5.2}$$

and where n is the effective propagation index of the modulator electrodes and c is the speed of light in a vacuum. As expected at low frequencies, where the distinction between copropagating and counterpropagating traveling waves degenerates into a lumped element electrode, the cancellation as expressed in (5.1) approaches unity, or 0 dB. As the frequency increases, the modeled cancellation continues to improve while the measured cancellation begins to plateau at about –20 dB. A careful examination of these two plots also reveals that the nulls in the plot of the modeled data using (5.1) are deeper than in the plot of the measured data. Both of these effects can be traced to the loss of the electrodes, which for the modeled case was assumed to be zero. The reason that the cancellation plateaus is that the electrode loss increases at the same rate that the cancellation is decreasing.

It is important to note that the cancellation occurs because of how inefficiently the transmit signal imparts phase modulation onto the optical carrier. Thus, this cancellation occurs before the modulation encounters its first nonlinearity, which is as desired.

5.2.1.2 Apodized Electrodes

There are applications in which additional cancellation is desired. The question is: what design parameters can we modify to achieve increased cancellation? A productive place to start answering this question is to understand why the cancellation follows a $\sin(x)/x$ function, as shown in (5.1). It is well-known that the Fourier transform of $\sin(x)/x$ is a single rectangular pulse, but what does this pulse represent here? The upper-left side of Figure 5.6 again shows a MZM with standard electrodes, which results in the $\sin(x)/x$ cancellation function that is repeated as the upper plot on the graph in the center of Figure 5.6. Neglecting the fringing fields at the ends of the electrodes, the E-O interaction is uniform along the length of the electrodes, as shown by the plot in the lower left of Figure 5.6, albeit with decreasing magnitude due to the electrode loss. Hence, this is the single rectangular pulse that results in the $\sin(x)/x$ shape of the transmit-signal modulation function. When maximizing the E-O modulation efficiency for the receive signal was the only MZM design criterion, the standard electrode layout, with uniform interaction, was the optimal choice.

We can modify the shape of the E-O interaction function, which, in turn, will modify the cancellation function. The upper right of Figure 5.6 shows an apodized electrode shape, which has the nonuniform E-O interaction function shown in the plot in the lower right of Figure 5.6 and which is expressed mathematically as

$$E\text{-}O\,Interaction \propto \frac{1}{2}\left[1-\cos\left(\frac{2\pi z}{L}\right)\right]. \tag{5.3}$$

The cancellation resulting from the apodized E-O interaction expressed by (5.3) is shown by the lower plot on the graph in the center of Figure 5.6, and with the assumption of low electrode loss is expressed approximately as

$$Transmit\,Signal\,Cancellation = \left[\frac{\pi^2}{\pi^2-x^2}\frac{\sin(x)}{x}\right]^2, \tag{5.4}$$

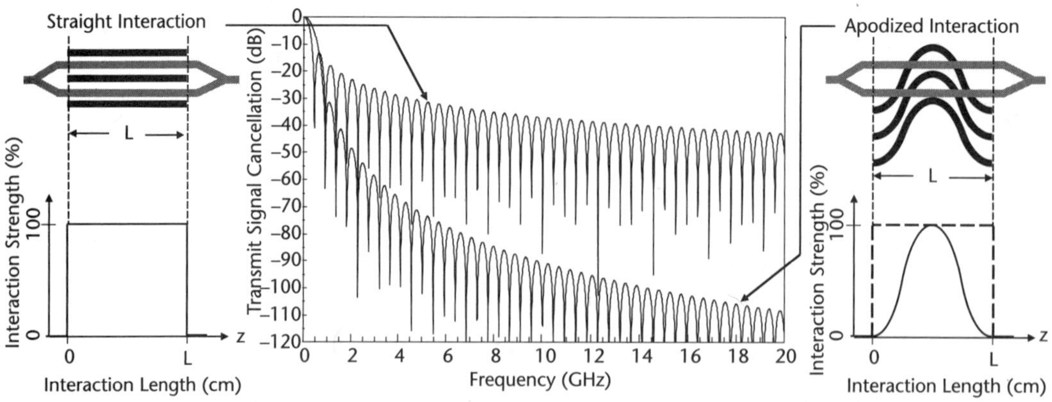

Figure 5.6 Comparison of modeled transmit signal cancellation versus frequency for a TIPRx in which the E-O interaction is constant, as shown in the left portion of the diagram and the upper curve on the plot, versus one in which the E-O interaction is apodized, as shown in the right portion of the diagram and the lower curve in the plot.

as was derived in [9]. At frequencies for which the electrodes are of long electrical length, x is large [see (5.2)], and thus (5.4) approaches its limit of

$$\lim_{x\to\infty} \textit{Transmit Signal Cancellation} = \left[\frac{\sin(x)}{x^3}\right]^2, \tag{5.5}$$

indicating that this degree of apodization causes the cancellation to improve much more rapidly as frequency (to which x is proportional) is increased than in the case of modulator electrodes configured for constant E-O interaction. The electrode loss of the apodized electrode is still proportional to length. Hence, the apodized modulator is able to asymptotically approach a greater level of cancellation.

The upper and lower curves on the graph in the center of Figure 5.6 are plots of (5.1) and (5.5) for a modulator with electrodes having length L = 14 cm and index n = 2.2. From these plots of modeled transmit-signal cancellation, it can be seen that apodization according to (5.3) can result in substantially greater cancellation of the transmit signal. The apodization also reduces the E-O interaction for the receive signal, which is generally undesirable [9]. Thus, for a single modulator that is to perform disparate functions for receive and transmit signals, there is now a design trade-off between the receive-signal sensitivity and the transmit-signal cancellation. Even with the extreme degree of apodization represented by (5.3) and depicted in the rightmost portion of Figure 5.6, which enables a roughly 10-dB improvement in measured transmit-signal cancellation, as evidenced by comparing the solid black curves in Figures 5.5 and 5.7, the E-O modulation efficiency decreases only by a factor of 2. Many designers of broadband radar or commu-

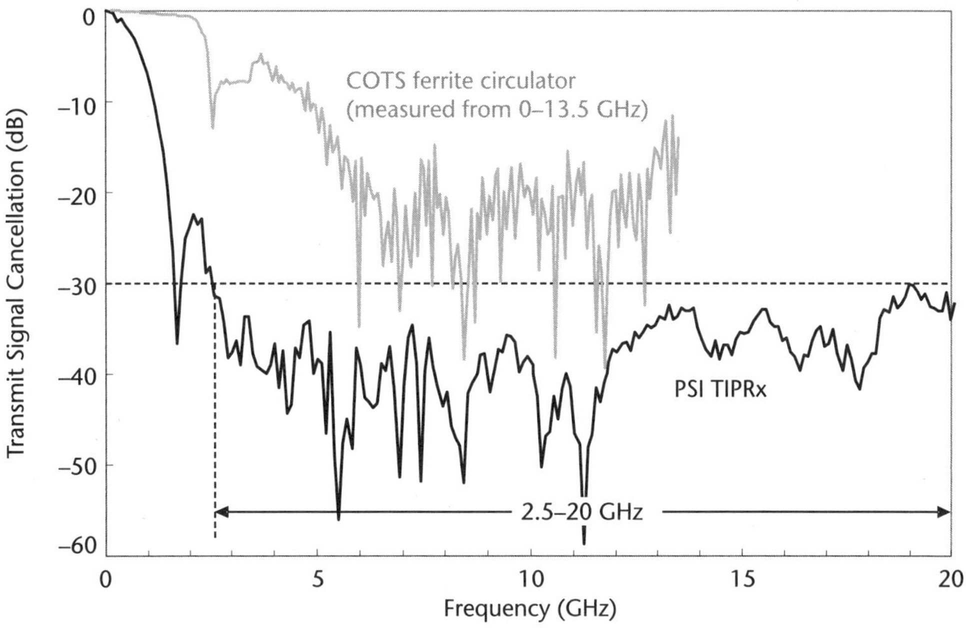

Figure 5.7 Measured transmit signal cancellation versus frequency for a TIPRx with apodized E-O interaction (black curve) versus measured cancellation for a COTS magnetic circulator (gray curve).

nications systems would consider this a worthwhile price to pay for the improved broadband transmit-signal suppression afforded by the apodized interaction.

Figure 5.7 shows the cancellation versus frequency that we measured using the same apodized modulator for which the lower curve in Figure 5.6 showed the cancellation predicted by (5.4) and compares it to the measured data for a commercially available RF magnetic-based circulator [10]. Across the octave-wide 6 to 12-GHz frequency band over which this RF circulator has approximately –20 dB of transmit-signal cancellation, the modulator with apodized electrodes has approximately –40 dB of cancellation. Moreover, as illustrated in the figure, the apodized modulator affords better than –30 dB of cancellation across the nearly decade-wide frequency range from 2.5 to 20 GHz.

5.2.1.3 Single-Sideband Filtering

An even greater degree of transmit-signal cancellation than described above can be achieved if one takes even further advantage of the fact that, in the counterpropagation cancellation architecture, the transmit and receive signals are fed to the modulator's traveling-wave electrodes from opposite ends and additionally incorporates a well-established method for accomplishing single-sideband (SSB) modulation of an optical carrier.

To understand how SSB modulation is achieved, it is first necessary to introduce an alternative modulator electrode topology to what has been shown in figures thus far. Whereas the modulators shown in Figure 5.3(a, b) enable a single input voltage applied to the center electrode to impose phase modulation of light in the upper optical waveguide of the interferometer that is equal in magnitude but opposite in sign with respect to modulation of light in the lower optical waveguide, the advantageous push-pull configuration, Figure 5.8(a) shows push-pull operation being achieved in a dual-drive modulator. With the aid of a 180° RF hybrid coupler, signal is applied to the upper electrode and a copy of this same signal with the opposite sign is applied to the lower electrode.

Push-pull modulation achieved either with the configurations of Figure 5.3 or the configuration of Figure 5.8(a) results in double-sideband (DSB) modulation of the optical carrier, as indicated by the spectral diagram at the output of the modulator in Figure 5.8(a). It has been shown [11] that replacing the 180° coupler in Figure 5.8(a) with a 90° coupler results in constructive interference at the modulator output for only the portion of the optical spectrum to one side of the optical carrier and destructive interference in the portion of the optical spectrum on the other side of the optical carrier. Hence, this configuration results in SSB modulation of the optical carrier. The original intent for inventing this method of SSB modulation was to extend the maximum distance over which an optical fiber link can convey high-frequency signals. In the case of DSB modulation, chromatic dispersion in the optical fiber causes each sideband to propagate at a slightly different velocity, such that, at some distance, the two sidebands will arrive at the photodetector 180° out of phase with one another, resulting in no optical modulation for the photodetector to detect. SSB modulation prevents chromatic dispersion from imposing this restriction on propagation length.

Combining this method of achieving SSB modulation with the counterpropagation cancellation architecture of Figure 5.3(b) can enable greater cancellation of the

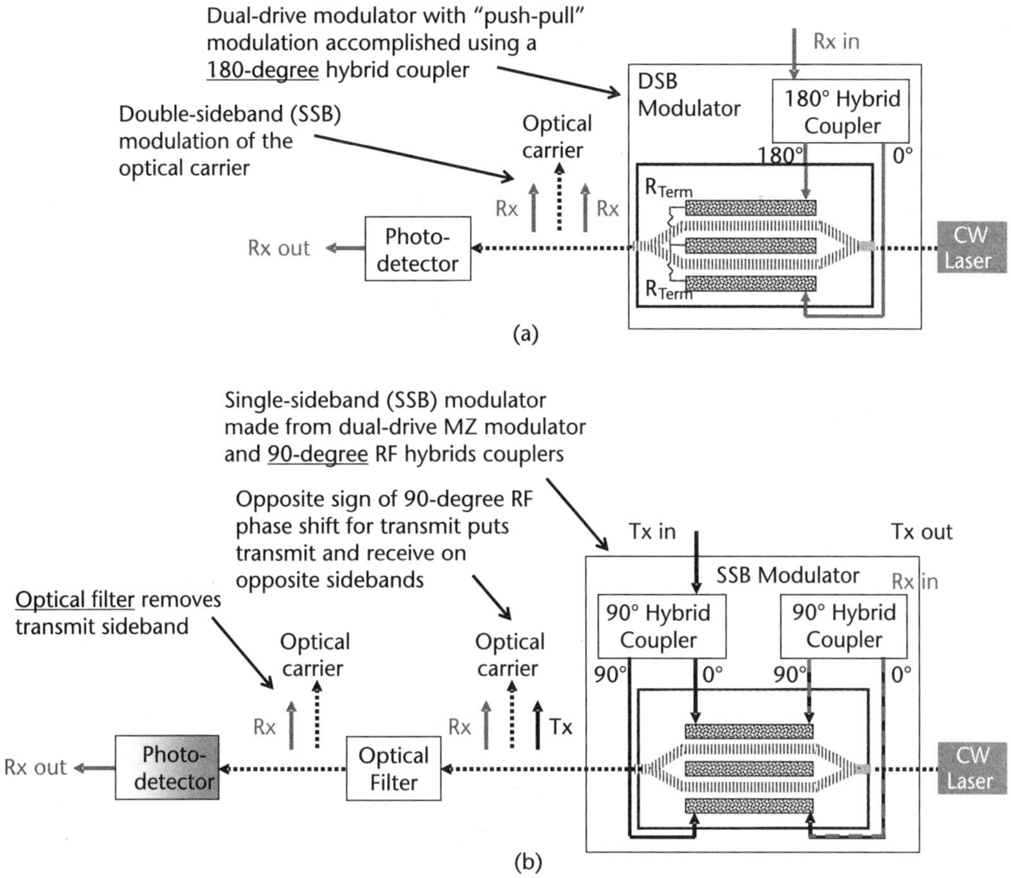

Figure 5.8 (a) Analog fiber-optic link using a dual-drive MZM and a 180° RF hybrid coupler to achieve the conventional DSB modulation of the optical carrier; and (b) method of transmit signal cancellation that uses a 90° RF hybrid coupler to impose the receive signal upon an SSB of the optical carrier and the transmit signal upon the opposite sideband where it can be removed by an optical filter.

transmit signal. Specifically, if the transmit and receive signals are supplied to the traveling-wave electrodes of a dual-drive MZM through a pair of 90° hybrid couplers as shown in Figure 5.8(b) (i.e., with the 0° and 90° ports of one coupler connected to the upper and lower electrodes, respectively, for one of the signals and in the reverse order for the other signal), then each of these signals will modulate only one of the two sidebands of the optical carrier supplied by the laser. This reversal of 90° coupler connections at each end of the electrodes also causes the transmit signal launched from one end to add back together in phase at the output of the 90° coupler at the other end, such that the only loss incurred by the transmit signal is the excess loss of the two couplers, typically less than 1 dB per coupler, plus the insertion loss of the modulator electrodes.

With the transmit and receive signals occupying opposite sidebands of the optical carrier, the addition of an optical filter retains the carrier and the one sideband occupied by the receive signal and rejects the opposite sideband occupied by the transmit signal [12]. The transmit signal is therefore cancelled in two ways: (1) via preferential modulation of the optical carrier by signals on the modulator's traveling-wave electrodes that copropagate rather than counterpropagate with the optical carrier in its optical waveguides, as in the case of all of the counterpropagation cancellation architectures described in this section; and (2) via preferential modulation of the optical carrier in the one sideband that is within the passband of the optical filter by the signal launched from port 2 and preferential modulation of the optical carrier in the one sideband that is outside the passband of the optical filter by the signal launched from port 1.

Figure 5.9 shows the measured transmit-signal cancellation afforded by the counterpropagation plus SSB filtering configuration shown in Figure 5.8(b), compared to the measured performance using the same modulator configured in the DSB counterpropagation configuration of Figure 5.3(b). At all frequencies above 3 GHz, adding the necessary hardware to realize SSB modulation and optical filtering to retain only one of the sidebands results in a 10 to 20-dB improvement in transmit-signal cancellation. Below 3 GHz, there is little improvement because, at these frequencies, the optical filter that retains the optical carrier has not yet rolled off sufficiently to reject the sideband that carries the transmit signal [13].

5.2.2 Subtraction Cancellation

Besides enabling the SSB filtering method of achieving transmit-signal cancellation as described earlier, the use of a dual-drive modulator also gives rise to an entirely different method of realizing a TIPRx. We call this method subtraction cancellation because it accomplishes cancellation of the transmit signal by subtracting the transmit signal from a second copy of this signal, similar to the electronic techniques discussed in Chapters 8 through 10. This chapter focuses on photonic methods for achieving cancellation at the antenna interface. Subtraction cancellation can also be achieved using methods that do not involve photonics technology, and one such method will be compared to its photonic counterpart at the end of this section.

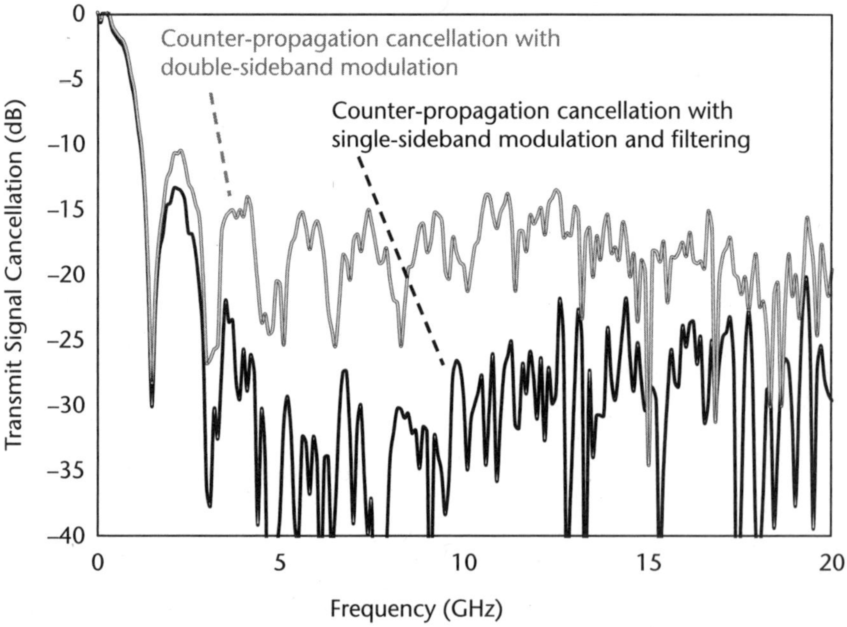

Figure 5.9 Measured performance of the SSB filtering method of achieving transmit signal cancellation (black curve) versus that of a TIPRx with conventional DSB modulation (gray curve).

5.2.2.1 Shared Antenna

The beginning of Section 5.2.1.3 explained how a dual-drive modulator efficiently intensity-modulates light if the two drive electrodes are driven out-of-phase with another by the modulation signal, such as is enabled by using a 180° RF coupler, as shown in Figure 5.8(a). What was not stated in Section 5.2.1.3 was the converse, that is, the fact that driving these two electrodes with in-phase signals would result in zero intensity modulation of the light, because the intensity modulation at the optical output of the MZM is proportional to the difference between the degrees of phase modulation imposed on the light in the two arms of the interferometer, as was explained in Section 5.2.1.1. Subtraction cancellation can therefore be achieved using a dual-drive MZM.

Figure 5.10 shows how a transmit-signal source and antenna can be connected to a dual-drive MZM to feed the transmit signal to the antenna while simultaneously recovering signals received by this same antenna in the same band at the photodetector. Applying the transmit signal to both electrodes results in the same optical phase modulation being applied to each arm of the modulator. Because the optical intensity at the output of the modulator is proportional to the difference between the phase modulation in the two arms, the result of imposing the transmit signal onto both electrodes is, ostensibly, zero intensity modulation of the light by the transmit signal. The receive signal modulates the light efficiently because only one of the electrodes is connected to the antenna.

Figure 5.11 is a plot of the measured transmit-signal cancellation versus frequency of the architecture shown in Figure 5.10. Although it is only at high frequencies that this architecture benefits from counterpropagation cancellation, the cancellation that it achieves by subtraction is effective at all frequencies where the

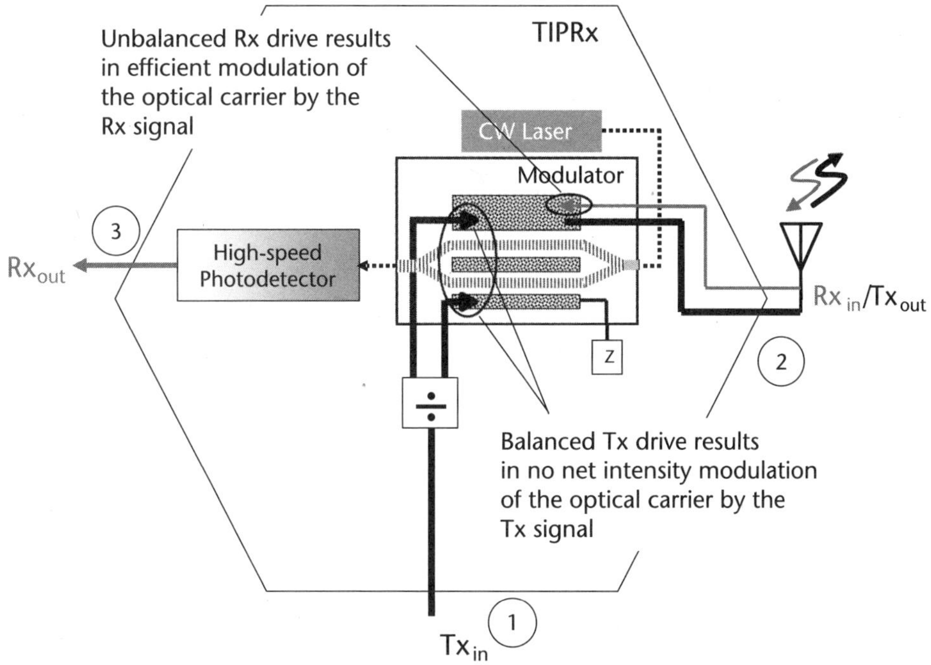

Figure 5.10 Illustration of the subtraction method of achieving transmit signal cancellation using a dual-drive MZM, for the case where a single antenna transmits and receives.

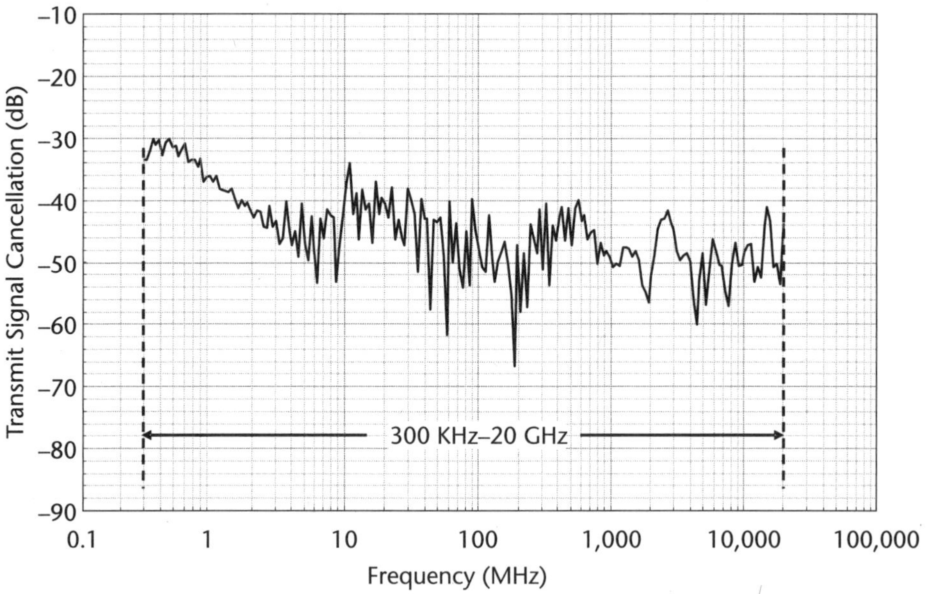

Figure 5.11 Measured transmit signal cancellation versus frequency for the subtraction cancellation architecture shown in Figure 5.10. Result is displayed for the first time on a logarithmic frequency scale to emphasize that this architecture imparts cancellation performance across decades of bandwidth that includes very low RF. The 300-kHz lower limit is imposed by the four-port network analyzer that performed this measurement.

transmit signal is fed to the two electrodes with equal magnitude and phase [14, 15]. As shown in Figure 5.11, the measured cancellation, which was limited by the instrumentation to the frequency range of 300 kHz to 20 GHz, afforded by the subtraction technique, is therefore better than –30 dB across this nearly 5-decade-wide band, and better than –40 dB at all frequencies in the band from 40 MHz to 20 GHz. The use of subtraction cancellation to extend to much lower frequencies than the already remarkably broadband cancellation performance enabled by the counterpropagation cancellation architecture does come at a price. Specifically, there are doubtlessly some system designers who would not be willing to throw away half of the transmit signal's power, as is required by this architecture and as indicated by the power divider shown in Figure 5.10.

It is important to note that, just as in the case of the counterpropagation cancellation architecture, when the cancellation is achieved, the modulator's sensitivity to small receive signals is not compromised. As explained at the beginning of Section 5.2, the two copies of the large transmit signal subtract from one another, and this subtraction occurs prior to the nonlinearity of the interferometric combining or any other nonlinear process.

Because the subtraction occurs linearly, we can again use the superposition principle described in Section 5.2.1.1 to establish the fundamental validity of the subtraction cancellation topologies described here. As described previously, we imagine a transmit signal source connected to port 1 and a receive signal source connected to port 2 in place of the antenna, and observe the output at port 3 with first one, and then the other, source voltage set to zero. Setting the transmit signal voltage source to zero results in there being only a passive impedance connected at port 1, such that the only voltage applied to the MZM is the receive signal voltage that is applied to one of the MZM's two drive electrodes. The intensity modulation at the output of the MZM, and incident on the photodetector, is proportional to the difference between the phase modulation on the first arm of the interferometer addressed by this one drive electrode and by the phase modulation in the second arm, the latter of which is zero because no voltage is applied to the second drive electrode. The receive signal applied by a source at port 2 therefore reaches port 3. Conversely, setting the receive signal voltage source to zero results in the antenna appearing only as an impedance at port 2, with the signal radiated by the antenna being proportional to any voltage imposed across this impedance. The voltage from the transmit signal source at port 1 encounters the power divider and is therefore applied in with equal amplitude and phase to both drive electrodes. It propagates to port 2 via the upper drive electrode and is imposed across the antenna's impedance, causing the antenna to radiate the transmit signal. The light reaching port 3 is not intensity-modulated by the transmit signal, because the phase modulation of the light by the transmit signal imposed on the first (upper) electrode is subtracted from the phase modulation of the light by the transmit signal imposed by the second (lower) electrode, to which it is equal.

5.2.2.2 Multiple Antennas

The subtraction cancellation architecture shown in Figure 5.10 is readily adaptable to the situation where the signal to be received overlaps in time and frequency with a signal or signals transmitted from other antennas. As long as a reference copy of the transmitted signal that is fed to the other antenna is available, it can be supplied to the second electrode of the modulator as shown in Figure 5.12, so that this copy of the transmit signal will be subtracted from the total signal sensed by the antenna connected to the first electrode of the modulator. The superposition principle can

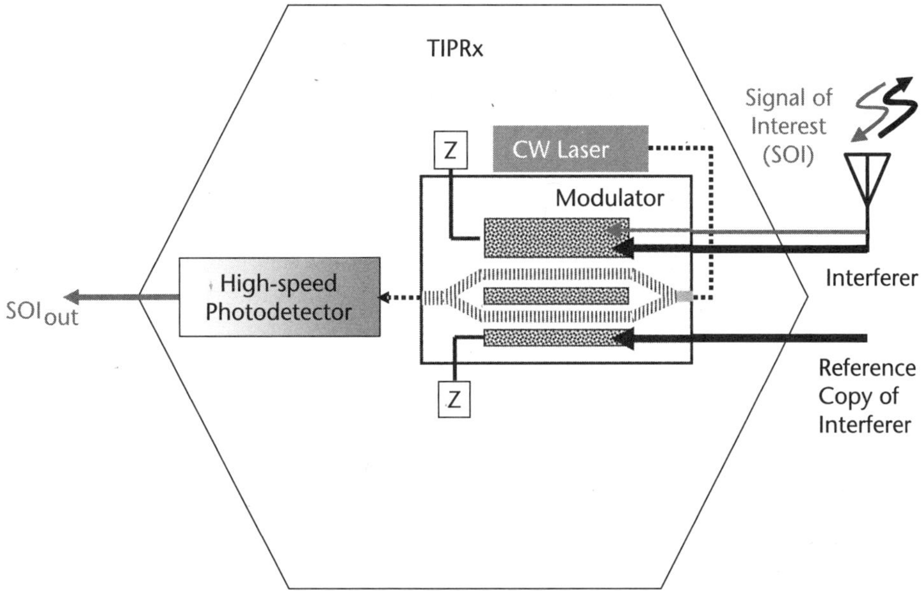

Figure 5.12 Illustration of the subtraction method of achieving transmit signal cancellation using a dual-drive MZM, for the case where a separate antenna is the source of the transmit signal.

be applied to convince ourselves of the validity of the subtraction cancellation architecture of Figure 5.12 exactly as was done for the single-antenna configuration of Figure 5.10 in the previous section. Rather than repeating that application here, this is left as an exercise for the reader.

It is again important to note that, when cancellation is achieved in the architecture shown in Figure 5.12, the modulator's sensitivity to small receive signals is not compromised. As explained at the beginning of Section 5.2, the large transmit signal sensed by the receive antenna connected to one electrode subtracts from the reference copy of this signal fed to the other electrode, and this subtraction occurs prior to the nonlinearity of the interferometric combining or any other nonlinear process. This is in stark contrast to how subtraction cancellation is achieved using some methods not involving photonics technology. For example, it is possible to substitute the differential amplifier shown in Figure 5.2 for the laser-modulator-photodetector combination shown in Figure 5.12. Connecting the antenna at port 2 and the reference copy of the interferer – with the interferer in this case being a signal transmitted from another antenna or antennas – at port 1, the difference between these two signals would appear at the output port of the differential amplifier. Whereas in the photonic subtraction architecture of Figure 5.10 this output signal is proportional to the difference between linear copies of the two input signals, the output of the differential amplifier is proportional to the difference between nonlinear copies of the two input signals. Subtraction cancellation using photonics technology can therefore be achieved with a much greater dynamic range (i.e., it enables cancellation of transmit signals having much higher power). For example, we have demonstrated simultaneous transmission of signals with up to 2W (+33 dBm) of power from one antenna while simultaneously receiving signals from a second, nearby antenna at 0.5 to 6 GHz [16], with this frequency range being limited only by the bandwidth of the two antennas.

5.3 Practical Considerations

A key advantage of photonic versus all-electronic implementations of IBFD is their wide instantaneous bandwidth, which can equal their operational bandwidth, spanning multiple decades as shown by the examples in this chapter. All-electronic implementations have only been able to achieve instantaneous bandwidths of, at most, hundreds of megahertz.

The approaches described in Section 5.2 can all be readily implemented today using mature, COTS components, except in the case of the apodized-electrode modulator of Section 5.2.1.2, which has been experimentally demonstrated but is only available as a custom part. The cost and size of discrete component implementations, while acceptable for proof-of-concept demonstrations, limit the range of applications. Photonic integrated circuits (PICs), as well as photonic-electronic integrated circuits, are presently under development. The tighter integration enabled by these integrated circuits (ICs) will reduce the cost of photonic-based antenna interfaces and expand their application to a broader range of systems.

5.4 Conclusion

5.4.1 Summary

With an increasingly congested spectral environment, methods for cancelling in-band high-power interference signals, whether transmitted from the same antenna as, or from different antennas than, the antenna that is listening for a signal to receive, will likely be in ever-increasing demand. The architectures described in this chapter comprise several ways that photonics technology can be leveraged to achieve a high degree of transmit-signal cancellation prior to the first nonlinear component in the receive path and across broad bandwidths that can include the receive-signal frequencies, enabling high-performance STAR and IBFD systems.

5.4.2 Future Directions

The photonic architectures described here include the earliest demonstrations of antenna interfaces that enabled same-frequency STAR or IBFD using a single antenna for both transmit and receive signals. At the time of these demonstrations, most, if not all, systems for which such capabilities were desired operated at frequencies below 20 GHz, which is why the measured results shown are all at frequencies below 20 GHz. With the exception of the photonic architecture described in Section 5.2.2.2 for achieving subtraction cancellation of a signal transmitted by a separate antenna, all of the techniques described here benefit from the fact that the transmit signal is supplied to the antenna in such a way that it modulates the light reaching a photodetector much less efficiently than the receive signal modulates this same light, and this phenomenon has the rare quality of actually imparting better performance at higher frequencies than at lower frequencies. Since the early demonstrations of STAR and IBFD enabled by photonics technology, components have come on the market that make it possible to construct analog fiber-optic links for operation at frequencies up to 110 GHz [17]. Therefore, when the photonic architectures described in this chapter are implemented at frequencies between 20 GHz and 110 GHz, we expect to see even more dramatic evidence of not only their feasibility but also their superiority to alternative (i.e., nonphotonic) techniques.

References

[1] Cox, C., and E. Ackerman, "Bi-Directional Signal Interface," U.S. Patent No. 7,555,219, June 30, 2009.

[2] Ackerman, E., and C. Cox, "Optimization of Photonic Transmit/Receive Module Performance," *Proc. IEEE International Topical Meeting on Microwave Photonics*, 2009.

[3] Cox, C., *Analog Optical Links: Theory and Practice*, New York: Cambridge University Press, 2004.

[4] Ackerman, E., and C. Cox, "Microwave Photonic Links with Gain and Low Noise Figure," *Proc. IEEE Lasers Electro-Optic Society Annual Meeting*, October 2007, pp. 38–39.

[5] Mach, L., "Ueber einen Interferenzrefraktor," *Zeitschrift für Instrumentenkunde,* Vol. 12, 1892, pp. 83–89.

[6] Zehnder, L., "Ein neuer Interferenzrefraktor," *Zeitschrift für Instrumentenkunde,* Vol. 11, 1891, pp. 275–285.

[7] Gopalakrishnan, G., et al., "Performance and Modeling of Broadband $LiNbO_3$ Traveling Wave Optical Intensity Modulators," *J. Lightwave Technol.*, Vol. 12, October 1994, pp. 1807–1819.

[8] Dertouzos, M., et al., *Systems, Networks, and Computation: Basic Concepts*, New York: McGraw-Hill, 1972, Section 3.6, pp. 60–65, 136–139.

[9] Cox, C., and E. Ackerman, "TIPRx: A Transmit-Isolating Photonic Receiver," *J. Lightwave Technology*, Vol. 32, October 2014, pp. 3630–3637.

[10] http://www.ditom.com/images/D3C6012.pdf.

[11] Smith, G., D. Novak, and Z. Ahmed, "Overcoming Chromatic-Dispersion Effects in Fiber-Wireless Systems Incorporating External Modulators," *IEEE Transactions on Microwave Theory and Techniques,* Vol. 45, August 1997, pp. 1410–1415.

[12] Devgan, P., "Isolation of RF Signals Using Optical Single Side Band Modulation Combined with Optical Filtering," U.S. Patent No. 9,240,842, January 19, 2016.

[13] Ackerman, E., et al., "Broadband Simultaneous Transmit and Receive from a Single Antenna Using Improved Photonic Architecture," *Proc. IEEE International Microwave Symp.*, 2019, p. 778–781.

[14] Cox, C., and E. Ackerman, "Transmit Isolating Photonic Receive Links: A New Capability for Antenna Remoting," *Proc. Conf. on Optical Fiber Communications*, 2011.

[15] Cox, C., and E. Ackerman, "Same-Aperture Any-Frequency Simultaneous Transmit and Receive Communication System," U.S. Patent No. 9,935,680, April 3, 2018.

[16] Cox, C., and E. Ackerman, "Maximizing RF Spectrum Utilization with Simultaneous Transmit and Receive," *Microwave Journal*, September 2014, pp. 114–126.

[17] www.photonicsinc.com, Billerica, MA.

CHAPTER 6

Code-Based RF Filtering

Ahmed Hamza, Cameron Hill, and James F. Buckwalter

6.1 Introduction

Direct-sequence spread spectrum (DSSS) techniques offer an alternative approach to IBFD wireless networks. DSSS has been utilized in code-division multiple access (CDMA) wireless networks, allowing multiple receivers to occupy the same bandwidth (BW) [1]. Commercial applications of CDMA operated with FDD such that the colocated Tx operated at a different band from the Rx. The relaxed dynamic range (DR) of the FDD system allowed pseudorandom noise (PN) code correlation using DSP after the ADC.

By utilizing DSSS for IBFD, the Tx and Rx signals overlap and the spectral efficiency is improved, as compared with FDD. The DR and BW of the Rx must be improved to tolerate Tx SI. Advances in RF CMOS process technology have demonstrated N-path techniques that couple RF-domain signal processing with high-speed, high-linearity switches to enable DSSS techniques to be implemented in the RF domain with low power and noise figure (NF) overhead. Moving signal correlation to the RF front end removes the additional DR and BW requirements in the Rx ADC while maintaining compatibility with other IBFD techniques. When combined with other Tx SIC techniques (such as circulators, RF cancellation, and digital SIC), the total SIC can theoretically exceed 120 dB, as discussed in Chapter 1. The principles of the RF code-domain rejection for IBFD are explored here, as highlighted within the IBFD transceiver diagram of Figure 6.1(a) and SIC techniques tree of Figure 6.1(b).

6.2 Code-Domain Principle

A code-domain IBFD transceiver, shown in Figure 6.2, operates by modulating the Tx signal by a PN code (PN_{Tx}) and demodulating the Rx signal by an orthogonal PN code (PN_{Rx}). Both codes are of length M, implying that they are modulated at a rate M times higher than the symbol rate of the original uncoded signals. The PN rate (i.e., the chip rate) and the symbol rate are determined by the number of

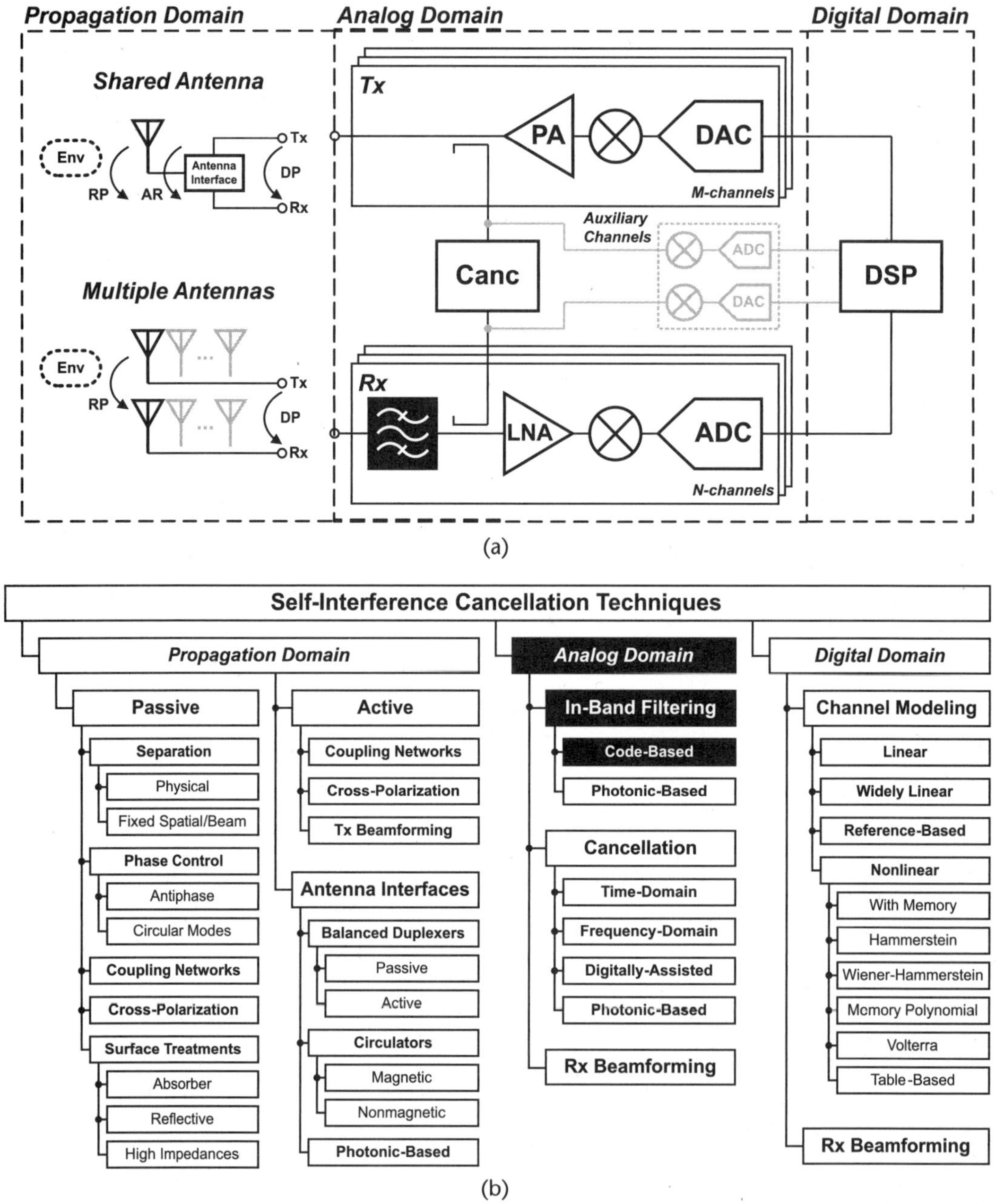

Figure 6.1 Code-based RF filtering and specific chapter focus highlighted within (a) a generic IBFD transceiver and (b) a SIC techniques tree diagram.

simultaneous channel users. The code space should provide M codes, ideally, an orthogonal code set with M-length codes. The PN code multiplication spreads the Tx signal over a spread BW_{spread} given by $M \cdot BW_{sig}$.

The RF signal after the antenna interface consists of the Tx signal spread by PN_{Tx} and the desired Rx signal spread by PN_{Rx}.

$$RF_{in}(t) = Rx(t)PN_{Rx}(t) + Tx(t)PN_{Tx}(t). \tag{6.1}$$

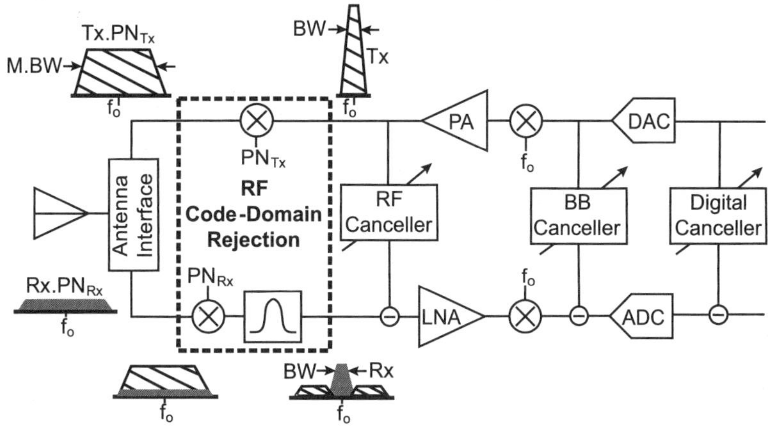

Figure 6.2 RF code-domain rejection and example signals shown within an IBFD transceiver.

To provide in-band SIC, the Tx signal is filtered with low insertion loss (IL) to the desired Rx signal. The incoming Rx signal is correlated with PN_{Rx} that is orthogonal to PN_{Tx}, that is,

$$\frac{1}{M}\int_0^{MT_C} PN_i(t)PN_j(t-\tau)dt = \begin{cases} 0 & \forall\tau,\ i \neq j \\ 1 & \tau = 0,\ i = j \end{cases}, \tag{6.2}$$

where T_c is the chip period. The resulting signal after code correlation is

$$RF_{out}(t) = \frac{1}{M}\int_0^{MT_C} PN_{Rx}(t) \times RF_{in}(t)dt \cong Rx(t). \tag{6.3}$$

This equation assumes ideal PN code orthogonality, eliminating Tx SI within the signal BW. In a real circuit, code correlation despreads the desired Rx signal into the signal BW while the Tx signal is spread outside the signal BW. Residual in-band Tx power exists due to nonideal orthogonality depending on the PN codes.

6.3 System-Level Discussion

6.3.1 Spectral Efficiency

RF code-domain DSSS techniques offer spectral efficiency in a multiuser full-duplex system as each user is assigned a specific PN code that is orthogonal to other users who concurrently operate within the same-spread BW. For synchronized transceivers, the capacity of a multiuser code-domain IBFD system is the same as or even exceeds the capacity of an FDD IBFD system, where each Rx operates in full-duplex mode with respect to the Tx SI but in FDD mode with respect to other transceivers [2]. However, DSSS IBFD requires network coordination to assign and synchronize codes for the simultaneous users.

6.3.2 Code-Domain Codes

6.3.2.1 Ideal Codes

An ideal PN code should have the following values for its normalized auto-correlation (R_{xx}) and cross-correlation (R_{xy}).

$$R_{xx}[i] = \frac{1}{M}\sum_{k=0}^{M-1} x[k]x[k+i] = \delta[i] \tag{6.4}$$

$$R_{xy}[i] = \frac{1}{M}\sum_{k=0}^{M-1} x[k]y[k+i] = 0,\ \forall\ x \neq y \tag{6.5}$$

The discrete interval, i, is the lag between the two versions of the code. A single auto-correlation peak allows synchronization while cross-correlation with other PN codes is zero for any lag for good rejection. Unfortunately, no PN code has these ideal properties, and Figure 6.3 summarizes the properties of different code families.

6.3.2.2 M-Sequences

M-sequences are odd-length sequences with code length $M = 2^n - 1$, where n is an integer. The auto-correlation is either 1 or $-1/M$, which guides synchronization. The cross-correlation takes large values, which degrades rejection.

6.3.2.3 Gold Codes

Gold codes are generated by XORing preferred m-sequences to bound the cross-correlation function to three values, improving rejection at the expense of worse auto-correlation. Figure 6.4 plots the best-case and worst-case rejection of m-sequence and Gold codes of different code length, indicating better rejection for Gold codes [2].

6.3.2.4 Walsh Codes

An M-length Walsh code is constructed from

$$W_2 = \begin{bmatrix} 1 & 1 \\ 1 & -1 \end{bmatrix}, W_{2^k} = \begin{bmatrix} W_{2^{k-1}} & W_{2^{k-1}} \\ W_{2^{k-1}} & -W_{2^{k-1}} \end{bmatrix}, \tag{6.6}$$

where $k = \log_2(M)$. Walsh codes have zero cross-correlation for any delay, making them ideal for rejection. The auto-correlation has many peaks due to their periodic properties, resulting in difficult synchronization. The repetitive pattern of Walsh codes spreads the signal at specific harmonically related frequency bins, resulting in reduced spreading efficiency.

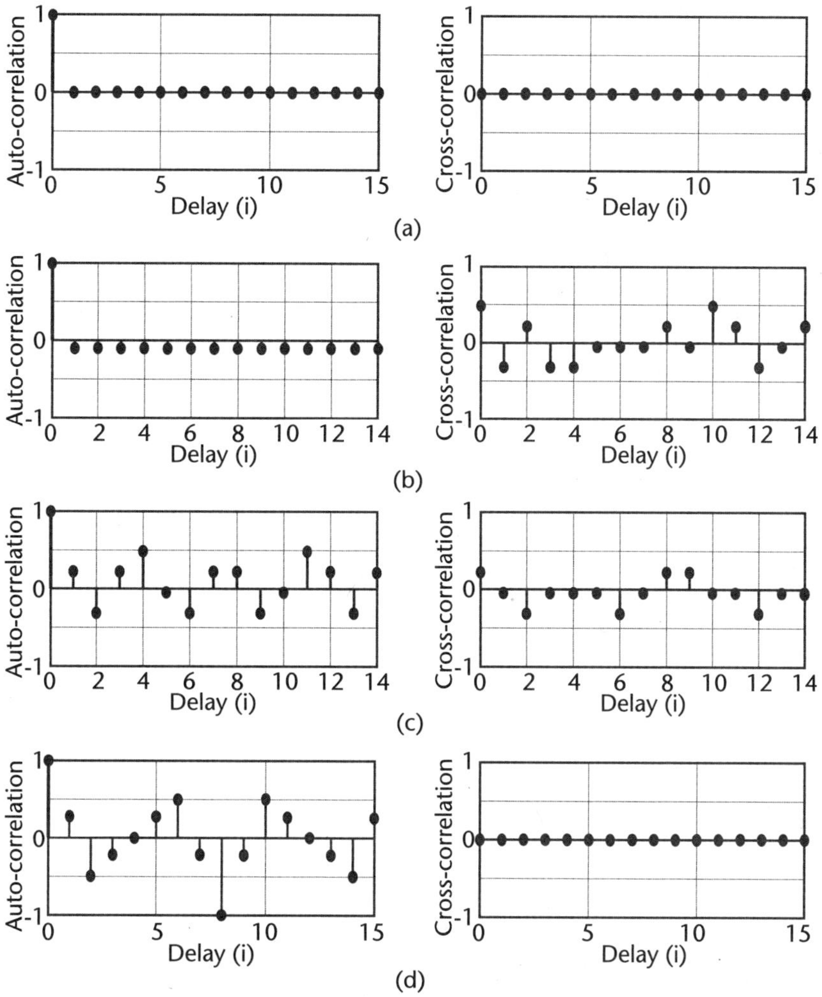

Figure 6.3 Auto-correlation and cross-correlation for different code families: (a) ideal code; (b) M-sequence code; (c) Gold code; and (d) Walsh-Hadamard code. (© 2020 IEEE. Reprinted, with permission, from [3].)

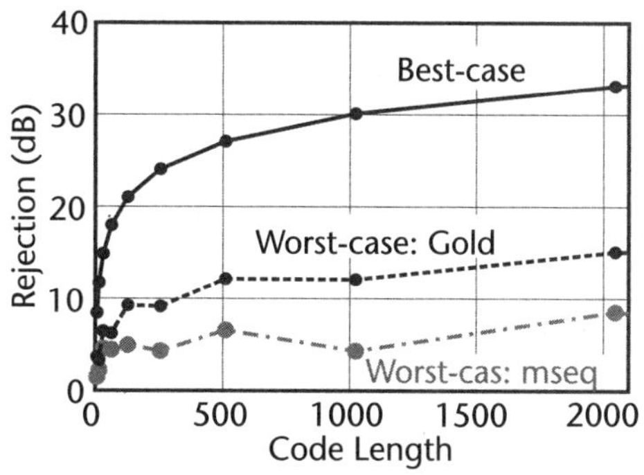

Figure 6.4 Best-case and worst-case rejection versus code length for m-sequence and Gold codes.

6.3.2.5 Composite Codes

To increase the code length for processing gain ($PG = 10 \times \log_{10}(M)$), synchronization and rejection improve at the expense of the symbol rate for the same spread BW. Instead of increasing the code length, longer codes are constructed through combining two different code families similar to long code scrambling techniques used in wideband CDMA systems [4].

Code multiplication is one approach to synthesize desirable properties [5]. For example, multiplying a 128-length random PN code with repeated 8-length Walsh codes results in multiple 128-length codes that combine the synchronization benefits of random codes with the superior rejection of Walsh codes. Figure 6.5(a) plots the auto-correlation and cross-correlation of the resulting code.

Another approach repetitively cycles through Walsh codes based on the phase of Gold codes or m-sequences [6]. For example, 8-length Walsh codes are repeated 15 times based on the phase of a 15-length Gold code, resulting in multiple 120-length codes with improved rejection and synchronization, as shown in Figure 6.5(b).

These composite codes only have good rejection when synchronized, placing stricter requirements compared to Walsh codes that are orthogonal at any time lag. Figure 6.6 plots the rejection and IL for orthogonal and matched codes constructed through code multiplication (a similar profile was observed for code repetition). The rejection exceeds 25 dB for a synchronization error under 10%, while the IL increases by less than 1 dB for 20% synchronization mismatch.

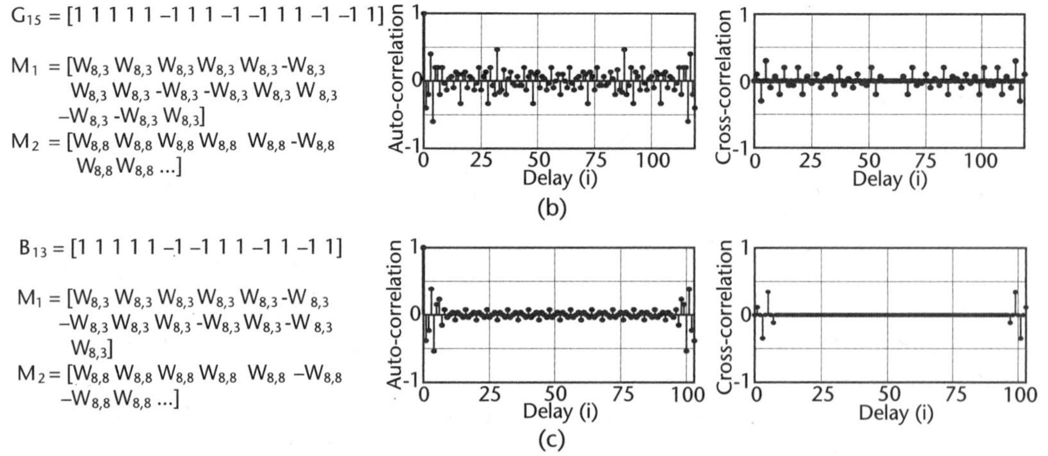

Figure 6.5 Constructing long codes from different code families: (a) code multiplication using 128-length random code and repeated 8-length Walsh code; (b) Walsh code repetition based on 15-length Gold code; and (c) Walsh code repetition based on a 13-length Barker code.

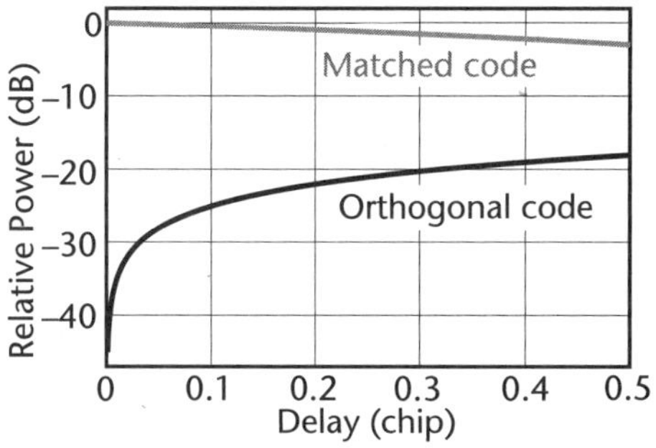

Figure 6.6 Effect of synchronization error on IL and rejection.

To avoid the synchronization constraints, Barker codes are constructed such that $\left|\sum_{i=1}^{M-k} b_i b_{i+k}\right| \le 1$ where b_i is the Barker code string of ±1. Barker codes have good auto-correlation, facilitating synchronization. Limited numbers of Barker codes prevent stand-alone usage in a multi-user system (for example, there is only one Barker code of length 13). Combining Barker codes with Walsh codes through code repetition retains the auto-correlation peak and high rejection over most lags, as shown in Figure 6.5(c) for a 104-length code constructed by combining a 13-length Barker code and 8-length Walsh code.

6.3.3 Jammer and Noise Tolerance

One of the advantages of code-domain IBFD is robustness against noise and jammers [7]. In addition to Tx SI tolerance, code-domain front ends (CDFEs) tolerate continuous-wave (CW) and modulated jammers. Also, the CDFE can receive signals below the noise floor due to the processing gain (PG) of the PN code. Figure 6.7 plots the simulated spectrum at the input of the CDFE, including the desired signal of 0.78125 MSymbol/sec spread by a 128-length PN code along with noise, two in-band CW jammers, and a 5 MSymbol/sec modulated jammer. After code correlation, the desired signal is despread at the Rx, which increases its power within the signal BW while jammers and noise are spread across the spread BW. The output spectrum indicates the correct reception of the desired signal even with the existence of noise and jammers at higher power levels. The improvement in

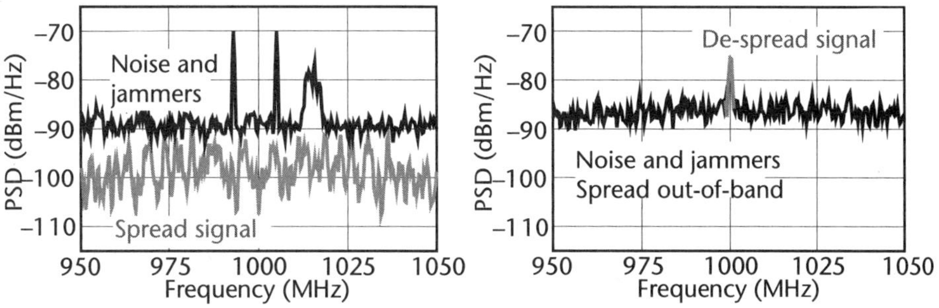

Figure 6.7 Noise and jammer tolerance of RF code-domain signal processing: (a) input spectrum including a desired spread signal below the noise and jammers power level; and (b) output spectrum after code correlation de-spreading the desired signal while rejecting noise and jammers. (© 2020 IEEE. Reprinted, with permission, from [3].)

the signal-to-interference-plus-noise ratio (*SINR*) is given by the ratio between the input and output *SINR*,

$$SINR_{imp} = 10\log_{10}\frac{SINR_o}{SINR_i} = 10\log_{10}\frac{BW_{spread}}{BW_{sig}} = PG. \tag{6.7}$$

6.3.4 Multipath Tolerance

The wireless channel between the Tx and Rx is characterized with multiple paths, as shown in Figure 6.8. These paths cause intersymbol interference (ISI) as the symbol period approaches the path delays, degrading the Rx signal's SNR.

As discussed in Chapter 1, limited Tx-Rx isolation at the antenna causes some of the Tx SI signal to leak to the Rx. The antenna reflection (AR) causes the Tx SI to be reflected to the Rx with a different attenuation and delay compared to the direct path (DP). Meanwhile, the Tx SI also generates reflection paths (RPs) from nearby objects that reappear at the Rx with significant power. The multipath nature of this SI channel increases the power and area overhead of RF cancellers to cover such large delay spreads in the order of tens of nanoseconds, as mentioned in Chapter 8.

The received signal at the full-duplex transceiver is expressed as

$$\begin{aligned} RF_{in}(t) &= m_{Rx}(t) * h_{CH}(t) + m_{SI}(t) * h_{SI}(t) \\ &= \sum_{i=1}^{I} h_{CH,i} \times m_{Rx}\left(t - t_{CH,i}\right) + \sum_{j=1}^{J} h_{SI,j} \times m_{SI}\left(t - t_{SI,j}\right), \end{aligned} \tag{6.8}$$

where m_{Rx} and m_{SI} are the Rx and Tx SI message signals; $h_{CH,i}$, $t_{CH,i}$ are the gain and delay of the different multipaths between the far Tx and Rx; $h_{SI,j}$, $t_{SI,j}$ are the gain and delay of the SI channel as illustrated in Figure 6.8; and * denotes the convolution operation. For an IBFD transceiver, the Rx and Tx SI signals are found from the multiplication of the corresponding signal by its PN code.

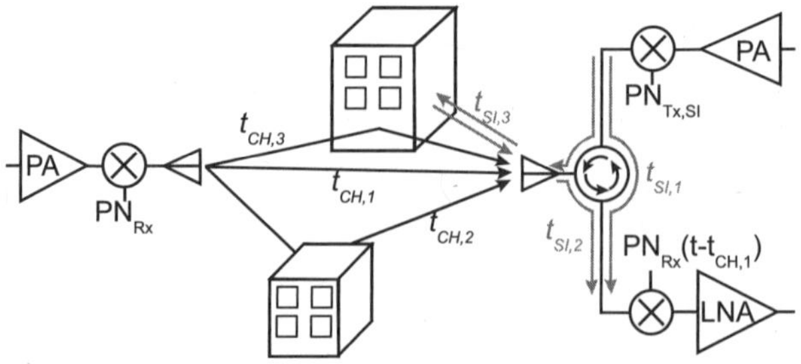

Figure 6.8 IBFD link in multipath environments. Multipath propagation between the Tx and Rx affect the quality of the received signal and multipath between the Tx SI and Rx complicates the design of an RF canceller.

$$
\begin{aligned}
RF_{in}(t) &= \sum_{i=1}^{I} h_{CH,i} \times Rx\left(t - t_{CH,i}\right) \times PN_{RX}\left(t - t_{CH,i}\right) \\
&+ \sum_{j=1}^{J} h_{SI,j} \times Tx_{SI}\left(t - t_{SI,j}\right) \times PN_{Tx,SI}\left(t - t_{SI,j}\right)
\end{aligned} \tag{6.9}
$$

One of the unique properties of PN codes is that they only correlate synchronized signals that are tagged with the same PN code in (6.2). This is useful for IBFD for two reasons. First, it allows operation in multipath environments. Moreover, it allows the rejection of the main Tx SI as well as mismatch and environment-generated reflections. Thus, the output of the CDFE after multiplication by the synchronized PN code of the main path ($PN_{Rx}(t - t_{CH,1})$) is

$$
RF_{out(t)} = PN_{Rx}\left(t - t_{CH,1}\right) \times RF_{in}(t) \approx h_{CH,1} \times Rx\left(t - t_{CH,1}\right). \tag{6.10}
$$

Only the desired signal is correlated at the Rx while rejecting multipath Rx components. Also, the Tx SI and its delayed reflections are rejected by code orthogonality. The CDFE can also be configured as a rake Rx with multiple fingers, where each finger correlates a specific multipath component by synchronizing the PN_{Rx} code delay to the path delays, increasing the Rx SNR [8].

6.4 CMOS Techniques for RF Code-Domain Rx Front Ends

6.4.1 Modulating N-Path Filters with Pseudorandom Noise Codes

The circuit implementation of the Rx code correlators offers several technical challenges. RF code-domain signal processing requires PN code correlation and filtering in the RF domain without penalizing the Rx linearity or NF. Advances in RF CMOS processes allow the implementation of tunable high-Q, high-linearity RF filters using N-path techniques, similar to those used to create the nonmagnetic circulators covered in Chapter 4. N-path filters utilize CMOS switches and capacitors to perform RF filtering using the impedance translation property of the 1/N duty cycle clock pulses of the N-path filter switches [9, 10], resulting in linear, low-loss, high-Q RF filtering that is compatible with chip integration. N-path filters can have more than a 10-dBm 1-dB compression point and their input third-order intercept point (IIP3) can exceed 30 dBm, which are sufficient to handle 30-dBm Tx SI, given that the CDFE is proceeded by an antenna interface offering 20 to 30 dB of rejection [11].

N-path filters can be modified to offer frequency and code selectivity. A code-selective filter (CSF) can be constructed by modulating the N-path filter switches with PN codes on top of the $1/N$ duty cycle clock pulses. Modulating the input RF signal by ±1 according to the PN code is done by adding an extra pair of switches to the N-path differential switches, as shown in Figure 6.9(a). The upper pair of switches are enabled when the PN code is 1 and the RF signal passes as it is, while the lower pair is enabled when the PN code is 0, reversing the input RF signal phase. Another option for modulating N-path filters is through selecting the clock

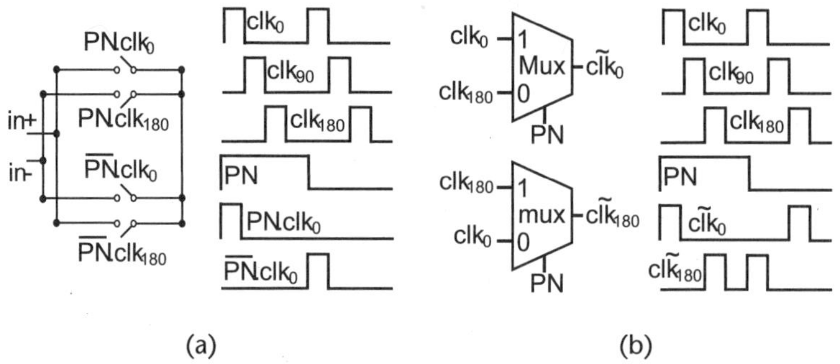

Figure 6.9 Modulating N-path filters by PN codes: (a) using extra switches to reverse the phase of the input RF signal; and (b) using multiplexers to select between two differential clock phases.

pulse between 2 differential phases (i.e., select either clk_0 or clk_{180}, clk_{90} or clk_{270}, ...) using a multiplexer controlled by the PN code, as shown in Figure 6.9(b).

Modulating N-path filters with a PN code results in three CSF architectures based on N-path architectures [9, 12, 13]. These are the code-selective bandpass filter (BPF), code-selective notch filter, and LNA with CSF feedback.

6.4.2 Code-Selective Bandpass Filters

An N-path BPF translates a lowpass resistor-capacitor (RC) filter at DC to the clock frequency of the N-path switches. A code-selective BPF can be implemented either in a shunt or a series configuration [3, 14]. A shunt filter is shown in Figure 6.10(a), where the correlated spectrum is available at baseband, making it unsuitable for an RF transparent approach (where the filter is inserted at RF before the LNA of a receiver). For an RF CSF, the shunt filter can be modified by performing PN modulation ahead of the shunt filter by adding chopping switches [6] or by using a series CSF. A series CSF is shown in Figure 6.10(b). Only the first N-path switches are modulated by PN codes for an RF transparent architecture. Figure 6.10(c) shows the simulated S21 of 8-path shunt and series CSFs with PN codes disabled. The switch resistance is 5Ω and the capacitance is 50 pF. Series filters have better out-of-band rejection at the expense of higher power consumption and IL due to extra switches. Modulating the filter switches with PN codes either correlates the input spectrum inside the BPF BW for a matched code or spreads it out-of-band for orthogonal codes. Figure 6.10(d) shows the simulated input spectrum of a 6.25-MSymbol/sec signal modulated by a 16-length Walsh code at 100 MChip/sec. The spectrum at the CSF RF output is shown in Figure 6.10(e) and at the baseband capacitors of both the shunt and series filters similarly in Figure 6.10(f) for matched and orthogonal PN codes, indicating how the CSF rejects the orthogonally coded spectrum.

The shunt and series CSFs can achieve up to 50 dB of rejection across the signal BW using Walsh codes. The rejection with other code families is limited to the PG of the code. This rejection can be improved using code-selective notch filters.

6.4.3 Code-Selective Notch Filters

Code-selective notch filters are implemented by modulating shunt or series N-path notch filters by the PN code of the Tx signal [3, 14]. A shunt notch filter uses

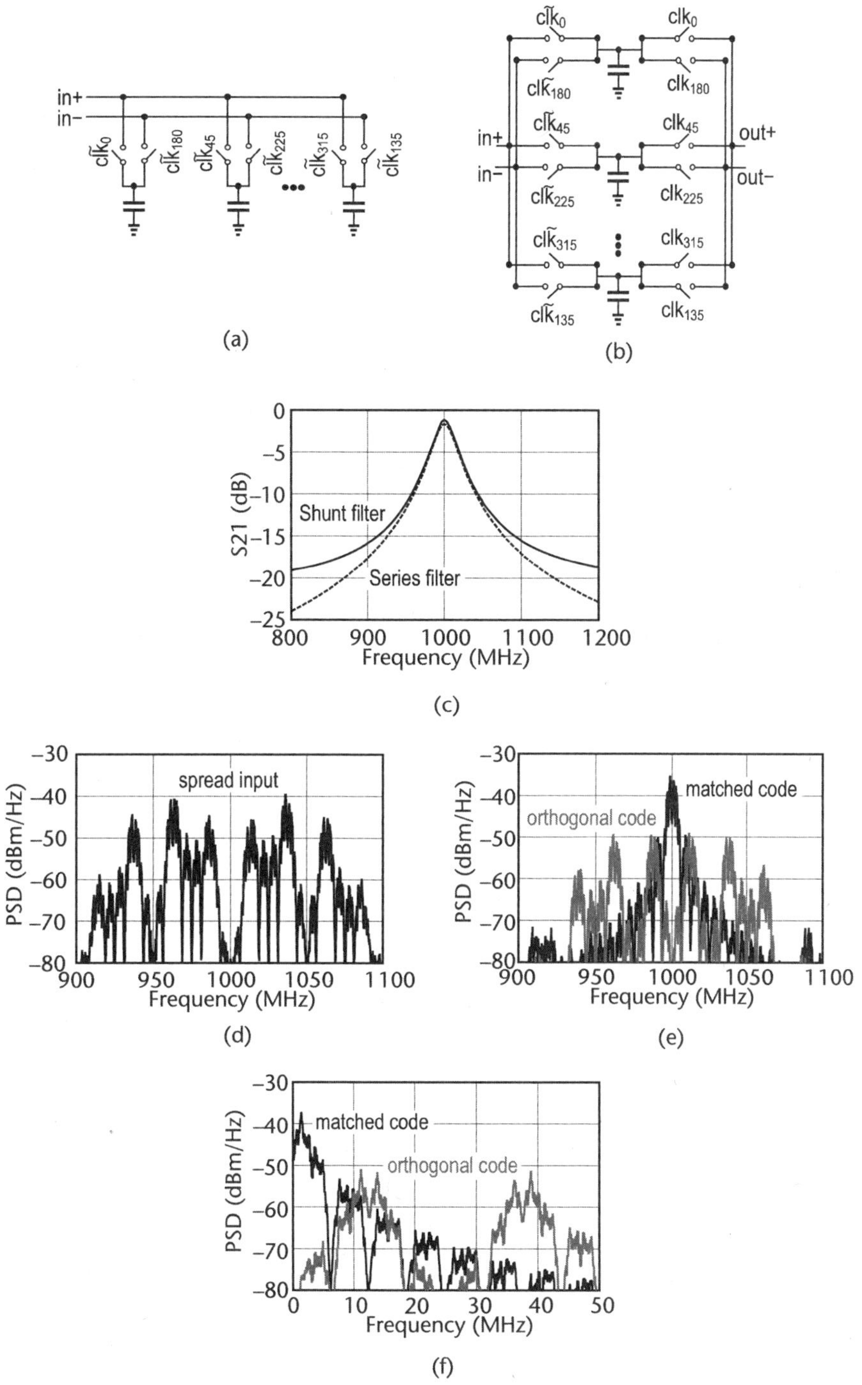

Figure 6.10 Code-selective BPFs: (a) shunt code-selective BPF; (b) series code-selective BPF; (c) simulated S21 of the shunt and series filters; (d) input RF spectrum modulated by Walsh code; (e) output RF spectrum of the series filter for matched and orthogonal codes; and (f) spectrum at baseband capacitors for matched and orthogonal codes.

impedance transformation with a quarter-wave transmission line to transform the bandpass response to a notch response. A series notch filter frequency translates a highpass RC filter at DC to the RF of the switches' clock pulses. Both filters are shown in Figure 6.11, along with simulated $S21$ for 8-path filters with a switch resistance of 5Ω and a capacitance of 50 pF and 10 pF for the shunt and series filters, respectively.

Shunt CSFs have lower IL and power consumption at the expense of larger area and limited tuning range (due to the $\lambda/4$ transmission line), compared to series CSFs. The simulated RF output spectrum of the code-selective notch filters is shown in Figure 6.11(d) for matched and orthogonal codes. For matched codes, the input spectrum is correlated inside the notch to be rejected. For orthogonal codes, the input spectrum is spread out of the notch and passes to the output with minimal IL. Both shunt and series filters are transparent at RF, given that the second switches of the series filter are modulated by PN codes. The iterative multiplication by the same PN code restores the spectrum of the input signal ($PN \times PN = 1$), which is required for the notch filter to be followed by a code-selective BPF and verified by comparing the input spectrum in Figure 6.10(d) with the notch filter orthogonal code output spectrum in Figure 6.11(d).

Figure 6.12 plots the rejection and IL of cascaded code-selective notch and bandpass filters using random PN codes with varying code length at a chip rate of 128 Mchips/sec. Without the notch filter, the rejection is 21.8 dB, 17.3 dB, and 14.2 dB for 128-length, 64-length, and 32-length codes, respectively (~PG). Increasing the notch BW increases the rejection at the expense of increased IL. Simulations indicate an optimal trade-off for a notch BW around 3 MHz, which increases the rejection to 36.8 dB, 31.7 dB, and 24.4 dB with an IL increase of

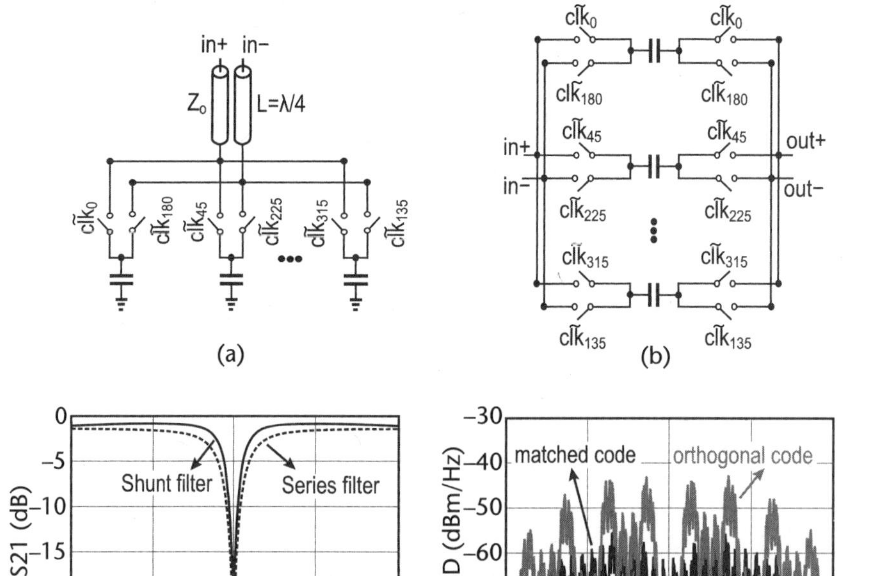

Figure 6.11 Code-selective notch filters: (a) shunt code-selective notch filter; (b) series code-selective notch filter; (c) simulated S21 of the shunt and series filters; and (d) output RF spectrum for matched and orthogonal codes.

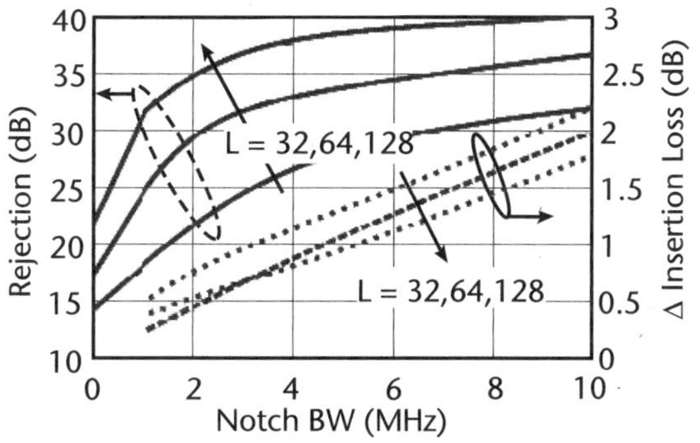

Figure 6.12 Effect of notch filter on rejection and IL for random PN codes with varying length. The notch filter increases the rejection beyond the codes' PG. (© 2020 IEEE. Reprinted, with permission, from [3].)

0.6 dB, 0.7 dB, and 1 dB for 128-length, 64-length, and 32-length codes, respectively. Evidently, the notch filter increases the rejection beyond the PG and is more profound with higher-length codes.

6.4.4 Low-Noise Amplifier with Code-Selective Feedback

An LNA with CSF feedback is another approach to implement CDFE with lower NF [5, 15]. An LNA with CSF feedback avoids the IL of the RF filters by implementing the filtering and code selectivity in the feedback path of the LNA, as shown in Figure 6.13(a). A notch CSF is added in the negative feedback path of the LNA such that the overall response is bandpass for matched codes.

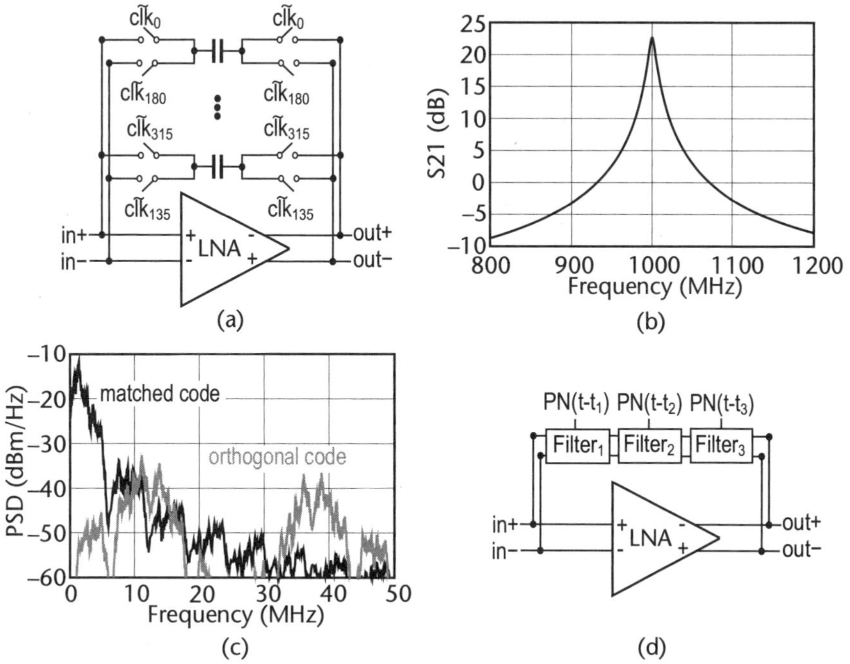

Figure 6.13 LNA with CSF feedback: (a) schematics; (b) simulated S21; (c) spectrum at the feedback capacitors for matched and orthogonal codes; and (d) rake Rx implemented as a LNA with feedback CSFs allowing multipath signal reception.

Besides the NF advantage, this architecture reduces the switch resistance and capacitance values required for performance similar to the BPF due to the Miller effect [13]. Figure 6.13(b) shows the simulated *S*21 of the LNA with 8-path feedback with a switch resistance of 5Ω and a capacitance of 10 pF. The LNA has narrower BW and higher rejection compared to the response in Figure 6.10(c). Figure 6.13(c) shows the simulated spectrum at the N-path capacitors for matched and orthogonal codes. Another advantage of the LNA with CSF feedback is that multiple filters can be connected in series in the feedback path [5]. This can be used as a rake Rx, where each filter is modulated by a delayed version of the Rx PN code allowing multipath reception, improving the overall SNR in a power-efficient architecture utilizing a single LNA, as shown in Figure 6.13(d) [15].

6.5 CMOS Techniques for Tx Code-Domain Modulation and Spectral Shaping

Complementary to the Rx code-domain architectures, the Tx PN modulation also plays an important role. Two types of modulators have been used for Tx modulation [16, 17]: a differential chopper configuration shown in Figure 6.14(a) and a reflection phase shifter (RPS) style shown in Figure 6.14(b). The chopper configuration generally allows for better phase and amplitude balance, while the RPS has lower losses by using a hybrid coupler. Since higher chip rates achieve better rejection, it is advantageous to place the code-domain modulator after the PA to avoid adding design constraints on the BW of other transmit components, such as the PA and the analog RF canceller.

6.5.1 High-Power, High-Modulation Bandwidth CMOS Switch Design

Because the modulator is after the PA, the signal power can be up to 40 dBm, and any IL incurred by the modulator has a high impact on system efficiency. Silicon-on-insulator (SOI) CMOS processes are desirable for Tx modulators due to their low cost and their ability to be integrated on the same chip as the Rx filters for code-synchronization simplicity. SOI also offers low parasitics at RF and the best performance in terms of loss and switching speed. CMOS typically lacks in power handling, where a single switch might have a compression point of 5 to 10 dBm in a 50Ω environment due to voltage compression.

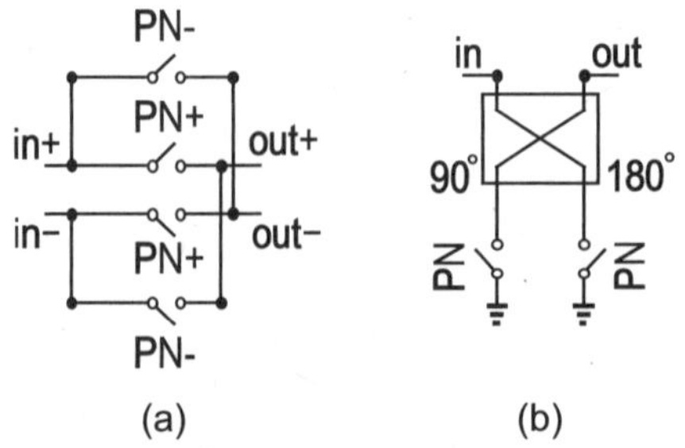

Figure 6.14 Types of PN modulators that can be used in the transmit chain.

To overcome the linearity shortcomings of SOI CMOS switches, RF switches incorporate large resistors at the gate of the device while combining several devices in series. The large gate resistor allows the gate voltage to swing in unison with the input RF voltage, preventing unwanted turn-off of the device when the switch is on. The series combination (or stacking) of the devices allow only a small voltage drop from drain to source when the device is off, preventing undesirable turn-on of the device, as illustrated in Figure 6.15. These types of switches are common in CMOS implementations of TDD systems. When applied to code-domain IBFD systems, resistive gate architectures lack switching speed due to RC rise times, limiting the achievable spreading BW. These rise times create a trade-off between power handling and modulation BW, illustrated in Figure 6.16(a), while the stacking creates a trade-off between power handling and insertion loss due to increased series resistance in the switch.

To circumvent the RC BW-power trade-off, different high-impedance elements can be considered. Figure 6.16(b) shows how the contours of maximum power handling can be moved upwards significantly on the fractional BW (FBW) axis by using a gate inductance rather than a gate resistance. The void through the middle of the contours represents a resonant region where the gate voltage will exceed the recommended process voltage. This can be dampened by a small series resistance on the gate, which has minimal impact on FBW.

Despite the benefit of inductive terminations for switching speed, stacking many switches requires a prohibitive number of inductors. A composite ladder switch (Figure 6.15(c)) can be used instead with an inductor terminating the control line. This type of switch allows the gates to see high capacitance when turned off but combines the series switches into one device with a gate inductance when turned on. Such a composite device reduces the number of large inductors since only one inductor per switch is needed.

These composite ladder switches also perform well as shunt switches without a high-impedance control. When the switch is on, the input terminal is shorted to ground, and when the switch is off, the capacitive division from the secondary devices is sufficient to prevent the device from compressing. An alternative PN

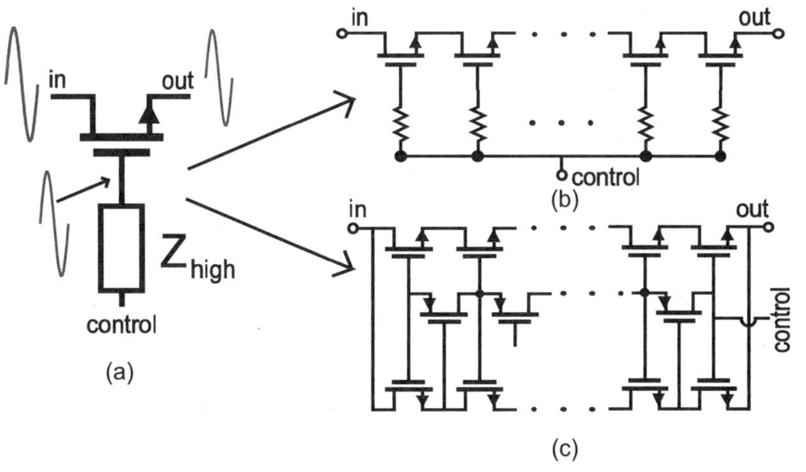

Figure 6.15 High-power CMOS switch implementations: (a) voltage division with a high-impedance element; (b) resistively stacked switch; and (c) ladder switch for faster modulation speeds.

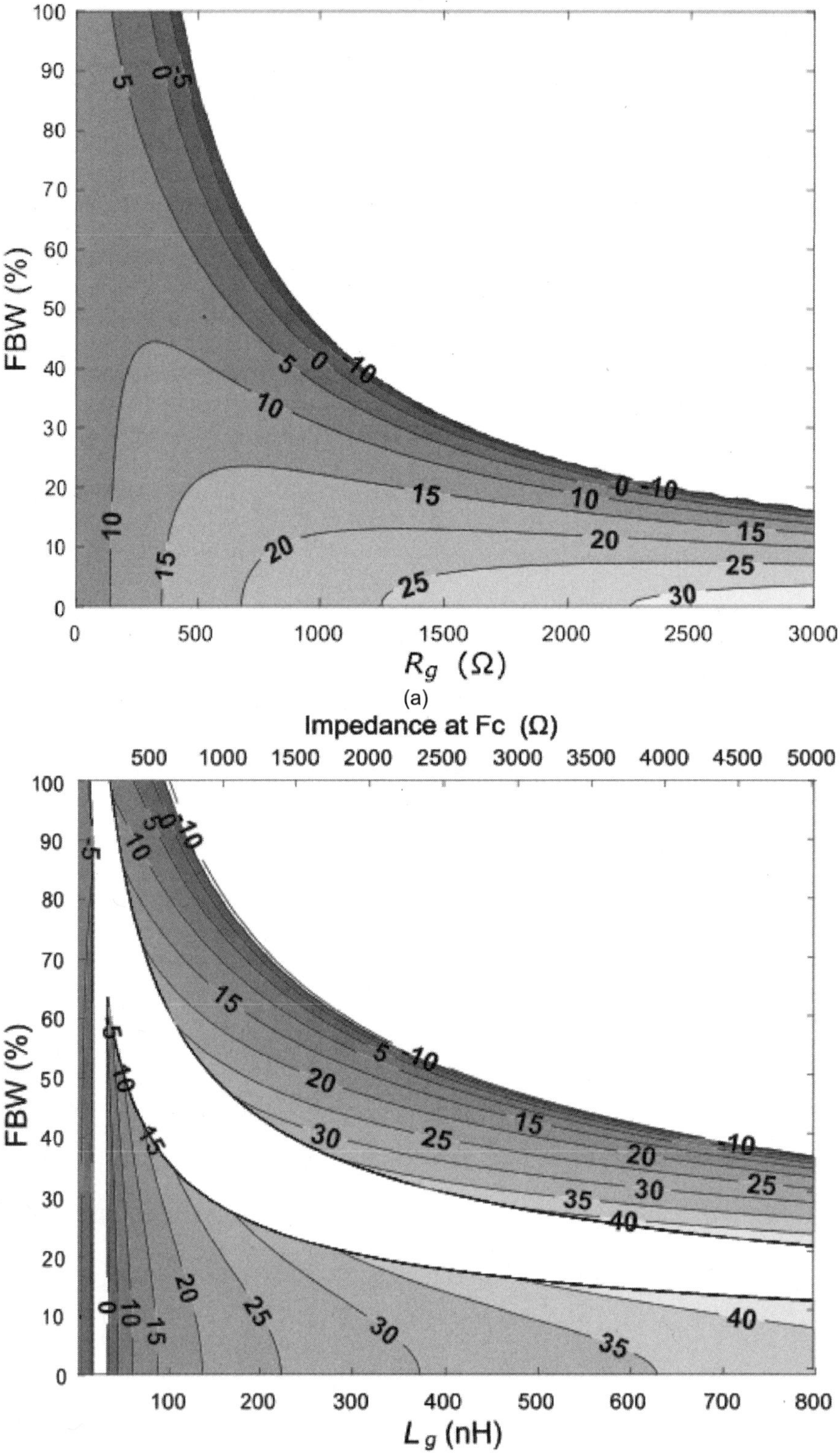

Figure 6.16 Contours of maximum power handling in dBm for various FBWs and gate impedances for (a) resistive gate impedance; and (b) inductive gate impedance. (© 2018 IEEE. Reprinted, with permission, from [16].)

modulator using only shunt switches in a reflection phase shifter is demonstrated in [17] and outlined in Figure 6.14(b). A comparison of the different PN architectures is shown in Figure 6.17 to demonstrate the impact of using nonresistive gate terminations. Each of the designs cited is capable of handling Tx powers > 1W.

6.5.2 Spectral Shaping Approaches

One central challenge of generating the PN sequences after the DAC is that there is no opportunity to pulse shape the coding waveforms. This leads to sharp edges in the time-domain waveforms and therefore a high level of out-of-band power. An example of a signal spectrum with pulse-shaped and square-wave codes is shown in Figure 6.18, where the harmonic power from the square-wave codes causes significant interference across a large BW.

If additional switching BW is available, it can be used to implement digital filtering at the Tx modulator in conjunction with a low-order BPF. One such technique is sigma-delta modulation ($\Sigma\Delta$M). An example of the signal chain for an N-order $\Sigma\Delta$M used for PN Tx modulation is shown in Figure 6.19. When designing for modulation after the PA, high-order $\Sigma\Delta$Ms can cause instability if the input signal is not backed off significantly, which leads to high IL [18].

Like $\Sigma\Delta$M approaches, it is possible to optimize the output bits and timing for both better noise-shaping properties and a reduced oversampling load for the PN modulators. Optimizing the output sequence leads to a pulse-encoded transitions (PET) technique outlined in Figure 6.20. Each transition in the PN code has two more transitions added to form $y(t)$, as in Figure 6.20(a). The resulting $y(t)$ can be decomposed into two signals: $x(t)$, the original PN code; and $p(t)$, the pulse encoding function, as in Figure 6.20(b, c) such that

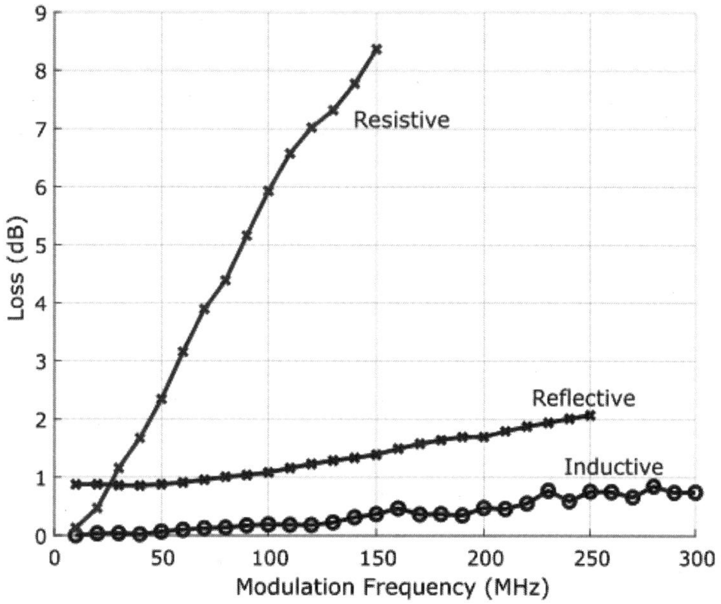

Figure 6.17 Comparison of the insertion loss with increasing modulation speed for inductive, resistive or reflective modulators. Test boards were de-embedded for resistive and inductive data but not for reflective data. Losses are normalized to the minimum of the inductive type switches.

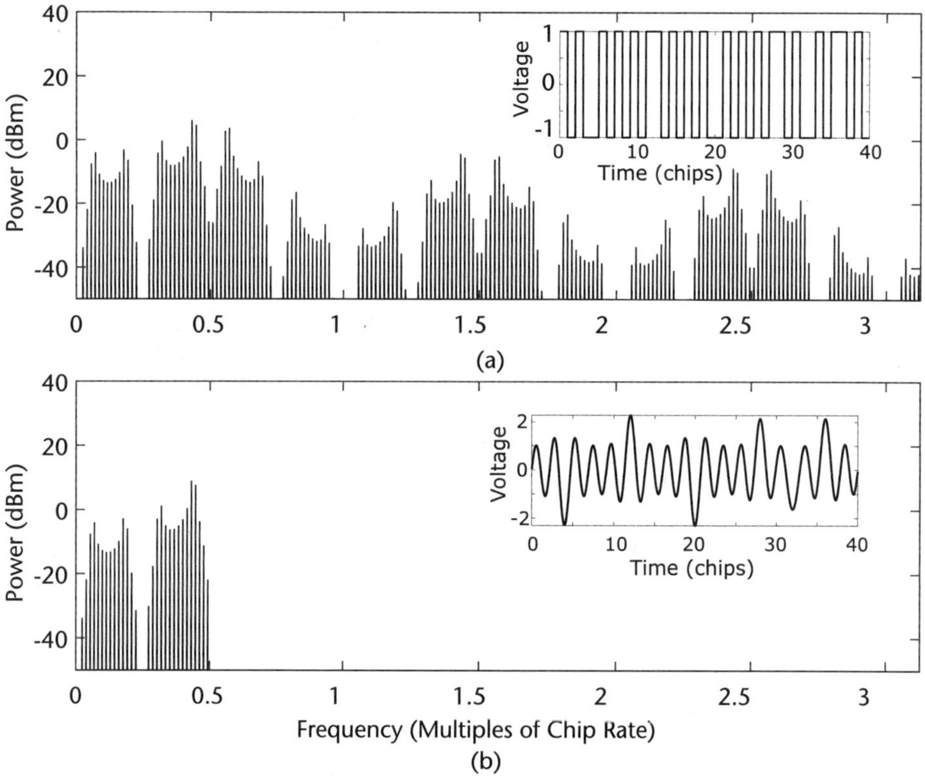

Figure 6.18 Comparison of code spectrum for a 128-length Walsh code (a) without; and (b) with pulse shaping as well as the corresponding time-domain signals.

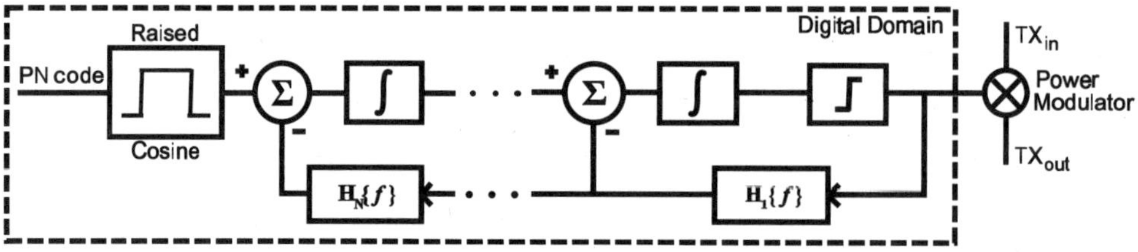

Figure 6.19 Example signal path for sigma-delta modulation (ΣΔM) for PN transmit codes.

$$y(t) = x(t) \cdot \prod_{n=1}^{N} p_n(t) x \rightarrow \infty \tag{6.11}$$

Taking the Fourier transform results in,

$$a_k = \frac{2}{\pi k} \sin\left(\frac{2\pi k T_1}{T}\right)\left[(-1)^N + 2\sum_{n=1}^{N}(-1)^{(N-n)} \cos\left(\frac{\pi k \Delta t_n}{T}\right)\right]. \tag{6.12}$$

For N added pulses, there are N equations with corresponding solutions for Δt_n to cancel N harmonics.

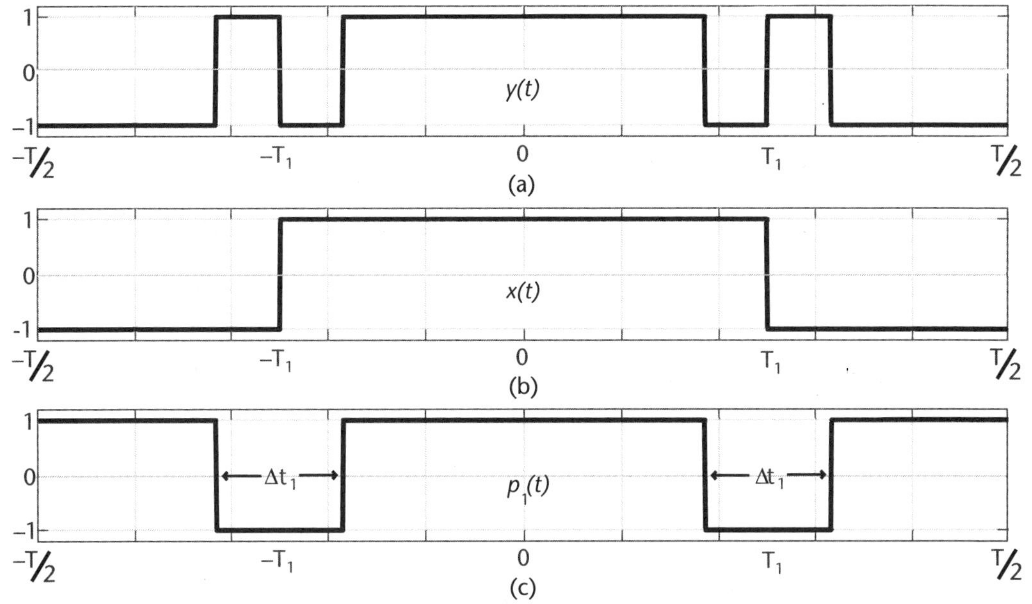

Figure 6.20 (a) Optimized pulse encoded transitions (PET) signal; (b) with the original pulse; and (c) the multiplying pulses. (© 2019 IEEE. Reprinted, with permission, from [17].)

The above analysis assumes periodic square waves for $y(t)$; however, the solutions for (6.12) do not depend on the pulse width of the original waveform, $x(t)$. This implies that the theory is valid for any arbitrary sequence of bits and that k does not need to be integer-valued. In the limit where a PN code is infinitely long, the spectrum approaches a random sequence of bits. Figure 6.21 shows how random spectrums can be nulled using different values of N and k. Measurements indicate that shaping the PN codes spectrum slightly affects the PN codes cross-correlation, reducing the CDFE rejection by 1 to 2 dB.

The primary advantage of PET over ΣΔM is a lower RF oversampling ratio (OSR), quantified by the minimum Δt of the output waveform. In terms of RF OSR, PET can outperform both first-order and second-order ΣΔM, as shown in Table 6.1, where the adjacent channel leakage ratio (ACLR) of the next adjacent channel is calculated for various values of PET N and for equivalent OSR in ΣΔM.

6.6 Practical Considerations

Code-domain techniques require a trade-off between code length and signal BW. For systems that demand large numbers of users, the codes should be long and higher data rates (>1 MS/s) will require wide RF BW. The large FBW will demand antennas with RF response variation, complicating the interaction of the filter and the antenna. An alternative would be applying code-domain approaches in millimeter-wave bands where available signal BW is large. The RF code-domain signal processing techniques are difficult to realize, even in CMOS SOI processes above 6 GHz.

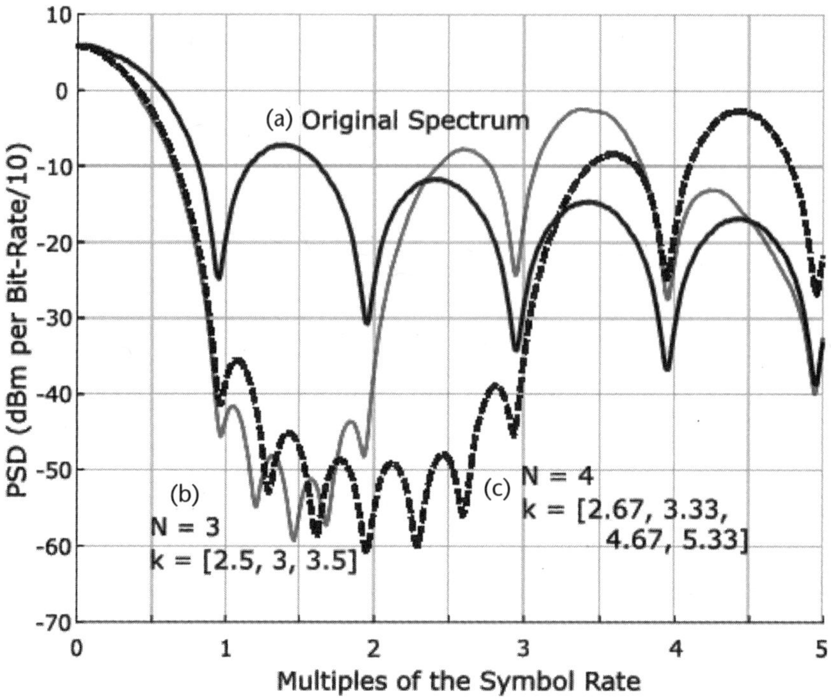

Figure 6.21 Examples of spectrum for combinations of N and k. (© 2019 IEEE. Reprinted, with permission, from [17].)

Table 6.1 Comparison of ACLR Values Achieved Using Different *N* with ΣΔMs Using Equivalent RF OSR

		Simulated ACLR (dBc)				
		PET N			*ΣΔM Order*	
PET N	*RF OSR*	*1*	*2*	*3*	*1*	*2*
1	18	25.8	—	—	16.1	10.7
2	32	—	38	—	25.5	25.9
3	58	—	—	49.1	34.4	39.2

6.7 Conclusion

6.7.1 Summary

This chapter introduced code-domain techniques for RF filtering and Tx SIC. Background on the use of PN codes for IBFD was introduced and an analysis of the IL, rejection, and power handling suggests that CMOS technologies are capable of RF signal processing. RF code-domain filters were demonstrated in a CMOS SOI process and applications to shunt and series filtering were demonstrated. RF code-domain filtering can support Tx SI rejection of around 50 dB while operating up to a Tx 1-dB compression point of 10 dBm. Additionally, RF code-domain modulation was discussed using a combination of high-power CMOS SOI RF switches supporting Tx power levels of >1W. Schemes to eliminate out-of-band switching emissions through wideband pulse transition encoding allows adjacent channel emissions to be suppressed by up to 40 dB.

6.7.2 Future Directions

Future efforts will include cointegration of the RF CDFEs into a common chip with an RF nonmagnetic circulator, similar to those discussed in Chapter 4. Additional research will also be directed towards applications of code-domain IBFD, including multi-element radar systems and MIMO channel estimation, as mentioned in Chapter 14.

References

[1] Salmasi, A., and K. S. Gilhousen, "On the System Design Aspects of Code Division Multiple Access Applied to Digital Cellular and Personal Communications Networks," *IEEE Vehicular Technology Conference*, May 1991, pp. 57–62.

[2] Proakis, J., and M. Salehi, *Digital Communications*, New York: McGraw-Hill, 2008.

[3] Hamza, A., et al., "High-Rejection RF Code Domain Receivers for Simultaneous Transmit and Receive Applications," *IEEE Journal of Solid-State Circuits*, Vol. 55, No. 7, July 2020, pp. 1909–1921.

[4] Holma, H., and A. Toskala, *WCDMA for UMTS*, New York: John Wiley & Sons, 2001.

[5] Agrawal, A., and A. Natarajan, "An Interferer-Tolerant CMOS Code-Domain Receiver Based on N-Path Filters," *IEEE Journal of Solid-State Circuits*, Vol. 53, No. 5, May 2018, pp. 1387–1397.

[6] AlShammary, H., et al., "A Code-Domain RF Signal Processing Front End with High Self-Interference Rejection and Power Handling for Simultaneous Transmit and Receive," *IEEE Journal of Solid-State Circuits*, Vol. 55, No. 5, May 2020, pp. 1199–1211.

[7] Abu-Rgheff, M., *Introduction to CDMA Wireless Communications*, Boston, MA: Academic Press, 2007.

[8] Price, R., and P. Green, "A Communication Technique for Multipath Channels," *Institute of Radio Engineers*, Vol. 46, No. 3, March 1958, pp. 555–570.

[9] Ghaffari, A., et al., "Tunable High-Q N-Path Band-Pass Filters: Modeling and Verification," *IEEE Journal of Solid-State Circuits*, Vol. 46, No. 5, May 2011, pp. 998–1010.

[10] Mirzaei, A., at al., "Architectural Evolution of Integrated M-Phase High-Q Bandpass Filters," *IEEE Transactions on Circuits and Systems-I*, Vol. 59, No. 1, January 2012, pp. 52–65.

[11] Luo, C., et al., "A 0.4-6-GHz 17-dBm B1dB 36-dBm IIP3 Channel-Selecting Low-Noise Amplifier for SAW-Less 3G/4G FDD Diversity Receivers," *IEEE Transactions on Microwave Theory and Techniques*, Vol. 64, No. 4, April 2016, pp. 1110–1121.

[12] Ghaffari, A., et al., "Tunable N-Path Notch Filters for Blocker Suppression: Modeling and Verification," *IEEE Journal of Solid-State Circuits*, Vol. 48, No. 6, June 2013, pp. 1370–1382.

[13] Park, J., and B. Razavi, "Channel Selection at RF Using Miller Bandpass Filters," *IEEE Journal of Solid-State Circuits*, Vol. 49, No. 12, December 2014, pp. 3063–3078.

[14] AlShammary, H., et al., "Code-Pass and Code-Reject Filters for Simultaneous Transmit and Receive in 45-nm CMOS SOI," *IEEE Transactions on Microwave Theory and Techniques*, Vol. 67, No. 7, July 2019, pp. 2730–2740.

[15] Hamza, A., et al., "A Self-Interference-Tolerant, Multipath Rake Receiver with More Than 40-dB Rejection and 9-dB SNR Multipath Gain in a Fading Channel," *IEEE Radio Frequency Integrated Circuits Symposium*, August 2020, pp. 51–54.

[16] Hill, C., et al., "RF Watt-Level Low-Insertion-Loss High-Bandwidth SOI CMOS Switches," *IEEE Transactions on Microwave Theory and Techniques*, Vol. 66, No. 12, December 2018, pp. 5724–5736.

[17] Hill, C., et al., "Watt-Level, Direct RF Modulation in CMOS SOI with Pulse-Encoded Transitions for Adjacent Channel Leakage Reduction," *IEEE Transactions on Microwave Theory and Techniques*, Vol. 67, No. 12, December 2019, pp. 5315–5328.

[18] Schreier, R., and G. C. Temes, *Understanding Delta–Sigma Data Converters*, New York: John Wiley & Sons, 2005.

CHAPTER 7

Photonic-Based RF Filtering

Vincent J. Urick

7.1 Introduction

The field of microwave photonics [1, 2] emerged decades ago from needs such as wideband point-to-point links, RF signal processing, and analog RF beamforming. The application of microwave photonics technologies to IBFD problems is relatively recent, only gaining traction since the early 2000s [3]. Similar to Chapter 6, this chapter will describe a novel RF filtering capability afforded, not by electronic devices, but by photonics [4–11] that are uniquely suited to suppress interfering signals using an in-band methodology, as highlighted within the IBFD transceiver diagram of Figure 7.1(a) and SIC techniques tree of Figure 7.1(b).

The technique described here filters signals based on their amplitudes, independent of their frequency. Typical microwave filters, photonics-based or otherwise, operate on signals in the frequency domain independent of their amplitudes. The method here is not a limiting function in that the amplitude-based filtering rejects strong signals only, while leaving small signals relatively unaffected. The technique is based on operating basic analog-photonic links in regimes where their RF responses are nonlinear, which is unique in that not all signals experience the same compression. This counterintuitive feature is exhibited by external intensity-, phase-, and polarization-modulated analog-optical links. A thorough explanation requires the theoretical development presented in the following section.

7.2 Concept and Theory

The three photonic architectures under consideration are shown in Figure 7.2. An intensity-modulated link [12] can be implemented by modulating a CW laser with an RF signal via an MZM, as discussed in Chapter 5 and shown in Figure 7.2(a). The MZM can have one or two outputs that would feed photodiodes to convert the signal back into the optical domain. A DC bias is applied to the MZM in order to control the amount of average optical power on each output. Figure 7.2(b) depicts a phase-modulated link [13], where an RF signal is imparted on the optical phase

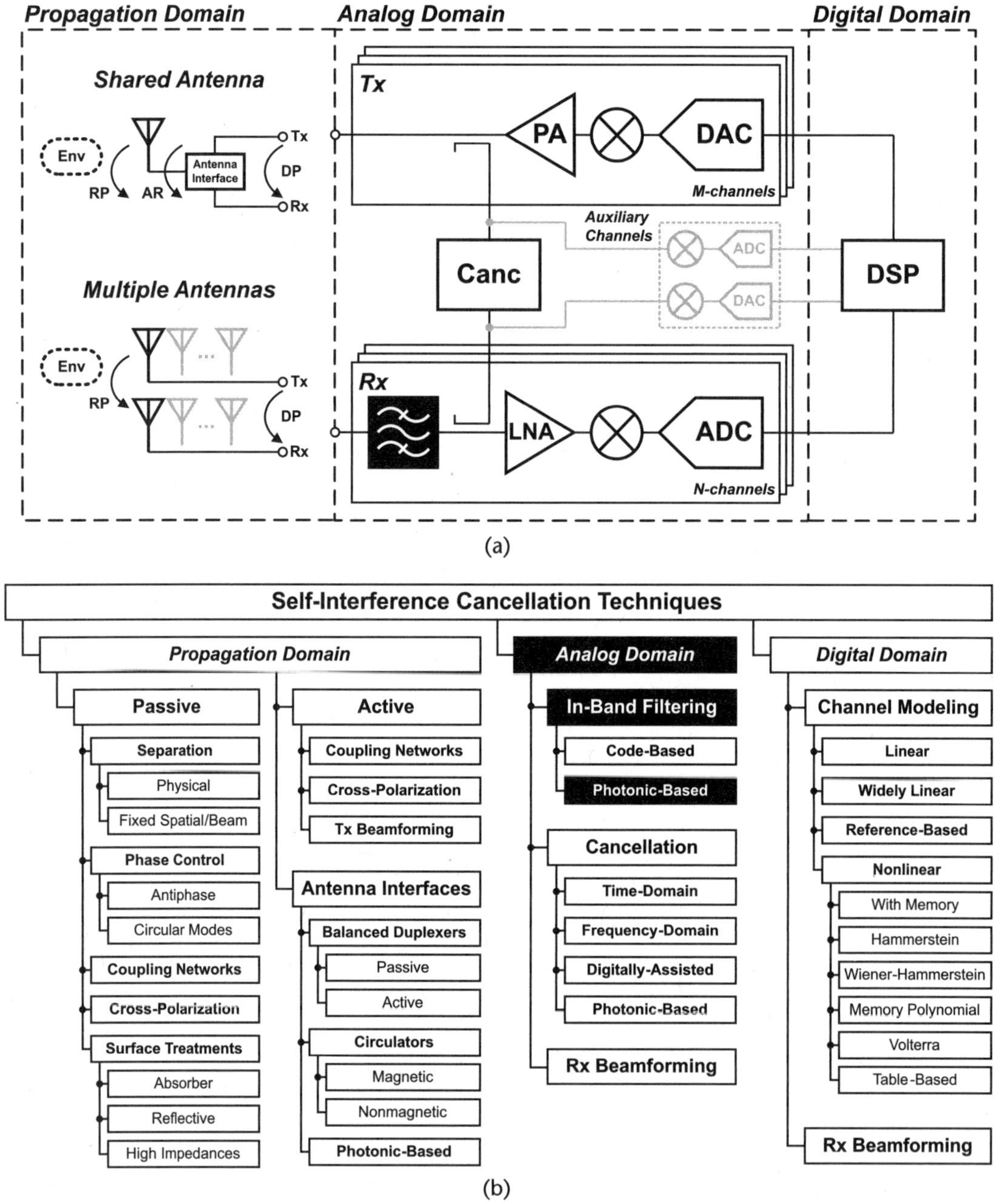

Figure 7.1 Photonic-based RF filtering and specific chapter focus highlighted within (a) a generic IBFD transceiver and (b) a SIC techniques tree diagram.

of a CW laser. The demodulation is achieved with an asymmetric Mach-Zehnder interferometer (MZI) with its outputs connected to two photodiodes. The average optical power on each MZI output is controlled with a DC bias. A polarization-modulated link [14] is shown in Figure 7.2(c), where an RF signal is encoded onto the polarization of a CW laser. Demodulation is achieved with a polarization controller, a half-wave plate (HWP) and a polarization beam splitter, the last of which

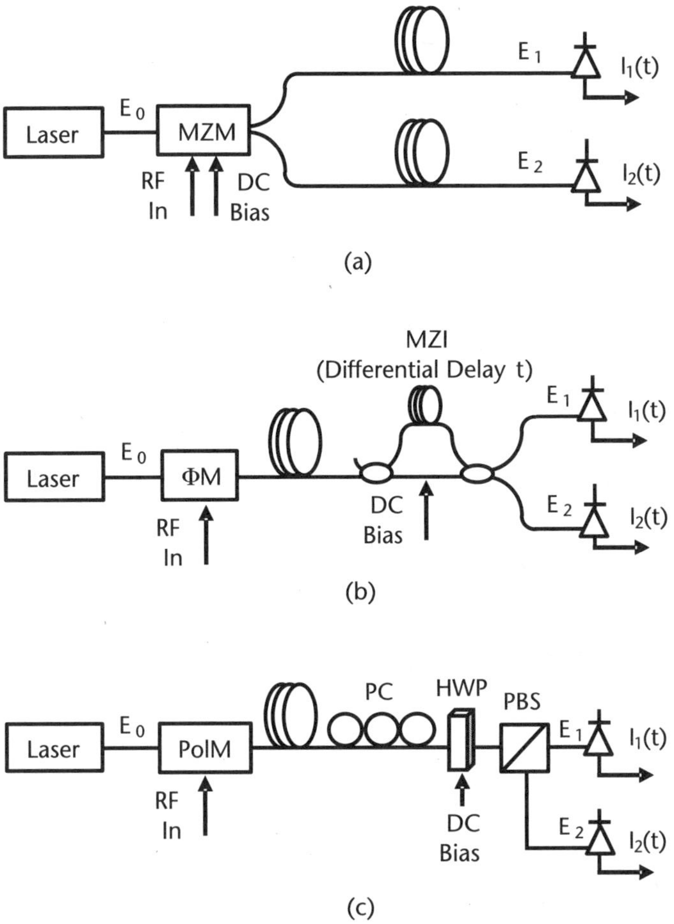

Figure 7.2 Block diagrams for three architectures that afford the novel filtering function. Shown are (a) an intensity-modulated link having a MZM; (b) a phase-modulated link using a phase modulator (ΦM) and MZI; and (c) a polarization-modulated link having a polarization modulator (PolM), a polarization controller (PC), an HWP, and a polarization beam splitter (PBS).

has two outputs. The average optical power in these two outputs is controlled by aligning the HWP.

The structures in Figure 7.2 can be analyzed using a previously developed [15] matrix technique. The transfer functions for each structure can be written as

$$\begin{bmatrix} E_1 \\ E_2 \end{bmatrix} = \frac{\sqrt{I_o}}{2}\begin{bmatrix} 1 & i \\ i & 1 \end{bmatrix}\begin{bmatrix} e^{i\varphi_{dc}/2+i\varphi/2} & 0 \\ 0 & e^{-i\varphi_{dc}/2-i\varphi/2} \end{bmatrix}\begin{bmatrix} 1 & i \\ i & 1 \end{bmatrix}\begin{bmatrix} E_0 \\ 0 \end{bmatrix} \tag{7.1a}$$

$$\begin{bmatrix} E_1 \\ E_2 \end{bmatrix} = \frac{\sqrt{I_o}}{2}\begin{bmatrix} 1 & i \\ i & 1 \end{bmatrix}\begin{bmatrix} T(\tau) & 0 \\ 0 & 1 \end{bmatrix}\begin{bmatrix} 1 & i \\ i & 1 \end{bmatrix}\begin{bmatrix} E_0 e^{i\varphi} \\ 0 \end{bmatrix} \tag{7.1b}$$

$$\begin{bmatrix} E_1 \\ E_2 \end{bmatrix} = \frac{\sqrt{l_o}}{2} \begin{bmatrix} \cos 2\theta & \sin 2\theta \\ \sin 2\theta & -\cos 2\theta \end{bmatrix} \begin{bmatrix} 1 & -i \\ -i & 1 \end{bmatrix} \begin{bmatrix} e^{i\pi/2+i\varphi} & 0 \\ 0 & e^{-i\pi/2-i\varphi} \end{bmatrix} \begin{bmatrix} 1 \\ 1 \end{bmatrix} E_0, \tag{7.1c}$$

where (7.1a) and (7.1b) utilize scalar fields to describe intensity and phase modulation, respectively, and (7.1c) is written in terms of Jones matrices to handle the polarization-modulated structure. In (7.1), l_o is the net optical power loss (or gain) factor, assumed to be the same for the paths to E_1 and E_2. The input optical field is given by

$$E_0(t) = \sqrt{\frac{2P_o}{A}} \left(\frac{\mu}{\varepsilon}\right)^{\frac{1}{4}} e^{i\omega_o t}, \tag{7.2}$$

where P_o is the average optical power, A is the cross-sectional area of the fiber core, μ is permeability of the fiber core, ε is the permittivity of the fiber core, and ω_o is the angular optical frequency of the laser. The phase shift waveform is

$$\varphi(t) = \frac{\pi V(t)}{V_\pi}, \tag{7.3}$$

where $V(t)$ is the voltage waveform applied to the modulator and V_π is the voltage required to shift an MZM from peak to null transmission, induce a π phase shift in a phase modulator, or rotate the polarization by 180° in a polarization modulator. In (7.1a), φ_{dc} is the phase shift induced by the DC bias voltage applied to the MZM. Note that the MZM described by (7.1a) is assumed to be perfectly symmetric with zero chirp; other MZMs are covered in [15]. The time delay operator in (7.1b) applies a time delay τ to a function $F(t)$ as $T(\tau)F(\tau) = F(t + \tau)$. Finally, θ in (7.1c) is the orientation of the HWP with respect to the x-axis.

The photocurrents shown in Figure 7.2 can be calculated from (7.1) as

$$I_{1,2}(t) = \frac{A\eta q}{2hf_o} \left(\frac{\varepsilon}{\mu}\right)^{\frac{1}{4}} E^*_{1,2} E_{1,2}, \tag{7.4}$$

where η is the photodetector quantum efficiency, q is the electron charge constant, h is Planck's constant, and f_o is the optical frequency. A two-tone drive signal will be used for this analysis:

$$V(t) = V_1 \sin(\omega_1 t) + V_2 \sin(\omega_2 t), \tag{7.5}$$

where ω is the angular RF. Assuming that there are equal optical powers incident on each photodetector controlled by φ_{dc}, $\omega_o\tau$, and θ, the photocurrent can be written as

$$I_1(t) = I_{dc} + 2I_{dc} \times \left\{ \begin{array}{c} J_0(x_2)\sum_{j=0}^{\infty} J_{2j+1}(x_1)\sin[(2j+1)\omega_1 t] \\ +J_0(x_1)\sum_{k=0}^{\infty} J_{2k+1}(x_2)\sin[(2k+1)\omega_2 t] \\ \pm\sum_{j=0}^{\infty}\sum_{m=1}^{\infty} J_{2j+1}(x_1)J_{2m}(x_2)\sin[(2m\omega_2 \pm (2j+1)\omega_1)t] \\ \pm\sum_{k=0}^{\infty}\sum_{n=1}^{\infty} J_{2k+1}(x_2)J_{2n}(x_1)\sin[(2n\omega_1 \pm (2k+1)\omega_2)t] \end{array} \right\}, \quad (7.6)$$

where I_{dc} is the average photocurrent and $J(\cdot)$ is a Bessel function of the first kind. Note that the expression for $I_2(t)$ is in the same form of (7.6) but with a 180° phase shift on all of the RF terms. The arguments of the Bessel functions depend on the structure from Figure 7.2 as

$$x_n = \begin{cases} \pi V_n / V_\pi(f_n) & \text{(intensity modulation)} \\ 2\pi V_n \sin(\pi f_n \tau)/V_\pi(f_n) & \text{(phase modualtion)}, \\ \pi V_n / V_\pi(f_n) & \text{(polariztion modulation)} \end{cases} \quad (7.7)$$

where $n = 1,2$ for the two-tone drive, $f = \omega/(2\pi)$, and the frequency dependence of V_π is explicitly stated. The power at each frequency in (7.6) can be calculated according to

$$P_\omega = \langle I_\omega^2 \rangle R |H_{pd}|^2, \quad (7.8)$$

where I_ω is the term in (7.6) at ω, $\langle\cdot\rangle$ denotes a time average, R is the output resistance, and $|H_{pd}|^2$ is the photodetector circuit frequency response.

The analysis can be focused on interference suppression by replacing the subscripts 1 and 2 with SOI and INT to correspond to a signal of interest and an interference signal, respectively. Inserting the fundamental terms from (7.6) into (7.8) yields

$$P_{soi} = 2I_{dc}^2 J_0^2(x_{int}) J_1^2(x_{soi}) R |H_{pd}|^2 \quad (7.9)$$

$$P_{int} = 2I_{dc}^2 J_0^2(x_{soi}) J_1^2(x_{int}) R |H_{pd}|^2. \quad (7.10)$$

Equations (7.9) and (7.10) are the basis for the technique. First consider the small-signal regime, where both x_{soi} and x_{int} are much less than 1. In this case, $J_0(x) \approx 1$ and $J_1(x) \approx x/2$, resulting in linear power responses for P_{soi} and P_{int} with no interdependence between the two signals. The situation under consideration is a small signal of interest and a large interference signal. Under these conditions, (7.9) and (7.10) can be rewritten as $P_{soi} = AJ_0^2(x_{int})x_{soi}^2/4$ and $P_{int} \pm AJ_1^2(x_{int})$, where $A = 2I_{dc}^2 R|H_{pd}|^2$. Therefore, P_{int} can be suppressed when $J_1(x_{int}) = 0$, the first zero of which occurs when $x_{int} = 3.8317$. The effect on P_{soi} under this condition is

$J_0^2(3.8317) = 0.1622$, or an 8-dB degradation from its small-signal response. This amplitude-based filtering process is demonstrated in Figure 7.3. Plotted there is a calculation of the RF gain for an interference signal (INT) and a signal of interest (SOI), as a function of the INT input power with the SOI power fixed. Equations (7.9) and (7.10) were used to calculate the RF gain by dividing by the input power, $V_\omega^2 / (2R)$, where V_ω is the input peak voltage at ω and R is the input resistance that is matched to the output resistance. The parameters for Figure 7.3 are $V_\pi = 1$V, I_{dc} = 10m A, $H_{pd} = 1/2$, and $R = 50\Omega$. The calculation yields the first null of the INT at 12 dBm, where the SOI experiences an 8-dB degradation of its small-signal gain.

The technique under a two-tone drive is fairly straightforward to understand. The analysis can be generalized to a series of M sinusoids by applying the input voltage

$$V(t) = \sum_{m=1}^{M} V_m \sin(\omega_m t). \tag{7.11}$$

Carrying through the calculation as above, the power at each frequency can be written

$$P_n = 2I_{dc}^2 R \left|H_{pd}\right|^2 J_1^2(x_n) \prod_{m\geq 1,\neq n}^{M} J_0^2(x_m). \tag{7.12}$$

The form of (7.12) suggests that the technique can be applied to a multitude of signals and/or signals with bandwidth, leveraging the unique interrelationships between different frequency components governed by Bessel functions. This feature will be demonstrated in the following section.

The tolerance on the interference signal amplitude can be analyzed by returning to the case of one dominant interference signal. An expression for the interference power suppression can be obtained by taking the ratio of the analytical power gain to the small-signal power gain:

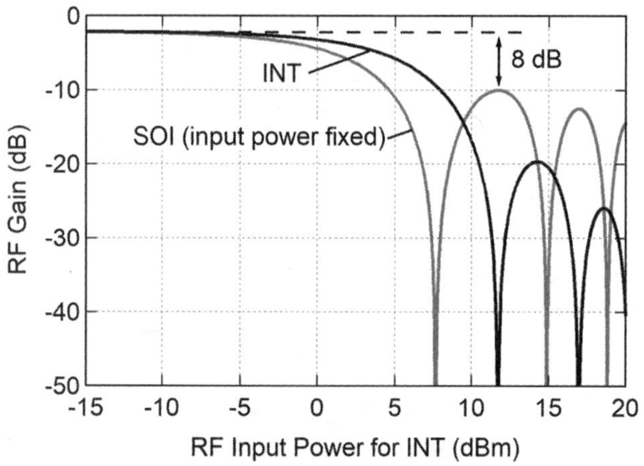

Figure 7.3 Calculated gain response for a SOI and INT, demonstrating the interference mitigation concept.

$$S = \frac{x_{int}^2}{4J_1^2\left(x_{int}\right)}. \tag{7.13}$$

In (7.13), it was assumed that the input and output resistances were matched at R and that $J_1(x) \approx x/2$ for the small-signal case. The zeros of $J_1(x_{int})$ are where S is mathematically infinite. Consider two solutions to (7.13) for a given value of S on either side of the first zero of $J_1(x_{int})$. The difference between two such values, the suppression region width, is plotted in Figure 7.4 for V_π = 1V. As an example of how to interpret Figure 7.4, 30-dB suppression of the INT requires control of the input power by ±0.35 dB about the null condition or a suppression width of 0.70 dB.

As a means to relax the tolerance on input power levels, consider the case of N parallel modulators, as shown in Figure 7.5. Following the analytical steps above for a two-tone drive results in the expressions

$$P_{soi} = 2R\left|H_{pd}\right|^2\left[\sum_{n=1}^{N} I_{dc,n} J_0\left(x_{int,n}\right) J_1\left(x_{soi,n}\right)\right]^2 \tag{7.14}$$

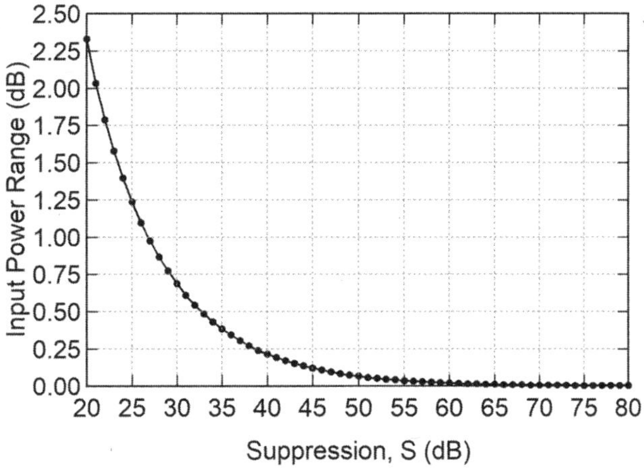

Figure 7.4 Calculation of the width of the first suppression region shown in Figure 7.3 as a function of the suppression level.

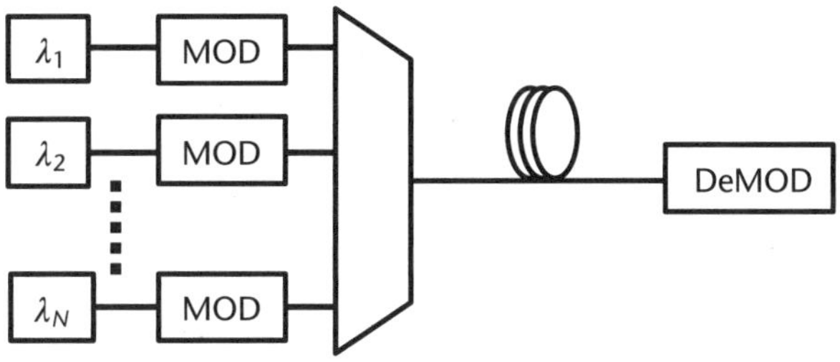

Figure 7.5 A wavelength multiplexed photonic link utilizing N lasers of different wavelengths, N modulators (MOD), and a single demodulation (DeMOD) stage.

$$P_{int} = 2R\left|H_{pd}\right|^2 \left[\sum_{n=1}^{N} I_{dc,n} J_0\left(x_{soi,n}\right) J_1\left(x_{int,n}\right)\right]^2. \tag{7.15}$$

According to (7.14) and (7.15), the total response of a parallel cascade of modulators can be tailored by controlling the voltage applied to each modulator and/or utilizing modulators with different V_π. To this end, (7.14) and (7.15) present a complex optimization problem beyond the scope of this presentation. Nonetheless, the validity of (7.14) and (7.15) will be demonstrated in the following section for $N = 2$.

7.3 Measured Data

There have been many experimental results validating the theory in the previous section [4–11], a representative sampling of which is presented here. The core functionality described by Figure 7.3 is demonstrated in Figure 7.6 with measured data. A phase-modulated structure such as shown in Figure 7.2(b) was used for this experiment. The SOI was at 5.001 GHz with a fixed input power of –30 dBm and the INT was at 5.000 GHz. A $LiNbO_3$ phase modulator was used having V_π = 1.5V near 5 GHz. Conversion from the optical to the electrical domain was achieved using an MZI with τ – 100 ps followed by a single photodiode operating at I_{dc} = 1.4 mA. The photodiode circuit exhibited H_{pd} = 1/2 and R = 50Ω. The measured results are compared to responses calculated using (7.9) and (7.10) with these parameters, showing very good agreement.

The concept of tailoring the suppression region via multiple parallel modulators is demonstrated in Figure 7.7. Shown there are data for a structure employing two phase modulators: a version of Figure 7.5 employing components from Figure 7.2(b). The SOI was at 5.001 GHz with a fixed input power of –14 dBm and the INT was at 5.000 GHz. The two $LiNbO_3$ phase modulators had $V_{\pi,1}$ = 1.5V and $V_{\pi,2}$ = 3.7V at 5 GHz. The two lasers were at wavelengths λ_1 = 1,552 nm and λ_2 = 1,565 nm. The MZI had τ = 100 ps and was connected to a single photodiode operating with a total photocurrent of 3.8 mA, containing contributions from each wavelength of $I_{dc,1}$ = 1.7 mA and $I_{dc,2}$ = 2.1 mA. The photodiode circuit parameters

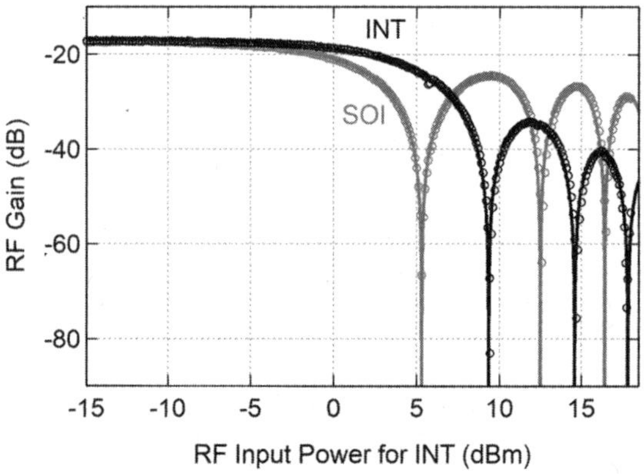

Figure 7.6 Measured (circles) and calculated (lines) gain responses for a SOI and INT through a phase-modulated link.

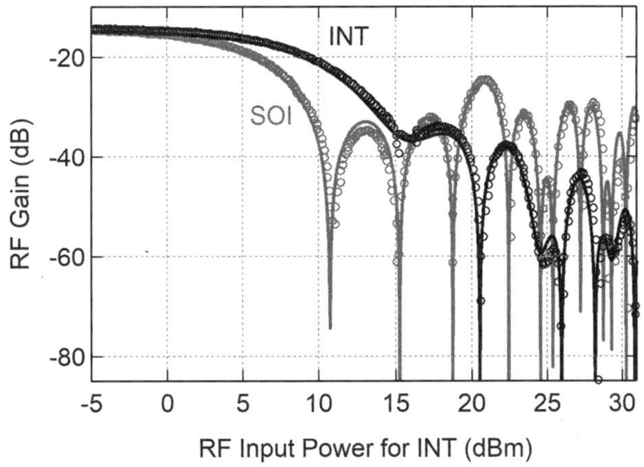

Figure 7.7 Measured (circles) and calculated (lines) gain responses for a SOI and INT through a phase-modulated link having two parallel modulators.

were $H_{pd} = 1/2$ and $R = 50\Omega$. These parameters were inserted into (7.14) and (7.15) to produce the calculated lines in Figure 7.7. There is excellent agreement between theory and experiment. A significantly different structure is shown in Figure 7.7 compared to the single-modulator case in Figure 7.6, with broader suppression regions for the two-modulator case. The data in Figure 7.7 validate the theory in (7.14) and (7.15); there was no attempt to optimize the response but (7.14) and (7.15) can be utilized to do so.

The technique can be further understood by considering frequency content beyond the SOI and INT fundamentals, as shown in Figure 7.8. Depicted there are measured results for an intensity-modulated structure like that in Figure 7.2(a). For these data, the SOI was at f_1 = 1.000 GHz with a power of –19 dBm, and the INT was at f_2 = 1.001 GHz. The single-output MZM exhibited V_π = 1.2V around 1 GHz, and the photodiode operated at I_{dc} = 2.7 mA, H_{pd} = 1/2, and $R = 50\Omega$. Equations (7.9) and (7.10) were used for the SOI and INT, respectively. Expressions for the remaining frequency components were obtained from (7.6). Two components

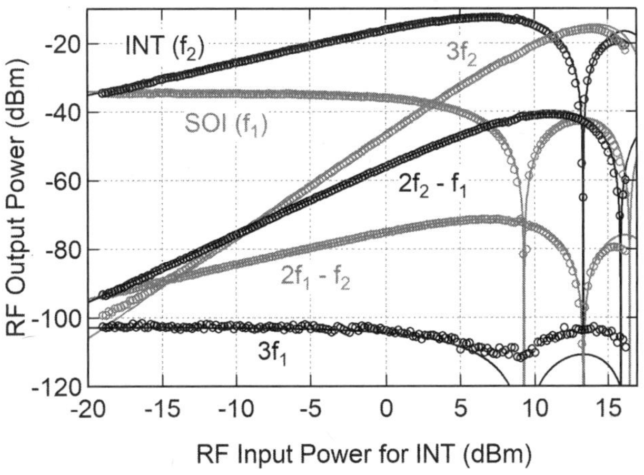

Figure 7.8 Measured (circles) and calculated (lines) power responses for a SOI and INT through an intensity-modulated link. In addition to the fundamental components, the third harmonics and third-order-intermodulation terms are shown.

of particular note are those at $(2f_2 - f_1)$ and $3f_2$ in Figure 7.8. At the suppression point near 13 dBm INT power, the term at $(2f_2 - f_1)$ is the same power as the SOI, which would severely limit the in-band spurious-free dynamic range. The term at $3f_2$ could be more problematic in that it approaches the level of the unsuppressed INT. These features must be considered and will be revisited in Section 7.4.

The bandwidth performance for a nonoptimized, MZM-based configuration is shown in Figure 7.9. These data were collected with a VNA and are cast in terms of isolation, which is defined experimentally as the inverse of (7.13). The configuration is nonoptimized over bandwidth in the sense that the INT power was fixed at 24.5 dBm, which met the suppression condition near 13.5 GHz only. The dominant factor in the response is the frequency-dependence of V_π, which varied from 4.0V at 6 GHz to 4.6V at 18 GHz. One might envision a conditioning circuit that would maintain the suppression condition as a function of frequency, a concept that will be discussed further in the following section. The remaining parameters for this experiment were an SOI power of −23.0 dBm, and a single photodiode with $I_{dc} = 5.3$ mA, $H_{pd} = 1/2$, and $R = 50\Omega$. Even without optimization, the level of suppression is impressive over the 12 GHz of bandwidth.

The results presented thus far have dealt with CW tones only. As implied by (7.12), the technique is also applicable to signals with appreciable bandwidth. Shown in Figure 7.10 are measured spectra demonstrating the suppression method for a pulsed signal at 450 MHz with a pulse width of 10 μs. For these data, an intensity-modulated structure having a single photodiode was employed. The structure parameters were $V_\pi = 1$V, $I_{dc} = 9$ mA, $H_{pd} = 1/2$, and $R = 50\Omega$. The expected 3-dB loss was observed under the small-signal condition. The loss increases to 48 dB across the bandwidth of the signal at the suppression condition, resulting in $S = 45$ dB. Data for a frequency-modulated CW signal are shown in Figure 7.11. The same structure as for the chirped experiment was used. The center frequency was 455 MHz, having a chirp of 25 MHz with a period of 50 μs. The expected 3-dB small-signal loss was again observed, with 33-dB loss under the suppression condition, resulting in $S = 30$ dB. Both of these results are impressive for signals having appreciable bandwidth.

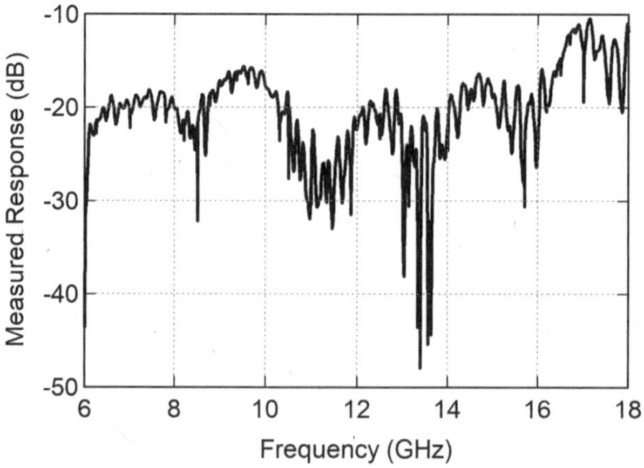

Figure 7.9 Measured isolation (S^{-1}) for an intensity modulated link set to the suppression condition near 13.5 GHz.

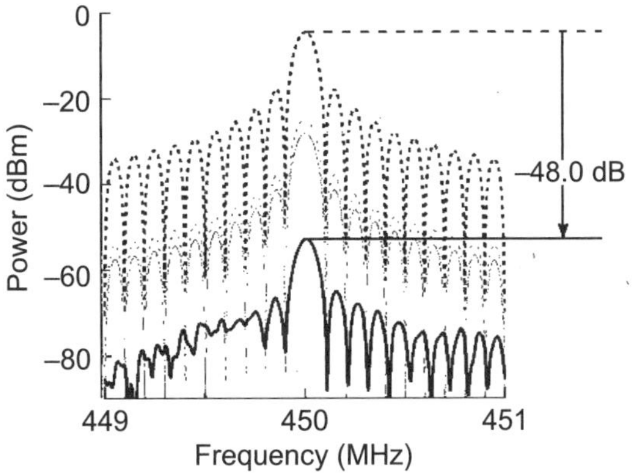

Figure 7.10 Measured power spectra for a pulsed 450-MHz signal with a 10-μs pulse width through an intensity-modulated link. Shown are the inputs (dashed) and outputs (solid) for the small-signal (gray) and suppression (black) conditions.

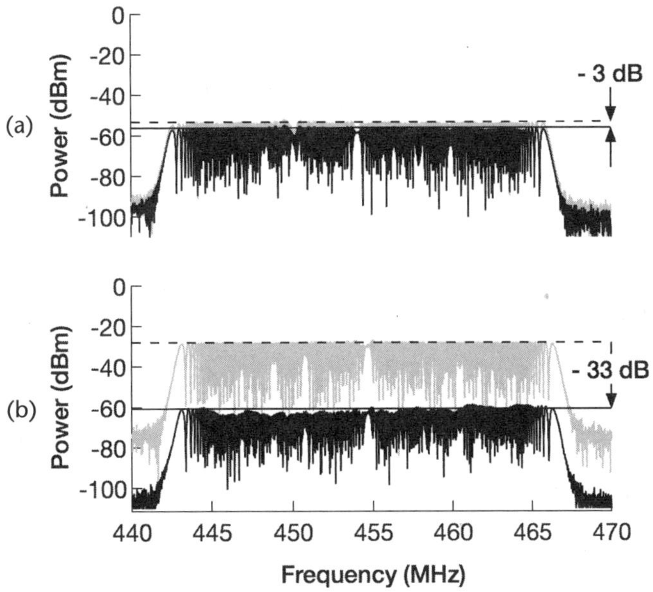

Figure 7.11 Measured power spectra for a 455-MHz signal with a linear 25-MHz chirp through an intensity-modulated link. Shown are the inputs (gray) and outputs (black) for the small-signal (top) and suppression (bottom) conditions.

7.4 Practical Considerations

The technique presented here would most likely be employed at the front end of a receiver system, perhaps connected directly to the antenna, as mentioned in Chapter 1. Indeed, photonic front ends have been considered because of the high RF power handling of $LiNbO_3$ [16, 17], an added benefit when employing this technique. On the other hand, the photonic structure (Figure 7.2) might be switched in at the time of need to avoid complications in the absence of strong interference. Working this method into a layered interference suppression strategy is an important consideration, where the photonic structure could be placed at various stages of a receive chain. In practice, the technique would be best paired with a control circuit, limiter, and/or equalizer in order to improve the performance over very large bandwidths. For example, the data in Figure 7.9 are for a free-running system without any conditioning ahead of the modulator. Multiple modulation stages (Figure 7.5) might be used to improve the performance, but trades with complexity must be considered in this case.

The unique suppression function and power handling of a photonic front end must be traded against the noise figure introduced by a photonic structure, as discussed in Chapters 3 and 15. Photonic links provide excellent loss and noise-figure performance for long-haul applications. However, high noise figure has historically plagued analog-photonic structures in standalone applications. For example, photonic architectures utilizing COTS components exhibit noise figures of about 10 to 20 dB in X-band, 25 dB in Ku-band, and 35 dB in W-band. Only a handful of "hero" experiments using state-of-the-art components have demonstrated noise figures below 10 dB in the 1–10-GHz range [18–20]. All of the low noise figure experiments employed $LiNbO_3$ modulators, which exhibit the high RF power handling mentioned above. The complexity issues with multiple modulation stages might be addressed with photonic integrated circuits (PICs), similar to those discussed for RF cancellation in Chapter 15. The future work concerning PICs as they relate to this technique will be addressed in Section 7.5.2. Regardless of the research direction, PICs will certainly have higher noise figures than photonic structures with state-of-the-art fiber-coupled components [21]. For example, today's PICs exhibit RF noise figures in the range of 40 to 50 dB. All of these noise figure considerations suggest that switching in a photonic structure to utilize the function described in this chapter may be the best approach. The reduced sensitivity when doing so would certainly be better than being blinded by interference or front-end damage.

While photonics does boast wide bandwidth, limitations on the operational frequency range when employing this technique must be considered. As described with respect to Figure 7.8, there are two frequency components under two-tone drive that are problematic. The third-order intermodulation term at $(2f_2 - f_1)$ rises to the level of the SOI at the suppression condition. This would not compress a receiver but would certainly waste processing resources. In the case of signals with bandwidth, this nonlinearity would cause asymmetric in-band spectral growth. The third harmonic of the INT, $3f_2$, would most likely compress the receiver as it rises to a level near that of the unsuppressed INT. In practice, this would limit the operational bandwidth of the technique from a minimum frequency f_{min} to $3f_{min}$. While this range could offer wide bandwidth, it would most likely limit the intrinsic bandwidth of the photonics. There are no even-order terms in (7.6) and this is a result of assuming that the bias terms φ_{dc}, $\omega_o\tau$, and θ are precisely set to achieve equal photocurrent on each photodiode. In practice, these bias conditions can be difficult to maintain [22], especially under a large signal drive, a condition on which this technique relies. The photodiodes themselves can present even-order and odd-order nonlinearities [22] that might impact the performance of this technique. The means to address spurious signals will be discussed in Section 7.5.2.

7.5 Conclusion

7.5.1 Summary

The technique presented in this chapter exhibits the typically advertised advantages of RF photonics but leverages a unique signal transfer to provide a novel signal processing function. This signal processing ability is intrinsic to the structures shown in Figure 7.2 and due to the interrelationships of multiple signals under optical angle modulation. Much interesting work remains in the application of this same-frequency, in-band, amplitude-based filtering technique, some of which will be discussed in the following section. The work to date is best summarized by collecting the relatively small number of measured results.

Figure 7.12 plots measured results from 11 experiments using this technique. Single-tone results are plotted as having 1 Hz of instantaneous bandwidth. The achieved suppression decreases as the signal bandwidth increases, but impressive suppression levels have been achieved approaching 1 GHz of signal bandwidth. The result in [11] was derived from a linearization technique that suppressed the $(2f_2 - f_1)$ term described above by 29 dB. Of the numerous photonic approaches for RF interference mitigation [3], the results shown in Figure 7.12 make one of the more compelling arguments to insert photonics into a RF receive chain.

7.5.2 Future Directions

Ongoing work related to the technique is largely related to the practical issues noted in Section 7.4. First and foremost, the technique must be integrated with microwave circuitry. As mentioned in Chapter 1, working this method into a layered approach is most likely the best strategy, where analog electronics, digital electronics, and

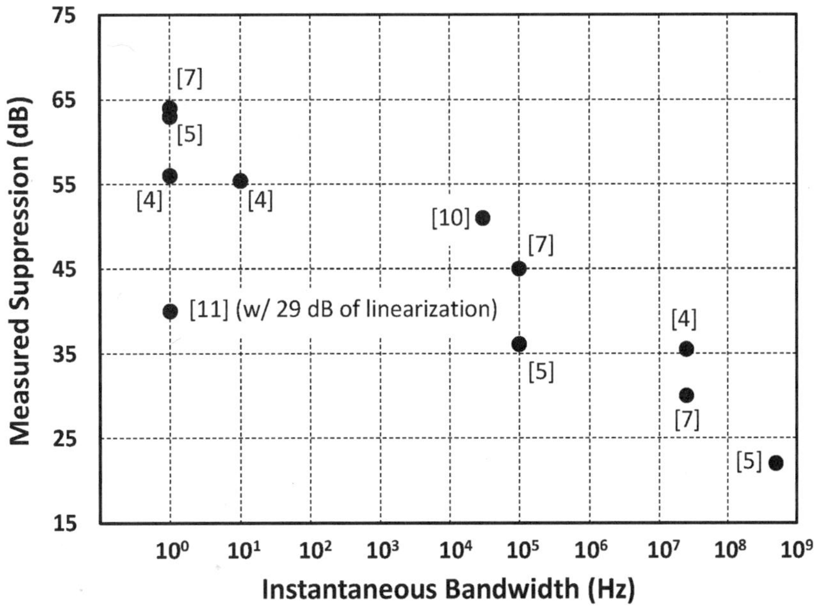

Figure 7.12 Measured results for the technique described in this chapter.

other analog-photonic approaches can be combined for a robust solution. The location of the photonics in the receive chain is critical. The technique as it stands is probably best implemented as a parallel path that can be switched to in the case of strong interference because of the associated noise figure.

As discussed previously, the high RF noise figure introduced by photonics would limit sensitivity if the structure was placed early in a receive chain, which is where an interference circuit is most needed. Reducing the RF noise figure is a focus in microwave photonics and many applications of the technology would benefit [20]. Most implementations of microwave photonics employ the structures in Figure 7.2 or slight variations thereof. The best noise figures in these structures have been achieved with low-noise solid-state lasers, $LiNbO_3$ modulators, and semiconductor photodetectors. Improving the performance of these components is the most direct path to reducing noise figure. Increasing laser power while maintaining shot-noise-limited performance has long been a concentration of research. Modulators would in turn need to handle the increased input power with low loss, while increasing the RF-to-optical conversion efficiency (i.e., reduce V_π). The photodetectors would then need to receive the increased power with enough linearity so as to not degrade the intrinsic spurious-free dynamic range of the modulation process. An alternative path to reducing the noise figure is to innovate new modulation techniques utilizing existing components. For this to be effective, any new technique must retain the functionality based on optical angle modulation to provide the interference suppression.

There is a large body of work on linearization of analog-photonic structures [23–33] that can be applied to this technique. As discussed in Section 7.4, there are numerous third-order nonlinearities that arise at the suppression condition. An existing linearization technique has already been adopted to reduce a problematic third-order intermodulation distortion term by 29 dB [11]. Investigations into other third-order linearization methods are an important line of research for this interference suppression technique. The impact of even-order distortion has not been considered in the experiments to date. As noted above, such terms could pose a problem to implementing this technique in practice. The characterization of even-order performance and, if needed, the development of mitigation techniques provide opportunity for future direction.

Integrated microwave photonics [34] is a topic under the broader PIC field that addresses analog applications of PICs. Transitions of the technique presented here to small platforms, such as handheld devices, would be limited by the size of bulk fiber-coupled lasers, modulators, and photodetectors. The size could become unmanageable for even larger platforms as multiple modulation stages are utilized to broaden the suppression region and/or linearize the response. Integrated microwave photonics therefore presents the most promising avenue of future research to improve the technique covered in this chapter. Such integration would be most impactful if it were heterogeneous because, as described above, the best microwave performance is achieved with different material platforms for lasers, modulators, and photodetectors. The end goal of such efforts would be optical microchips that provide the unique interference suppression functionality described in this chapter.

References

[1] Cox III, C. H., *Analog Optical Links*, New York: Cambridge University Press, 2004.

[2] Urick, V. J., J. D. McKinney, and K. J. Williams, *Fundamentals of Microwave Photonics*, New York: John Wiley & Sons, 2015.

[3] Urick, V. J., M. E. Godinez, and D. C. Mikeska, "Photonic-Assisted Radio-Frequency Interference Mitigation," *J. Lightwave Technol.*, Vol. 38, No. 6, March 2020, pp. 1268–1274.

[4] Loh, W., et al., "A Nonlinear Optoelectronic Filter for Electronic Signal Processing," *Scientific Reports*, Vol. 4, January 2014, p. 3613.

[5] Urick, V. J., et al. "An Optical Technique for Radio Frequency Interference Mitigation," *IEEE Photonics Technol. Lett.*, Vol. 27, No. 12, June 2015, pp. 1333–1336.

[6] McKinney, J. D., et al., "Interference Mitigation Via the Nonlinear Transfer Function of Angle-Modulated Optical Links," *IEEE Photonics Conf. Dig.*, paper TuC1.2, October 2015.

[7] Urick, V. J., et al., "Nonlinear Optical Angle Modulation for Suppression of RF Interference," *IEEE Transactions on Microwave Theory and Techniques*, Vol. 64, No. 7, July 2016, pp. 2198–2204.

[8] Dailey, J. M., et al., "Photonics-Based Same-Frequency RF Interference Mitigation," *IEEE Intl. Mtg. Microwave Photonics*, October 2016, pp. 83–86.

[9] McKinney, J. D., et al., "Interference Suppression Based on Power-Dependent Transfer Function of Angle-Modulated Optical Links," U.S. Patent No. 9,654,217, May 2017.

[10] Loh, W., et al., "Optical Unmasking of Spectrally Overlapping RF Signals," *Opt. Exp.*, Vol. 25, No. 22, October 2017, pp. 26581–26590.

[11] Ackerman, E., "Interference Suppression with Mitigation of Intermodulation Distortion," *IEEE AVFOP Conf.*, November 2019, pp. 1–2.

[12] Ackerman, E., et al., "Maximum Dynamic Range Operation of a Microwave External Modulation Fiber-Optic Link," *IEEE Transactions on Microwave Theory and Techniques*, Vol. 41, No. 8, August 1993, pp. 1299–1306.

[13] Urick, V. J., et al., "Equations for Two-Tone Analog Optical Phase Modulation with an Asymmetric Interferometer," *IEEE Photonics Technol. Lett.*, Vol. 25, No. 15, August 2013, pp. 1527–1530.

[14] Campillo, A. L., and F. Bucholtz, "Chromatic Dispersion Effects in Analog Polarization-Modulated Links," *Appl. Opt.*, Vol. 45, No. 12, April 2006, pp. 2742–2748.

[15] Diehl, J. F., et al., "Microwave Photonic Delay Line Signal Processing," *Appl. Opt.*, Vol. 54, No. 31, November 2015, pp. 35–41.

[16] Bucholtz, F., et al., "Susceptibility of Lithium-Niobate Modulator to High-Power Microwave Pulses," *Electron. Lett.*, Vol. 45, No. 5, February 2009, pp. 272–273.

[17] Schermer, R. T., et al., "Investigation of Electrooptic Modulator Disruption by Microwave-Induced Transients," *Opt. Exp.*, Vol. 17, No. 25, December 2009, pp. 22586–22602.

[18] Cox III, C. H., et al., "Limits on the Performance of RF-over-Fiber Links and Their Impact on Device Design," *IEEE Transactions on Microwave Theory and Techniques*, Vol. 54, No. 2, February 2006, pp. 906–920.

[19] Urick, V. J., J. D. McKinney, and K. J. Williams, "Fiber-Optic Links with All-Photonic RF Gain and Low RF Noise Figure," *IEEE Intl. Microwave Symp.*, paper WE2C-1, June 2011.

[20] Ackerman, E. I., and C. H. Cox III, "Optimization of Analog Optical Link Performance: First Minimize the Noise Figure," *IEEE Intl. Mtg. Microwave Photonics*, October 2014, pp. 430–433.

[21] Urick, V. J., "Requirements and Application Opportunities for Integrated Microwave Photonics," *OFC Tech. Dig.*, paper M2B.1, March 2016.

[22] Urick, V. J., et al., "Wideband Analog Photonic Links: Some Performance Limits and Considerations for Multi-Octave Implementations," *Proc. SPIE*, Vol. 8259, 2012.

[23] Ackerman, E. I., "Broad-Band Linearization of a Mach-Zehnder Electrooptic Modulator," *IEEE Transactions on Microwave Theory and Techniques*, Vol. 47, No. 12, December 1999, pp. 2271–2279.

[24] Urick, V. J., et al., "Wideband Predistortion Linearization for Externally-Modulated Long-Haul Analog Fiber Optic-Links," *IEEE Transactions on Microwave Theory and Techniques*, Vol. 54, No. 4, April 2006, pp. 1458–1463.

[25] Ismail, T., et al., "High-Dynamic-Range Wireless-over-Fiber Link Using Feedforward Linearization," *J. Lightwave Technol.*, Vol. 25, No. 11, November 2007, pp. 3274–3282.

[26] Haas, B. M., et al., "Dual-Wavelength Linearization of Optically Phase-Modulated Analog Microwave Signals," *J. Lightwave Technol.*, Vol. 26, No. 15, August 2008, pp. 2748–2753.

[27] O'Connor, S. R., T. R. Clark Jr., and D. Novak, "Wideband Adaptive Feedforward Photonic Link," *J. Lightwave Technol.*, Vol. 26, No. 15, August 2008, pp. 2810–2816.

[28] McKinney, J. D., K. Colladay, and K. J. Williams, "Linearization of Phase-Modulated Analog Optical Links Employing Interferometric Demodulation," *J. Lightwave Technol.*, Vol. 27, No. 9, May 2009, pp. 1212–1220.

[29] Urick, V. J., et al., "Suppression of Even-Order Photodiode Distortions Via Predistortion Linearization with a Bias-Shifted Mach-Zehnder Modulator," *Opt. Exp.*, Vol. 21, No. 12, June 2013, pp. 14368–14376.

[30] Ackerman, E. I., G. E. Betts, and C. H. Cox III, "Inherently Broadband Linearized Modulator for High-SFDR, Low-NF Microwave Photonic Links," *IEEE Intl. Mtg. Microwave Photonics*, October 2016, pp. 265–268.

[31] Bhatia, A., H. F. Ting, and M. A. Foster, "All-Optical Multiorder Distortion Elimination in a Phase-Modulated Microwave Photonic Link," *J. Lightwave Technol.*, Vol. 35, No. 4, February 2017, pp. 855–861.

[32] Li, Y., et al., "Coherent PM RF Photonic Link Achieving Record SFDR Performance," *IEEE Intl. Mtg. Microwave Photonics*, October 2019, pp. 1–4.

[33] Wang, F., S. Shi, and D. W. Prather, "Microwave Photonic Link with Improved SFDR Using Two Parallel MZMs and a Polarization Beam Combiner," *J. Lightwave Technol.*, Vol. 37, No. 24, December 2019, pp. 6156–6264.

[34] Marpaung, D., J. Yao, and J. Capmany, "Integrated Microwave Photonics," *Nat. Photonics*, Vol. 13, No. 2, February 2019, pp. 80–90.

CHAPTER 8

Time-Domain RF Cancellation

Kenneth E. Kolodziej

8.1 Introduction

RF cancellation within the analog domain of an IBFD system is a common technique to augment SI suppression in the preceding propagation domain [1]. Similar to the other SIC techniques presented throughout the book, general analog cancellers attempt to mitigate the transmitter leakage by injecting a cancellation signal within the receiver chain, which suppresses the SI and helps to avoid receiver saturation. While Chapters 6 and 7 presented alternative analog filtering methods, the cancellation approaches described here can be independently incorporated into a transceiver or designed to follow the code-based and/or photonic-based filters, respectively.

Within the receiver, these analog cancellation techniques can be implemented at RF (the transmitted/received frequency), an intermediate frequency (IF), and/or baseband (BB) before the ADC [2]. For any of these locations, a canceller's architecture can be categorized into one of three types: time-domain-based, frequency-domain-based, or digitally-assisted cancellation. This chapter focuses on the former, while the latter two are discussed in Chapters 9 and 10, respectively.

Time-domain-based cancellation architectures sample the transmit signal and split it into different analog paths with varying amounts of time delay along with amplitude and phase control. These branches are referred to as canceller taps and are analogous to digital tapped delay lines [3].

DISTRIBUTION STATEMENT A. Approved for public release. Distribution is unlimited. This material is based upon work supported by the Under Secretary of Defense for Research and Engineering under Air Force Contract No. FA8702-15-D-0001. Any opinions, findings, conclusions or recommendations expressed in this material are those of the author(s) and do not necessarily reflect the views of the Under Secretary of Defense for Research and Engineering.

It is the independent control and weighting of these taps that provides the canceller with the ability to modify the transmit signal before mitigating the SI by injecting it within a receiver. This chapter will present the motivation and critical design considerations for time-domain-based cancellers as well as the performance of several different adaptive prototypes implemented at RF, as highlighted within the IBFD transceiver diagram of Figure 8.1(a) and the SIC techniques tree of Figure 8.1(b).

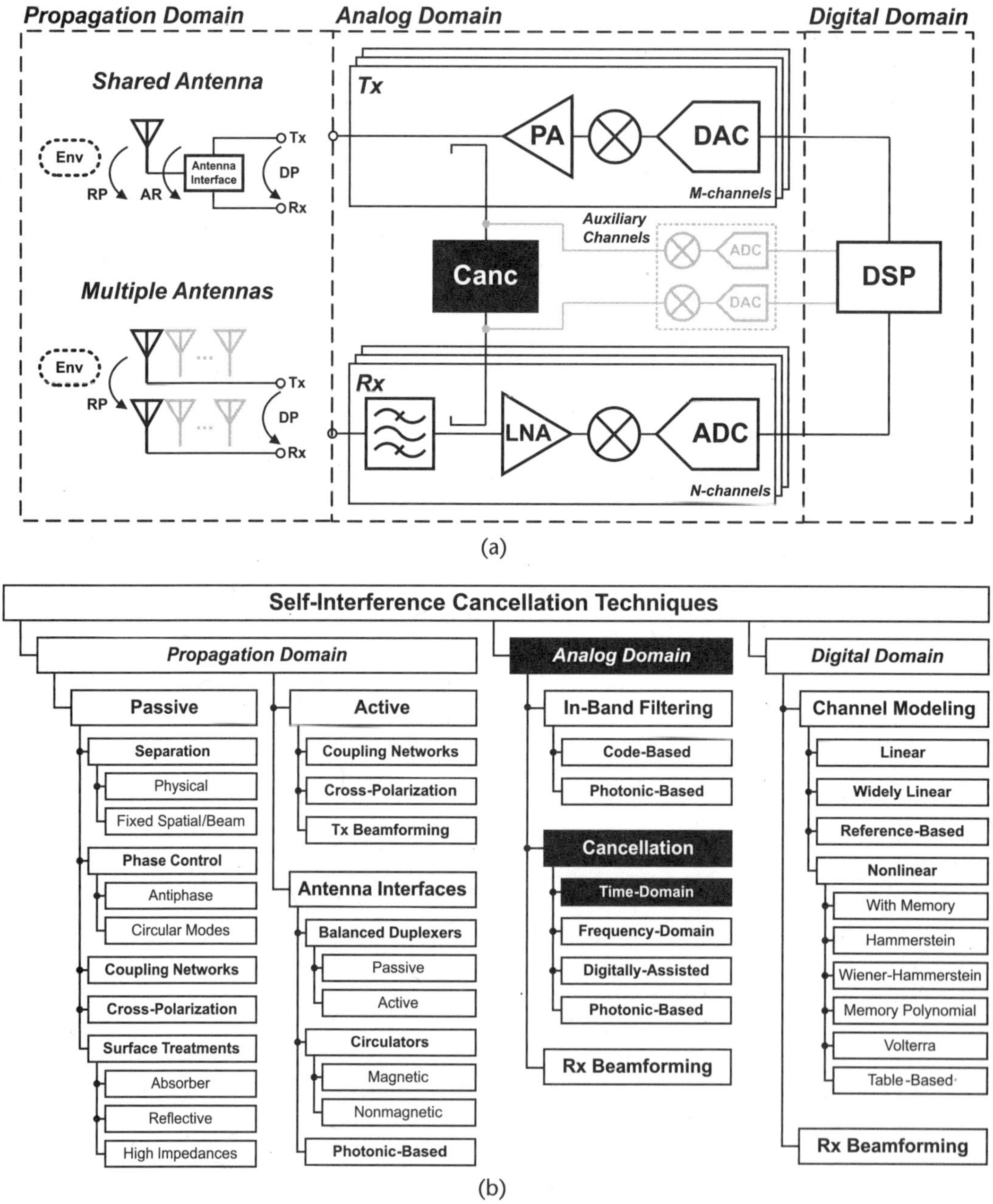

Figure 8.1 Time-domain RF cancellation and specific chapter focus highlighted within (a) a generic IBFD transceiver and (b) a SIC techniques tree diagram.

8.2 Cancellation Objective

The objective of RF cancellation within IBFD systems is simply to reduce the transmit SI power before the receiver input [4]. This would ideally occur before the LNA and helps to ensure that the SI is within the linear range of each component within the receiver, including below the full-scale level of the ADC. This technique is most effective when combined with one or more of the propagation-domain methods described in Chapters 2 through 5.

Typically, the resulting high-isolation antennas (with either shared or multiple elements) provide SIC for only the direct path (DP) SI signal because these structures are designed irrespective of their operating location. As mentioned in Chapter 1, the environment around the IBFD node creates reflection path (RP) SI signals that combine with the DP, as seen in Figure 8.2(a) for a system with two antennas. This composite SI channel can be modeled as a collection of impulse responses, such that

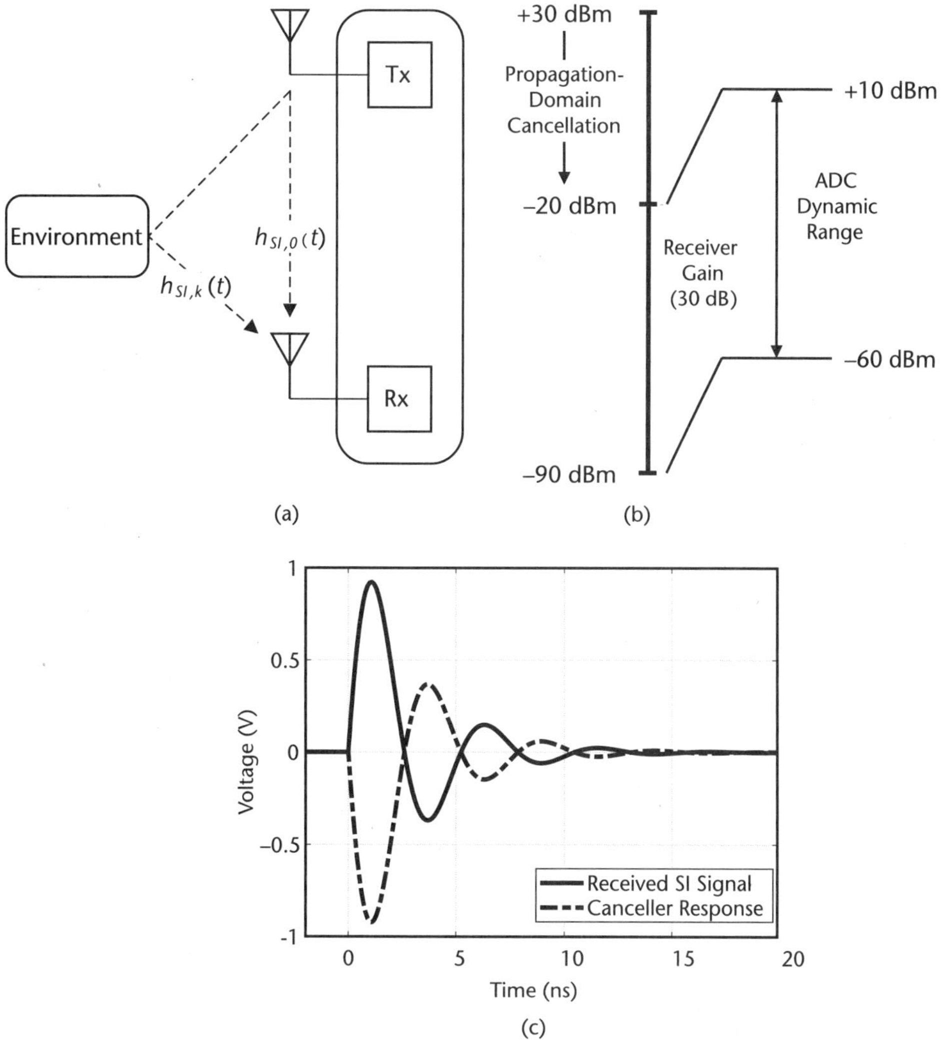

Figure 8.2 (a) Two-antenna node diagram illustrating the SI that results from direct and reflected path signals; (b) example transceiver power levels, highlighted the need for RF cancellation; and (c) an example cancellation response matching the anti-phased received SI signal.

$$h_{SI}(t) = h_{SI,0}(t) + \sum_{k=1}^{K} h_{SI,k}(t), \tag{8.1}$$

where $h_{SI,0}(t)$ represents the time-dependent DP signal and $h_{SI,k}(t)$ captures K environmentally-dependent RP signals. These SI components can be expanded with delayed impulses, described by

$$h_{SI}(t) = \alpha_0 \delta(t - \tau_0) * h_{\varphi_0}(t) + \sum_{k=1}^{K} \alpha_k \delta(t - \tau_k) * h_{\varphi_k}(t), \tag{8.2}$$

where α_0 is the attenuation coefficient for the DP signal that is delayed by τ_0 and convolved with a phase shift response, $h_{\varphi_0}(t)$. The latter multipath signals experience different attenuations and phase shifts and are also delayed by different amounts, denoted by τ_k, which are dependent on the surrounding environment.

The critical attenuation, phase shift, and delay variables can be clearly visualized in the frequency-domain representation of (8.2), such that

$$H_{SI}(j\omega) = \alpha_0 e^{-j\varphi_0} e^{-j\omega\tau_0} + \sum_{k=1}^{K} \alpha_k e^{-j\varphi_k} e^{-j\omega\tau_k}, \tag{8.3}$$

where φ represents the phase shift in radians, and (8.3) is valid for $\omega \geq 0$.

By focusing on the DP SI component of these equations, Figure 8.2(b) can be utilized to gain an understanding of the sample power levels within an IBFD transceiver as well as why an RF canceller is critical. The diagram depicts a node with +30 dBm (1W) of output power, which is suppressed with one or more propagation-domain methods by 50 dB, such that the receiver input power is –20 dBm. Typically, receivers contain signal gain (30 dB is illustrated in the figure) to reduce the noise contributions of the analog components through to the ADC. In this example, the transmit SI would experience this same amount of gain (assuming linear operation), which would increase it to +10 dBm, which is a common full-scale ADC level (2V peak-to-peak in 50Ω). This simple example illustrates the need for RF cancellation or more generally analog cancellation to ensure linear receiver operation and avoid saturation before the reflected multipath SI components are considered.

While the following sections will describe this concept in more detail, the canceller attempts to accomplish the aforementioned objective by closely matching the SI response of (8.2) in an anti-phase fashion, as depicted in Figure 8.2(c). This plot illustrates an example time-varying voltage of the composite SI at the receiver input and shows an amplitude-matched canceller response that is 180° offset. When these two signals are combined within the receiver, the desired signal cancellation occurs. The quality of this cancellation depends on several canceller design parameters that will be discussed in the following two sections.

8.3 Time-Domain Approach

Figure 8.3(a) illustrates the general architecture of a time-domain approach to RF cancellation. Similar to their frequency-domain counterparts (presented in Chapter 9), these designs obtain a copy of the transmit signal, which is typically accomplished through the use of a directional coupler. If this coupler is located after the transmit PA, the canceller input signal will contain not only the linear transmitted waveform, but also the consequential transmit noise and distortion. This subtle fact

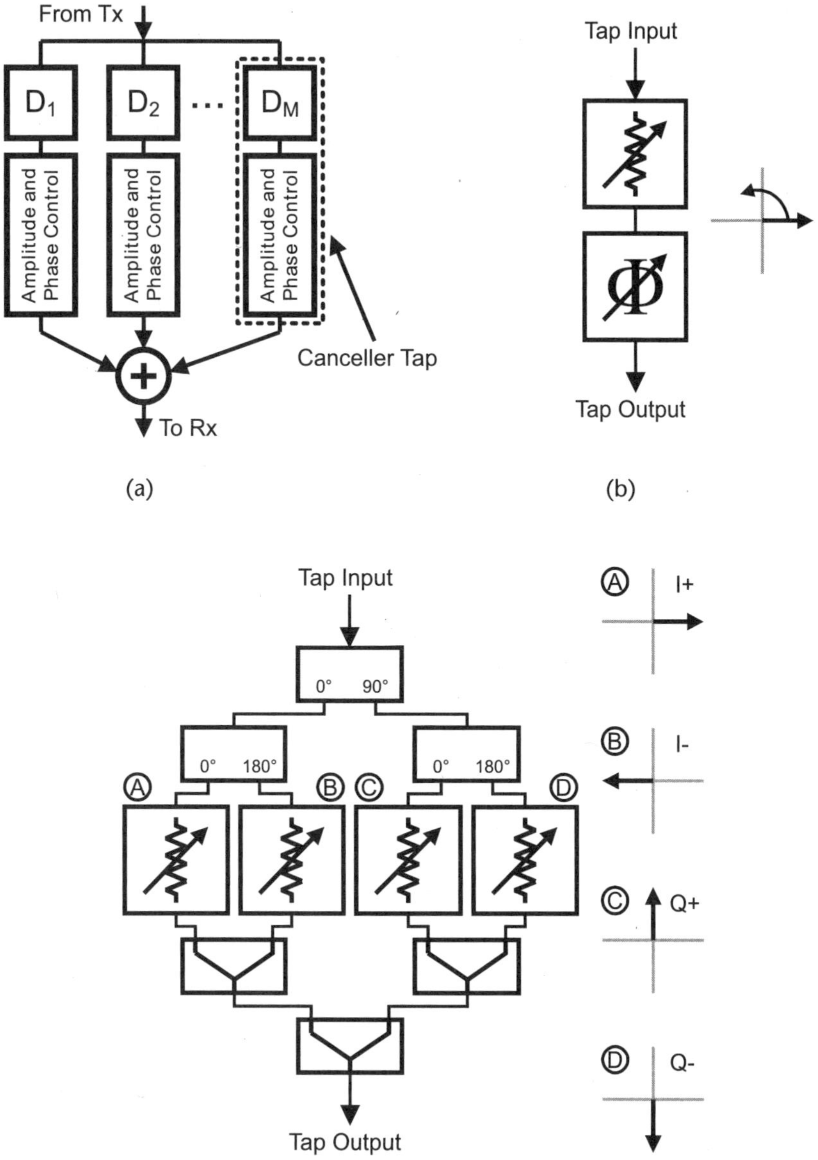

Figure 8.3 (a) Time-domain-based canceller structure with delay blocks D_M for a design with M taps; (b) tap amplitude and phase control using a variable attenuator and phase shifter; and (c) tap amplitude and phase control using a vector modulator architecture.

allows nonlinear components to also be cancelled within the analog domain and is a major advantage compared to digitally-assisted RF cancellers, as discussed in Chapter 10.

The diagram shows that the canceller input is subsequently split among several canceller taps that each have a different amount of time delay, denoted by D. These delay blocks are often fixed lengths of transmission line, but can also be tunable using integrated switched-capacitor circuits [5]. In addition to a delay block, each of these taps contains the ability to independently weight the signal, via amplitude and phase control elements, before being combined to form a single canceller output.

While some other variations exist, the tap amplitude and phase control are generally accomplished in one of two ways. First, variable attenuators and phase shifters can be combined to attenuate and rotate the tap input signal, as shown in Figure 8.3(b), with examples presented in [4, 6–10]. For this set of components, the impulse response of a canceller with M taps can be modeled as

$$h_c(t) = \sum_{m=1}^{M} \beta_m w_m \delta(t - \tau_m) * h_{\varphi_m}(t), \tag{8.4}$$

where β_m represents the fixed tap loss that captures the splitting/combing effects, w_m is the taps' tunable attenuation weight, and τ_m is the tap delay. Similar to as in (8.2), $h_{\varphi_m}(t)$ signifies the response of the taps' variable phase shifter, whose effect is more readily visualized in the equivalent frequency-domain response

$$H_c(j\omega) = \sum_{m=1}^{M} \beta_m w_m e^{-j\varphi_m} e^{-j\omega\tau_m}, \tag{8.5}$$

where φ represents the phase shift in radians and is valid for $\omega \geq 0$. It should be noted that the canceller response of (8.5) is very similar to the SI response of (8.3), which is intentional and provides the canceller with the ability to effectively match the SI response, as depicted in Figure 8.2(c).

While this tap construction works well, it is difficult to design variable phase shifters that operate over a wide range of input frequencies. The second amplitude and phase control approach overcomes this challenge by eliminating the tunable phase elements to create a vector modulator architecture, as depicted in Figure 8.3(c), with examples provided in [11–16]. Fixed phase shifts of 90° and 180° translate a tap input into four vectors (I+, I–, Q+ and Q–), as indicated with corresponding letters A through D, and are illustrated on the right of Figure 8.3(c). Similar to the previous control scheme, these vectors are then weighted using variable attenuators and combined to form a single tap output. This vector modulator approach can generate arbitrary phase shifts of the input signal by appropriately combining the weighted vectors.

Cancellers can be constructed using this tap structure, such that the resulting impulse response can be modeled as

$$h_c(t) = \sum_{m=1}^{M} \beta_m \left(w_{I,m} + jw_{Q,m}\right) \delta(t - \tau_m), \tag{8.6}$$

and contains parameters that are similar to those in (8.4), but replaces the phase shifter response with in-phase ($w_{I,m}$) and quadrature ($w_{Q,m}$) weight controls that can vary between –1 and 1 to represent the four attenuators shown in Figure 8.3(c). This model can also be converted into the frequency domain, such that

$$H_c(j\omega) = \sum_{m=1}^{M} \beta_m \left(w_{I,m} e^{-j\omega\tau_m} + w_{Q,m} e^{j(\pi - \omega\tau_m)} \right). \tag{8.7}$$

Both of these tap amplitude and phase control approaches work well, and canceller prototypes that were designed using each will be presented in Section 8.5. For either option, the benefit of incorporating multiple taps into the canceller design is highlighted in Figure 8.4, which is an expanded version of Figure 8.2(a). The diagram illustrates a time-domain-based RF canceller incorporated into an IBFD node with an FPGA used for the control of M canceller taps. The environment around the node is shown to contain two different objects, which reflect the transmit signal back towards the receiver.

The DP coupling between the transmit and receive antennas is represented by $h_{SI,0}(t_1)$, which indicates that this SI arrives at the receiver at time t_1. Because this path is fixed for the antenna structure, the signal in the first canceller tap (T_1) should be delayed by the same amount of time so they arrive at the receiver input together. Following a similar pattern for the two objects in this example, the second and third canceller taps (T_2 and T_3) can be assigned delays that match $h_{SI,1}(t_2)$ and $h_{SI,2}(t_3)$, respectively. Unfortunately, the environment around an IBFD node is often unknown and dynamically changing, and thus, the tap delay assignment is not as straightforward as in this example. Cancellers typically account for these conditions by utilizing multiple taps that increase their flexibility around a targeted delay spread and bandwidth capability, as discussed in the following section.

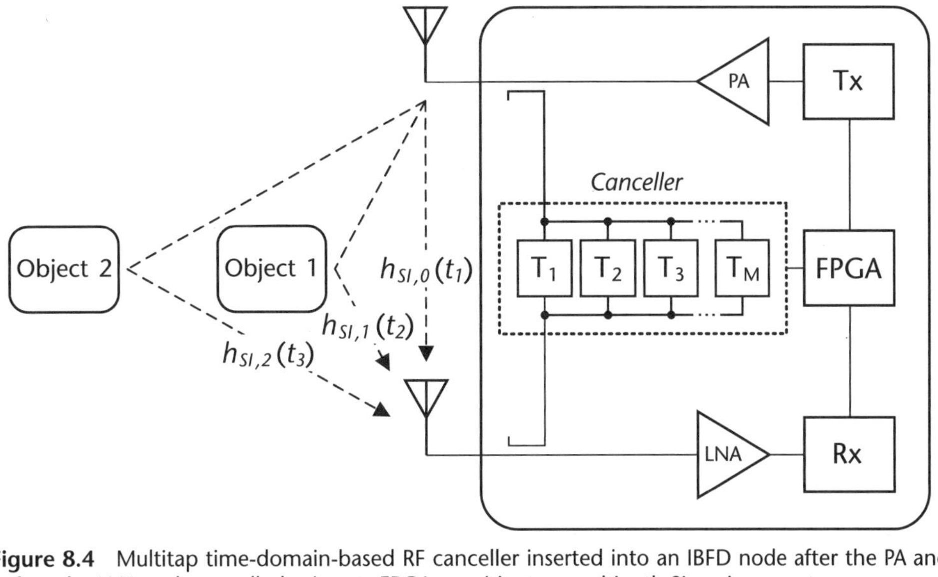

Figure 8.4 Multitap time-domain-based RF canceller inserted into an IBFD node after the PA and before the LNA and controlled using an FPGA to mitigate a multipath SI environment.

8.4 Design Considerations

As previously described, RF cancellers are constructed with multiple taps to provide robustness in unknown operating environments. The effectiveness of these designs can be enhanced by considering both the canceller delay spread and instantaneous bandwidth, as discussed in the following sections.

8.4.1 Delay Spread

The delay spread of a general wireless channel is defined as the time difference between the first and last significant multipath signal within a receiver [4]. For an IBFD node, the first signal is simply the DP coupling and the last multipath component depends on both the environment as well as an application-specific power threshold. Figure 8.5 shows example impulse responses for an IBFD system that was measured in isolated (anechoic chamber) and realistic (indoor) operating environments. While both of these cases include the DP signal delay, seen around 2 ns, the realistic indoor location exhibits much more multipath richness. If a threshold of –60 dB is considered, the delay spread of the isolated case would be very small (only the DP signal is above that level). For the realistic scenario, multipath above –60 dB occurs out to approximately 22 ns, which would result in a SI channel delay spread of 20 ns (22 ns minus 2 ns).

RF cancellers have a corresponding delay spread that is defined as the time difference between their maximum and minimum tap delays, $\tau_M - \tau_1$. As highlighted in Figure 8.4, the canceller delays would ideally match those of the multipath components in the SI channel, and thus, their delay spreads would be the same [17]. As previously mentioned, the SI channel is often unknown and requires that the canceller be designed with a delay spread that anticipates the maximum expected in the location of operation. Selecting a canceller delay spread that is significantly

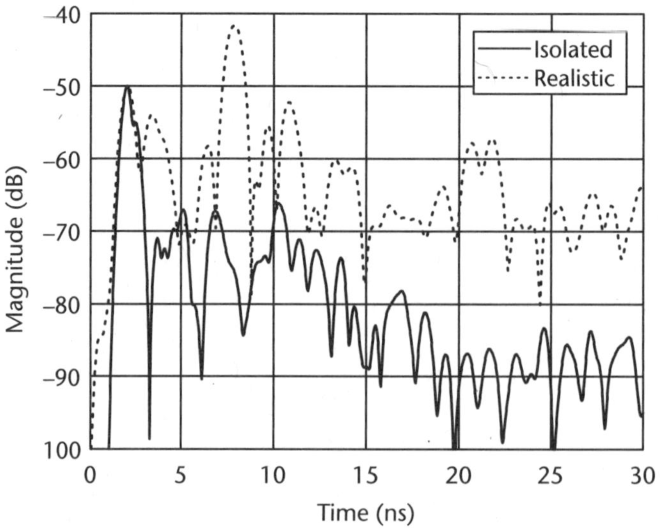

Figure 8.5 Measured channel impulse responses for an IBFD system in both an isolated and realistic (indoor) operating environment, highlighting the delay spread differences.

lower than that of the channel can severely limit its ability to mitigate multipath SI components.

8.4.2 Instantaneous Bandwidth

In addition to considering the canceller delay spread with respect to the channel, the number of canceller taps between those delay extremes greatly influences its ability to perform over a desired instantaneous bandwidth. As seen in Figure 8.6(a), for a simulated canceller with two taps, multitap cancellers have a periodic frequency response, which is also evident in both (8.5) and (8.7). While the top part of Figure 8.6(a) indicates the individual tap contributions, the bottom half depicts their

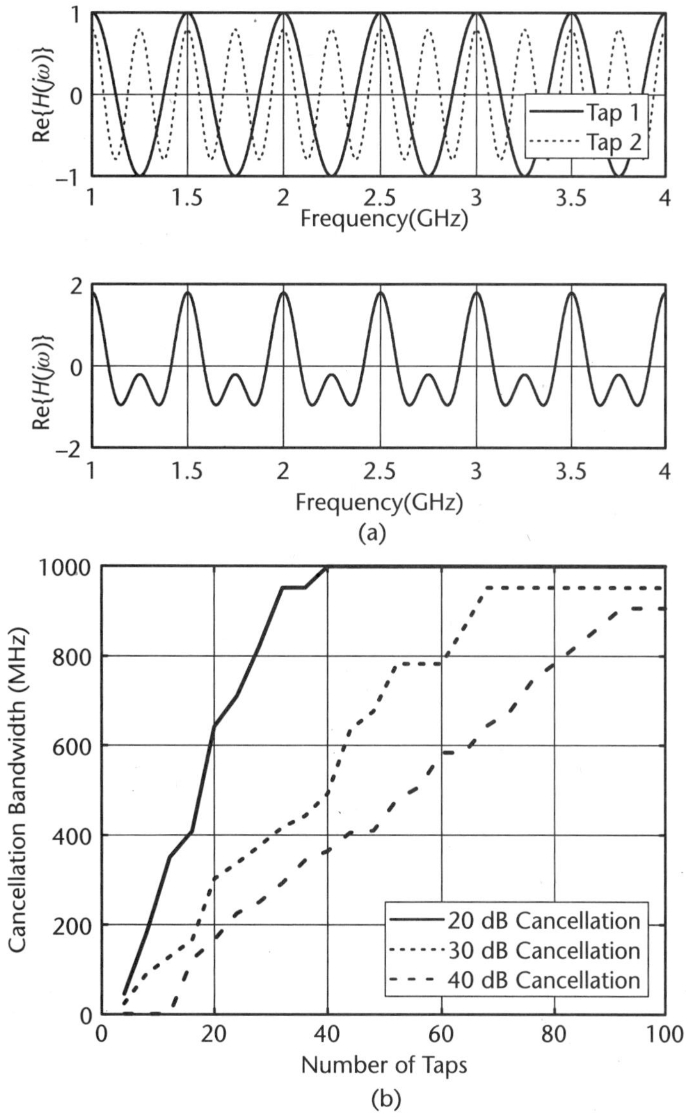

Figure 8.6 (a) Simulated two-tap canceller showing the periodic nature of the individual taps (top) and composite response (bottom); and (b) RF cancellation performance simulation over instantaneous bandwidth with a varying number of canceller taps for a measured self-interference channel, highlighting the need for large numbers of canceller taps to operate over wide bandwidths.

combined response, which has a period that is dependent on both the delay spread and number of taps.

The canceller's instantaneous bandwidth can be optimized by selecting the tap delays with some knowledge of the SI channel since the canceller is effectively sampling the channel using a limited number of taps. These taps would ideally provide closely spaced time samples of the channel, and match or exceed the channel's delay spread, which would allow it to replicate frequency-selective fading scenarios that result from multipath-rich environments.

Achieving wideband cancellation requires a large number of canceller taps with a large delay spread, which generates a combined response with a longer period. Figure 8.6(b) illustrates this through the simulation of the number of canceller taps required to achieve between 20 and 40 dB of cancellation for a given instantaneous bandwidth target. As evidenced in the plot for this example scenario, cancellers with a small number of taps can only provide cancellation over narrow bandwidths, while cancellers created with multiple taps are required to increase the bandwidth of operation [18]. Designing effective cancellers with both low and high numbers of taps will be discussed in the following section.

8.5 RF Canceller Examples

8.5.1 A 4-Tap Design

8.5.1.1 Canceller Architecture

Since the strength of reflected multipath SI signals decreases quickly away from an IBFD node, many narrowband systems can provide a sufficient amount of cancellation with only a few taps. The block diagram of a standard 4-tap canceller prototype is shown in Figure 8.7(a), where D represents the delay elements that were implemented with various lengths of coaxial cable [4]. The individual taps are clearly seen to replicate the amplitude and phase control scheme using variable attenuators and phase shifters, as shown in Figure 8.3(b). The tap weights (w_1 through w_8) were configured using on-board DACs to adjust the analog voltages required for tuning. It should be noted that this architecture exhibited a nonuniform tap weighting that was established using the directional couplers and splitters as shown. This weighting was created to match the SI channel response of an IBFD system employing a typical high-isolation antenna [19].

Figure 8.7(b) depicts a photo of the two-channel prototype board with the delay cables removed. Surface mount components were integrated onto this board to provide minimal size impact to the transceiver system, which was designed to focus on the 30-MHz bandwidth centered at 2.45 GHz. The measured impulse response of the canceller is shown in Figure 8.7(c) for the minimum and maximum attenuation states. The nonuniform tap weighting is evident in the minimum attenuation response with peaks that vary in magnitude. This plot also captures the canceller delay spread for this cable configuration as 8 ns (10 ns minus 2 ns for the last and first tap delays, respectively).

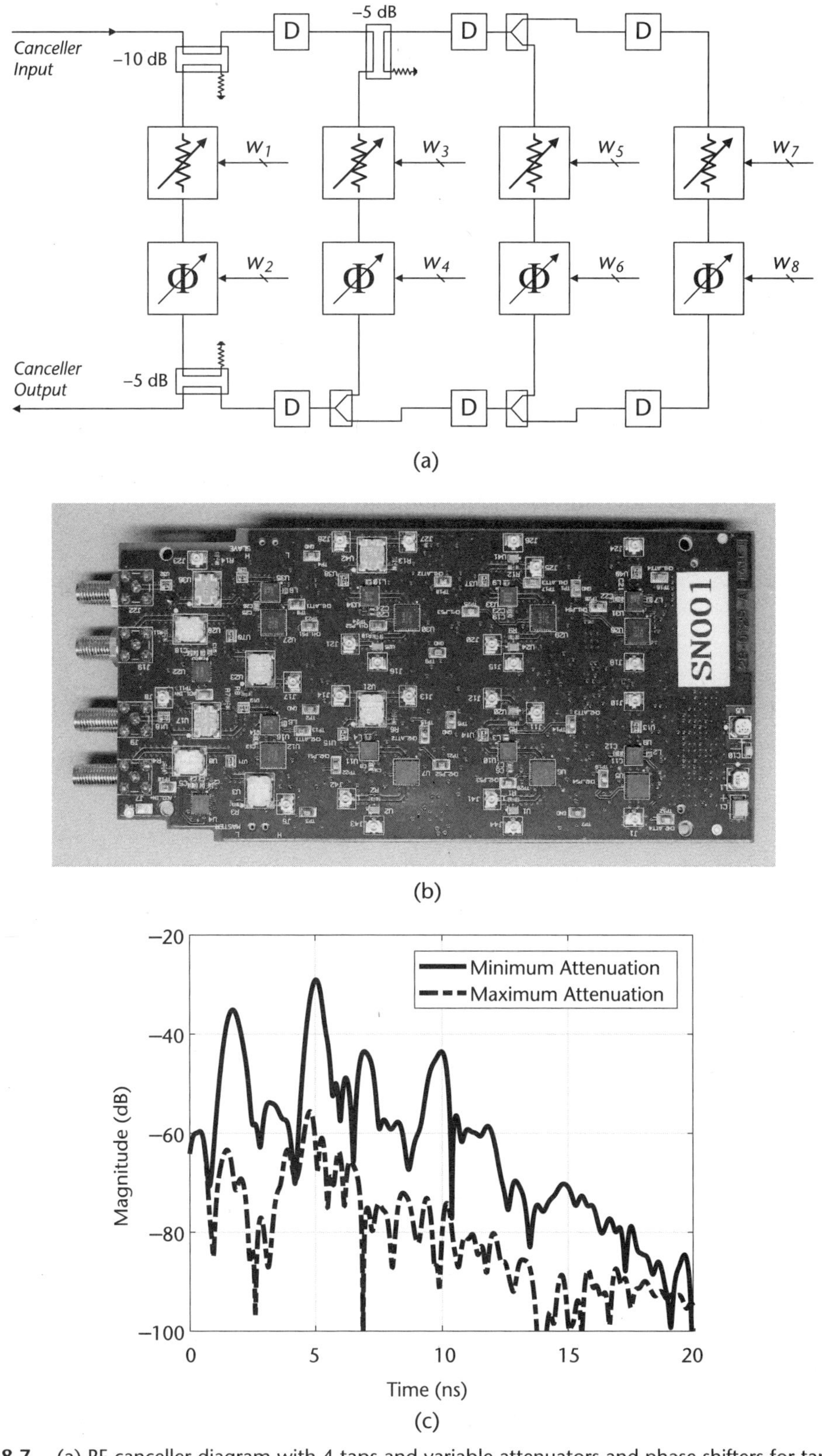

Figure 8.7 (a) RF canceller diagram with 4 taps and variable attenuators and phase shifters for tap amplitude and phase control; (b) prototype board photo with two channels, where the delay cables were removed for clarity; and (c) measured canceller impulse response for minimum and maximum attenuation states.

8.5.1.2 Measured Results

This 4-tap prototype was measured in both an anechoic chamber with a network analyzer as well as a standard indoor operating location, both with a realistic high-isolation antenna that suppressed the DP SI coupling [19]. Figure 8.8(a) illustrates the magnitude responses of the SI channel without the canceller active as well as the cancellation performance that results from tuning to bandwidths of 30 and 100 MHz centered at 2.45 GHz. Over those bandwidths, the average cancellations were measured to be 34.4 and 23.0 dB, respectively. This indicates that the

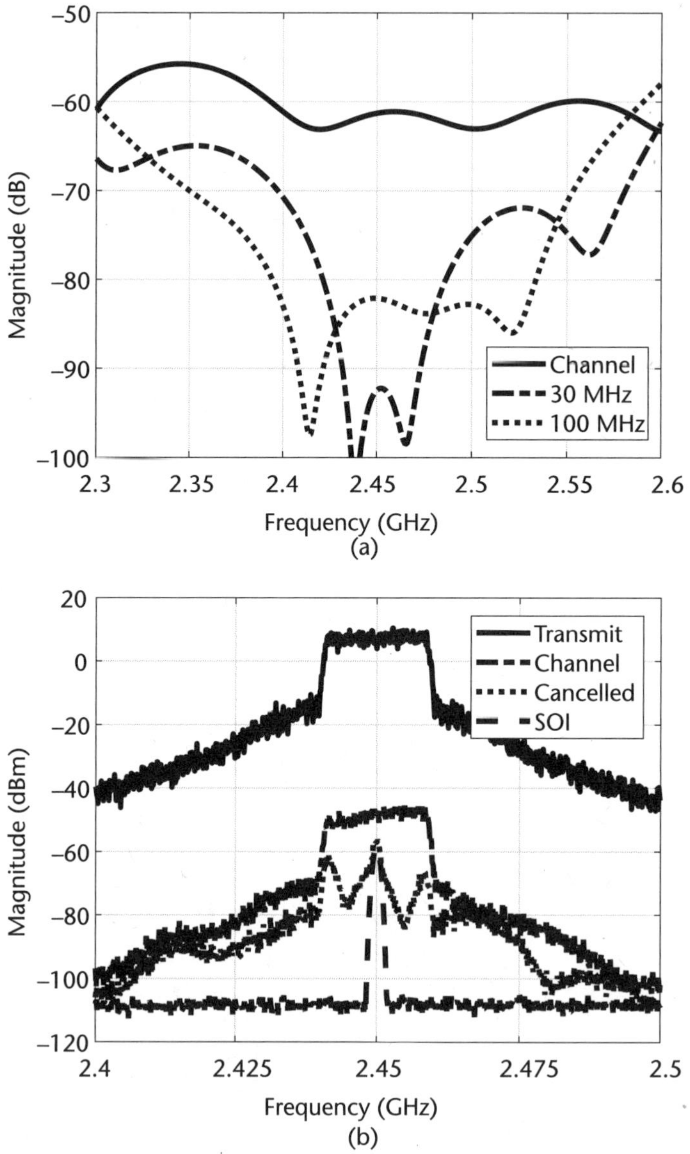

Figure 8.8 (a) Measured magnitude responses of self-interference channel and suppression by a 4-tap canceller targeting different instantaneous bandwidths in an anechoic chamber; and (b) measured magnitude responses of transmit, channel, cancelled and signal-of-interest (SOI) for a +30-dBm OFDM signal in an indoor operating environment. (© 2016 IEEE. Reprinted, with permission, from [4].)

prototype is capable of generating deep and narrow, or slightly shallower but wider, cancellations when comparing different bandwidth targets [20].

The results of the indoor evaluation are shown in Figure 8.8(b) and include both noise and nonlinear distortion products that were generated by an OFDM transmit waveform passing through a typical PA. Similar to the isolated measurement, this plot indicates that the +30-dBm transmit signal is first suppressed by the antenna and that the RF canceller additionally provides 22 dB of cancellation over the 20-MHz bandwidth. Additionally, the canceller allows a representative SOI to be recovered, which was not possible before tuning and underscores the need for RF cancellation in this scenario.

8.5.2 An 80-Tap Design

8.5.2.1 Canceller Architecture

The ability of photonics to enable the use of wide instantaneous bandwidths in IBFD systems was highlighted in Chapters 5 and 7, which each discussed unique properties of MZMs. These components are analogous to analog mixers and can be utilized to create a wideband time-domain-based RF canceller that processes signals in the optical domain after upconverting an RF input with the help of laser drivers. The block diagram of an RF-photonic canceller structure is shown in Figure 8.9(a) [16]. Although it is not straightforward to visualize the canceller taps, the overall design maintains a vector modulator architecture for the tap amplitude and phase control, as represented in Figure 8.3(c).

The unique aspect of this canceller is the use of fiber Bragg gratings (FBGs) that act as time delay elements, which are tunable based on the wavelength of an optical carrier, as depicted at the bottom of Figure 8.9(a) [18]. The design allows for the selection of various optical wavelengths using the tunable lasers for the I- and Q-channels, indicated on the top and bottom on the diagram, respectively. These lasers are independently weighted using variable optical attenuators (VOAs), modulated with the RF input signal, and subsequently delayed according to the desired wavelength and FBG structure before removing the optical carriers with photodiodes. This weighting and delay process allows the canceller to alter the RF input signal based on the node's environment and provide the desired SI mitigation.

This photonic-enabled design has RF inputs and outputs, and was demonstrated over 0.5 to 5.5 GHz using a prototype that was constructed using commercial components. While these parts were organized into two rack-mount chassis (as indicated in Figure 8.9(b)), the canceller architecture can be miniaturized using photonic integrated circuits (PICs), as discussed in Chapter 15, for optical-based canceller systems.

8.5.2.2 Measured Results

This photonic-RF canceller prototype utilized FBGs that generated up to 80 delay settings, which can be considered the number of canceller taps. These delays were spaced roughly every 1.3 ns and contained several gaps that were a result of manufacturing limitations. The canceller impulse response is depicted in Figure 8.10(a) for the 80 tunable locations with a delay spread of approximately 115 ns.

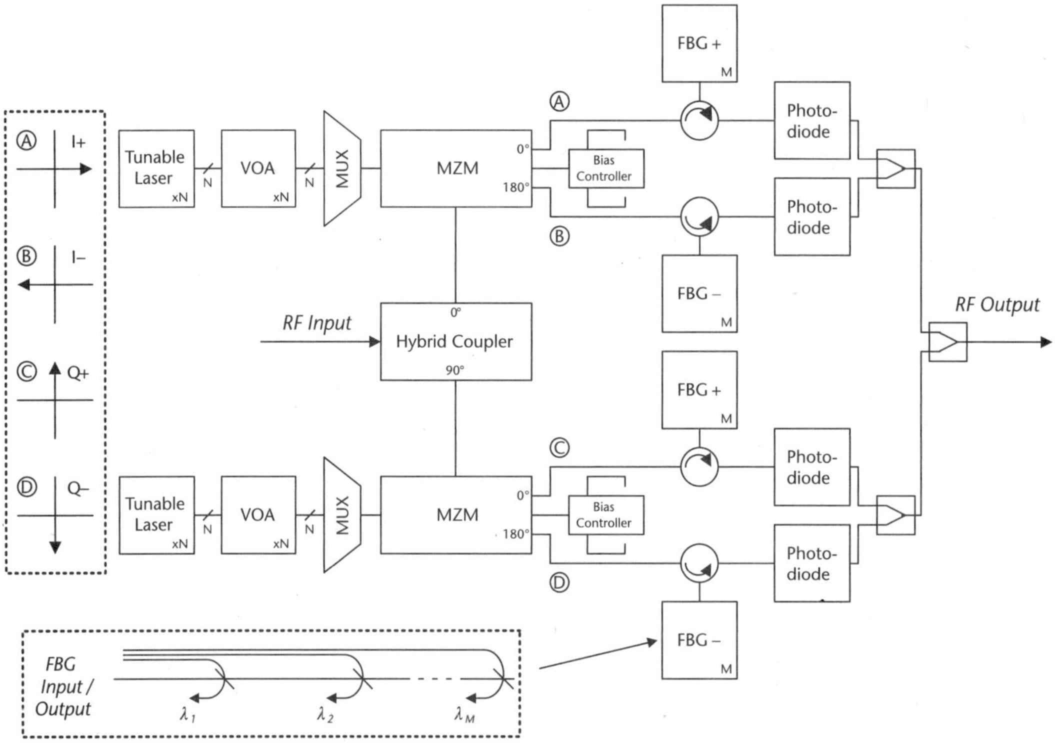

(a)

(b)

Figure 8.9 (a) RF-photonic canceller diagram with up to 80 taps and a vector modulator architecture for tap amplitude and phase control; and (b) prototype photo with two rack-mount chassis.

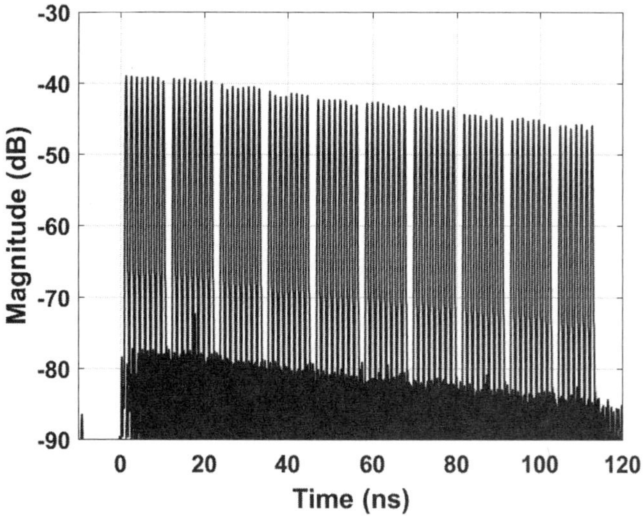

(a)

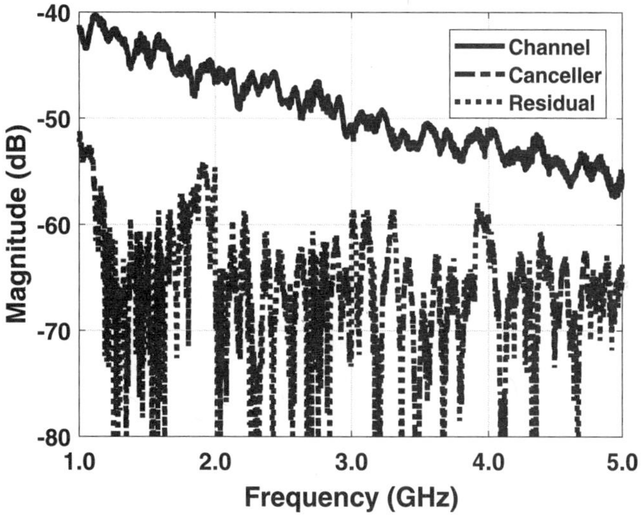

(b)

Figure 8.10 (a) Measured canceller impulse response for FBGs with 80 delay settings; and (b) measured magnitude responses of self-interference channel and suppression by 80-tap canceller targeting 1-GHz instantaneous bandwidths tuned between 1.0 and 5.0 GHz in four segments. (© 2019 IEEE. Reprinted, with permission, from [16].)

Comparing this plot to Figure 8.7(c) provides a sense of the increased number of taps and delay spread provided by photonic architectures as opposed to RF-only ones.

The performance capability of the canceller was evaluated by inserting it into a realistic multipath environment that was created using a channel emulator to generate an SI channel response for a node mounted on a base station tower. The channel was comprised of DP SI as well as strong reflections located at 0.75m, 1.85m, and 2.65m from the installation location. Figure 8.10(b) indicates the results of the canceller being tuned four times over the 4 GHz of bandwidth from 1.0 to 5.0 GHz (1.0 to 2.0, 2.0 to 3.0, 3.0 to 4.0, and 4.0 to 5.0 GHz). The average cancellation for these four 1-GHz bands was 18.6, 20.0, 15.5, and 13.3 dB, respectively, and yielded a total average of 16.9-dB cancellation over the 1.0 to 5.0-GHz bandwidth when combined [16]. As compared to the previously discussed all-RF approaches, this RF-photonic architecture can provide cancellation over a wide range of operating frequencies and instantaneous bandwidths, which significantly increases its flexibility.

8.6 Advanced Tuning Approaches

Equations (8.4) through (8.7) capture the fact that time-domain-based RF cancellers have tunable tap weights that must be adjusted to minimize the undesired SI. This can be formulated into an error metric, such that

$$E(w) = H_{SI}(j\omega) + H_c(j\omega), \tag{8.8}$$

where $E(w)$ is a frequency-domain representation of the residual SI at the receiver after cancellation. This error is dependent on the canceller weights, w, and is often minimized using standard gradient descent (GD) algorithms that provide weight updates according to

$$w_{n+1} = w_n - \mu \nabla E(w), \tag{8.9}$$

where μ is the step size, or learning rate, $\nabla E(w)$ is the gradient of the error surface, and the algorithm iterations are denoted by the subscript n. While many RF cancellers utilize this tuning approach, it can provide suboptimal results for the nonconvex error surfaces generated by multitap designs.

This challenge is illustrated in Figure 8.11(a), which represents a simplified one-dimensional surface for a single tuning weight. When a canceller is enabled for an unknown SI channel, it selects initial weight settings that are nonoptimal (such as point A). A successful tuning algorithm must then avoid local minima (such as point B), investigate interesting features on the surface (such as point C), and ultimately locate the global minimum (such as point D). While the standard GD algorithm of (8.9) can struggle with local minimum and provide slow convergence speeds, adaptive GD approaches can improve upon these issues by dynamically adjusting the algorithm's learning rate.

This is evident in the plot of Figure 8.11(b) that illustrates measured cancellation versus tuning iterations using several different algorithms for a 2-tap vector modulation canceller [21]. The adaptive GD is seen to converge to more than 30 dB of cancellation in less than 30 iterations [17], while the standard GD achieves less than 25 dB quickly, but cannot improve its performance due to its fixed parameters. Finally, the graph indicates that the canceller tuning can be further accelerated with the help of artificial neural networks (NNs), which utilize training datasets to construct models that provide canceller weight predictions for a given SI channel [22]. The use of these NN models can initialize the cancellation to higher values and require fewer tuning steps as the weights are starting near their optimal locations.

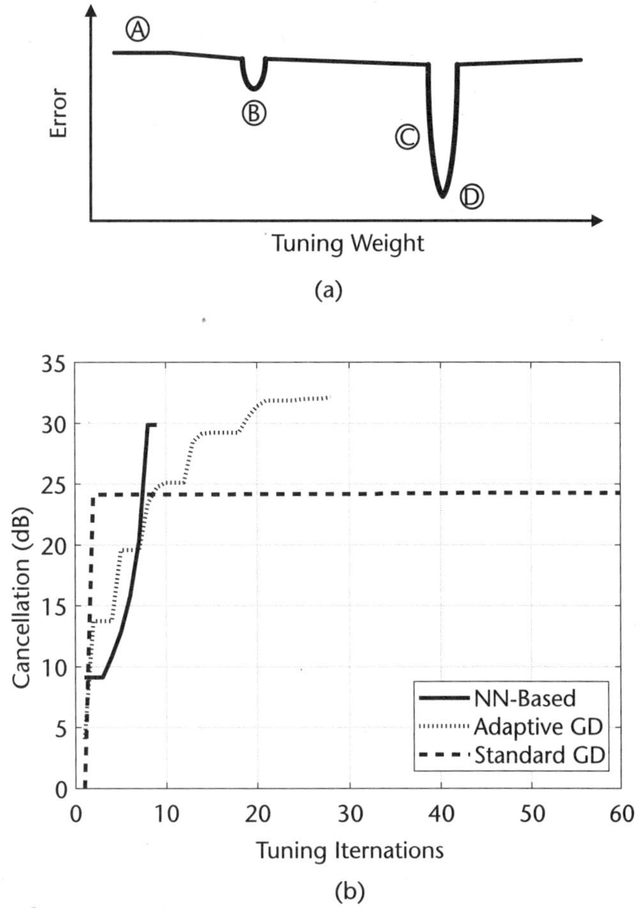

Figure 8.11 (a) Example cancellation error surface that is a function of a single tuning weight for a nonconvex scenario, and (b) average cancellation performance changes versus time, measured over 20 MHz centered at 2.5 GHz, for different tuning algorithms.

8.7 Practical Considerations

While this chapter discussed various time-domain-based RF canceller architectures, their incorporation into a large IBFD system requires careful planning to maximize SIC performance, specifically in three areas. First, the absolute delays in the canceller and SI should be calibrated so that the first canceller tap aligns in time with the fixed DP coupling, as previously mentioned. This will ensure that the canceller is provided the best opportunity to address delayed multipath signals. Second, the linearity and noise performance of the canceller should be considered during the design process. If the canceller generates nonlinearities that are independent of the transmitter components, they could become more difficult to remove with digital approaches, such as those discussed in Chapters 11 and 12. Finally, as the number of canceller taps increases, the canceller-tuning surface becomes more complex, which makes finding the optimal settings difficult. While both analog and digital adaptation approaches can be considered, hardware-specific tuning algorithms should be codesigned along with the canceller to improve robustness and reduce convergence time [23–25].

8.8 Conclusion

8.8.1 Summary

This chapter presented the motivation and design considerations for time-domain RF cancellers in IBFD systems. Similar to other types of RF cancellers, the goal is to reduce SI in the input of the receiver before nonlinearities are generated and/or saturation occurs. The instantaneous bandwidth capability and general flexibility of these cancellers depends on both their number of taps as well as their delay spread, which was defined as the time difference between the tap extremes. These time-domain-based taps can adjust their amplitude and phase characteristics using either variable attenuators and phase shifter or vector modulator architectures.

Increasing both the tap number and delay spread within a canceller allows it to mitigate SI in realistic environments with rich multipath characteristics. Two different prototype cancellers have demonstrated the effectiveness of these multitap designs with traditional RF components as well as an RF-photonic structure. These cancellers were created with 4 and 80 taps to target 30 and 1,000-MHz instantaneous bandwidths, respectively. Finally, the weights within the various taps need to be optimally configured to simultaneously maximum performance and minimize the time in doing so. Several advanced tuning approaches, including adaptive learning rate algorithms and the use of neural networks, were presented to address both of these goals.

8.8.2 Future Directions

The two prototype cancellers discussed in this chapter were constructed using commercially available parts, which significantly increased their overall size. Future efforts in this area will include research into integrated circuit (IC) approaches to time-domain-based cancellers that can be miniaturized similar to their frequency-domain counterparts, later discussed in Chapter 9. In addition to reducing the canceller form factor, architecture investigations should be conducted on approaches to extend the bandwidth of operation within both the instantaneous and tunable range. Future designs should be capable of maintaining sufficient linearity for the high transmit output powers that may exist in base stations and military systems. Finally, the multitap canceller tuning approaches should be further developed to improve both cancellation and settling time performance, as discussed.

References

[1] Jain, M., et al., "Practical, Real-Time, Full Duplex Wireless," *Proc. of the Annual Intl. Conf. on Mobile Computing and Networking (MobiCom '11)*, New York, 2011, pp. 301–312.

[2] Kolodziej, K. E., B. T. Perry, and J. S. Herd, "In-Band Full-Duplex Technology: Techniques and Systems Survey," *IEEE Transactions on Microwave Theory and Techniques*, Vol. 67, No. 7, July 2019, pp. 3025–3041.

[3] Bharadia, D., E. McMilin, and S. Katti, "Full Duplex Radios," *2013 ACM Conf. (SIGCOMM), SIGCOMM '13*, New York, 2013, pp. 375–386.

[4] Kolodziej, K. E., J. G. McMichael, and B. T. Perry, "Multitap RF Canceller for In-Band Full-Duplex Wireless Communications," *IEEE Transactions on Wireless Communications*, Vol. 15, No. 6, June 2016, pp. 4321–4334.

[5] Nagulu, A., et al., "A Full-Duplex Receiver Leveraging Multiphase Switched-Capacitor-Delay Based Multi-Domain FIR Filter Cancelers," *2020 IEEE Radio Frequency Integrated Circuits Symposium (RFIC)*, Los Angeles, CA, 2020, pp. 43–46.

[6] Zhang, Z., et al., "Full Duplex 2x2 MIMO Radios," *2014 Sixth International Conference on Wireless Communications and Signal Processing (WCSP)*, Hefei, 2014, pp. 1–6.

[7] Liu, Y., et al., "A Full-Duplex Transceiver with Two-Stage Analog Cancellations for Multipath Self-Interference," *IEEE Transactions on Microwave Theory and Techniques*, Vol. 65, No. 12, December 2017, pp. 5263–5273.

[8] Keogh, B., and A. Zhu, "Wideband Self-Interference Cancellation for 5G Full-Duplex Radio Using a Near-Field Sensor Array," *2018 IEEE MTT-S International Microwave Workshop Series on 5G Hardware and System Technologies (IMWS-5G)*, Dublin, 2018, pp. 1–3.

[9] Zhang, L., M. Ma, and B. Jiao, "Design and Implementation of Adaptive Multi-Tap Analog Interference Canceller," *IEEE Transactions on Wireless Communications*, Vol. 18, No. 3, March 2019, pp. 1698–1706.

[10] Ahn, I., J. Kim, and H. Song, "Adaptive Analog Self-Interference Cancellation for In-Band Full-Duplex Wireless Communication," *2019 IEEE Asia-Pacific Microwave Conference (APMC)*, Singapore, 2019, pp. 414–416.

[11] Huusari, T., et al., "Wideband Self-Adaptive RF Cancellation Circuit for Full-Duplex Radio: Operating Principle and Measurements," *2015 IEEE 81st Vehicular Technology Conference (VTC Spring)*, Glasgow, 2015, pp. 1–7.

[12] Tamminen, J., et al., "Digitally-Controlled RF Self-Interference Canceller for Full-Duplex Radios," *2016 24th European Signal Processing Conference (EUSIPCO)*, Budapest, 2016, pp. 783–787.

[13] Ershadi, A., and K. Entesari, "A 0.5-to-3.5 GHz Self-Interference-Canceling Receiver for In-Band Full-Duplex Wireless," *2019 IEEE Radio Frequency Integrated Circuits Symposium (RFIC)*, Boston, MA, 2019, pp. 151–154.

[14] Palaniappan, R., V. Gurumurthy, and S. Aniruddhan, "A Spectral Shaper Based Two-Tap RF Self-Interference Canceller for Full-Duplex Radios," *2019 IEEE MTT-S International Microwave Symposium (IMS)*, Boston, MA, 2019, pp. 614–617.

[15] Sharma, P. K., and N. Nallam, "A 0.1–0.95 GHz Full-Duplex Receiver with < 1 dB NF Degradation Using a Passive Continuous-Mode Charge-Sharing Vector Modulator," *IEEE Transactions on Microwave Theory and Techniques*, Vol. 67, No. 7, July 2019, pp. 3042–3052.

[16] Kolodziej, K. E., S. Yegnanarayanan, and B. T. Perry, "Photonic-Enabled RF Canceller for Wideband In-Band Full-Duplex Wireless Systems," *IEEE Transactions on Microwave Theory and Techniques*, Vol. 67, No. 5, May 2019, pp. 2076–2086.

[17] Kolodziej, K. E., A. U. Cookson, and B. T. Perry, "Adaptive Learning Rate Tuning Algorithm for RF Self-Interference Cancellation," *IEEE Transactions on Microwave Theory and Techniques*, 2020.

[18] Kolodziej, K. E., S. Yegnanarayanan, and B. T. Perry, "Fiber Bragg Grating Delay Lines for Wideband Self-Interference Cancellation," *IEEE Transactions on Microwave Theory and Techniques*, Vol. 67, No. 10, October 2019, pp. 4005–4014.

[19] Kolodziej, K. E., et al., "Ring Array Antenna with Optimized Beamformer for Simultaneous Transmit and Receive," *Proceedings of the 2012 IEEE International Symposium on Antennas and Propagation*, Chicago, IL, 2012, pp. 1–2.

[20] McMichael, J. G., and K. E. Kolodziej, "Optimal Tuning of Analog Self-Interference Cancellers for Full-Duplex Wireless Communication," *2012 50th Annual Allerton Conference on Communication, Control, and Computing (Allerton)*, Monticello, IL, 2012, pp. 246–251.

[21] Kolodziej, K. E., and B. T. Perry, "Wideband Vector Modulator for RF Cancellers in STAR Systems," *2018 IEEE Radio and Wireless Symposium (RWS)*, Anaheim, CA, 2018, pp. 64–67.

[22] Kolodziej, K. E., A. U. Cookson, and B. T. Perry, "Machine Learning for Accelerated IBFD Tuning in 5G Flexible Duplex Networks," *2020 IEEE/MTT-S International Microwave Symposium (IMS)*, Los Angeles, CA, 2020, pp. 691–694.

[23] Mikhael, M., et al., "An In-Band Full-Duplex Transceiver Prototype with an In-System Automated Tuning for RF Self-Interference Cancellation," *1st International Conference on 5G for Ubiquitous Connectivity*, Akaslompolo, 2014, pp. 110–115.

[24] Huang, X., and Y. J. Guo, "Radio Frequency Self-Interference Cancellation with Analog Least Mean-Square Loop," *IEEE Transactions on Microwave Theory and Techniques*, Vol. 65, No. 9, September 2017, pp. 3336–3350.

[25] Tapio, V., M. Sonkki, and M. Juntti, "Analog Self-Interference Cancellation with Automatic Gain Control for Full-Duplex Transceivers," *2018 IEEE 87th Vehicular Technology Conference (VTC Spring)*, Porto, 2018, pp. 1–5.

CHAPTER 9

Frequency-Domain RF Cancellation

Jin Zhou, Tingjun Chen, Yuhe Cao, Gil Zussman, and Harish Krishnaswamy

9.1 Introduction

Equalization-based RF SIC has the advantage of being able to emulate frequency-selective SI channel responses over a wide frequency range for wideband cancellation. For time-domain-based RF SIC such as the ones discussed in Chapter 8 and [1, 2], however, the generation of significant (nanosecond-scale) true time delay on silicon is challenging at its core due to both the length of the transmission lines required and the lossy nature of the silicon substrate. For example, to generate a 1-ns delay, a transmission line with a length of 15 cm is needed, which is too bulky and lossy to be integrated on silicon.

In this chapter, we address wideband SIC in the analog domain based on frequency-domain equalization (FDE) of the wireless SI channel [3], as highlighted within the IBFD transceiver diagram of Figure 9.1(a) and SIC techniques tree of Figure 9.1(b). The applicability of the FDE-based technique to integrated wideband SIC in the RF domain is made possible by recent advances in the implementation of tunable, reconfigurable, high-Q RF bandpass filters (BPFs) within nanoscale, CMOS, namely N-path filters, as mentioned in Chapters 4 and 6 [4–6].

9.2 Integrated N-Path-Filter-Based Wideband RF SIC

9.2.1 Concept of Frequency-Domain Equalization

Many RF SI cancellers lack equalization functionality and purely rely on amplitude-and-phase scaling. Consider a wireless transceiver with a frequency-selective antenna interface (such as an antenna pair, a duplexer, or a circulator, as discussed in Chapters 2, 3, and 4, respectively) and a conventional frequency-flat amplitude-based and phase-based RF canceller depicted in Figure 9.2(a). For simplicity, we will model the antenna interface response [$H_{SI,model}(j\omega)$] with a flat magnitude response and a constant group delay as $H_{SI,model}(j\omega) = A_0\exp(-j(\omega - \omega_0)\tau_{SI} + j\varphi_0)$. In particular, A_0 is the frequency-flat magnitude response, τ_{SI} is the isolation group delay, and φ_0 is the phase at frequency ω_0. For the frequency-flat RF canceller, the

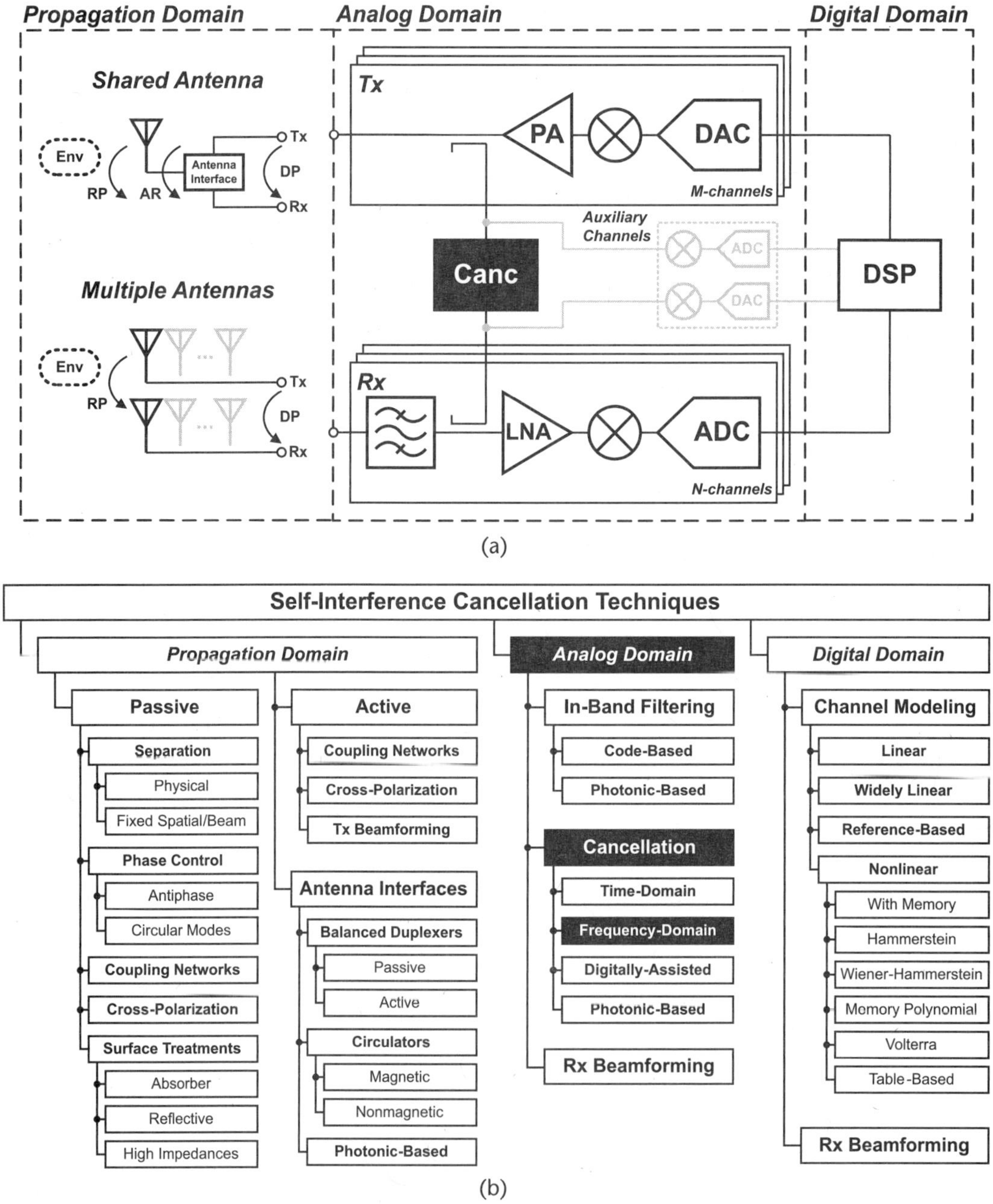

Figure 9.1 Frequency-domain RF cancellation and specific chapter focus highlighted within (a) a generic IBFD transceiver and (b) a SIC techniques tree diagram.

transfer function $\hat{H}_{flat}$ can be written as $\hat{H}_{flat}A_C\exp(+j\varphi_C)$, where A_C and φ_C are the frequency-flat magnitude and phase responses of the canceller, respectively. To achieve RF SIC centered at ω_0, one needs to set $A_C = A_0$ and $\varphi_C = \varphi_0$. An equation can then be written for the residual SI across ω, and assuming $(\omega - \omega_0)\,\tau_{SI} << 1$ rad, create one for the RF bandwidth ω_{BW} within which the worst-case SIC is SIC_{RF}. We obtain:

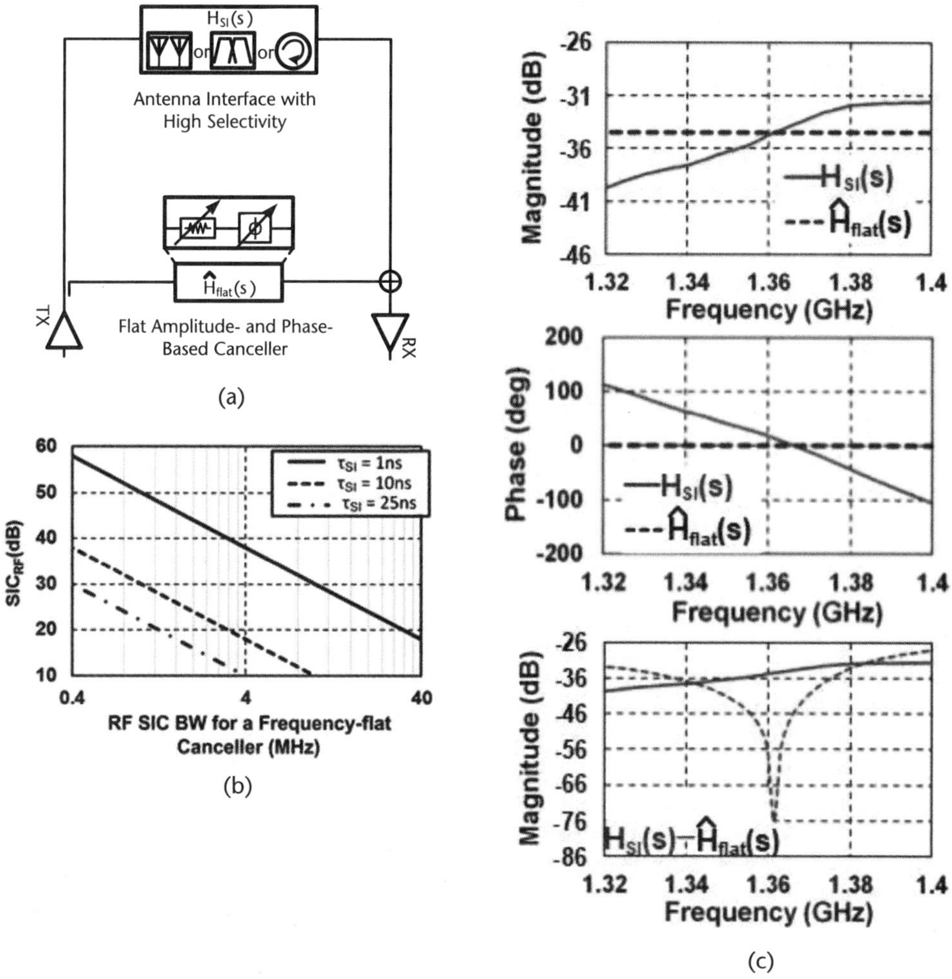

Figure 9.2 Conventional frequency-flat RF canceller based on amplitude and phase scaling. (a) Block diagram; (b) calculated RF SIC bandwidth for a given worst-case SIC for a frequency-flat amplitude-based and phase-based canceller using (9.1) across various SI channel delays; and (c) frequency responses of an antenna interface, the RF canceller, and the resultant SIC. (© 2015 IEEE. Reprinted, with permission, from [3].)

$$\omega_{BW} = \frac{2}{\tau_{SI}\sqrt{SIC_{RF}}}. \tag{9.1}$$

SIC_{RF} is plotted versus f_{BW} in Figure 9.2(b) for SI = 1 ns, 10 ns, and 25 ns. For SIC_{RF} = 20 dB, the supported SI RF bandwidths are 31 MHz, 3 MHz, and 1.2 MHz, respectively.

To verify the theoretical analysis in (9.1), Figure 9.2(c) plots the measured isolation transfer function of a 1.4-GHz antenna pair $H_{SI}(j\omega)$ with 32-dB peak Tx/Rx isolation, 9-ns peak isolation group delay, and 3-dB isolation magnitude variation over 1.36 to 1.38 GHz. The transfer function of a frequency-flat amplitude-and-phase-based RF canceller is also plotted in Figure 9.2(c). The conventional RF canceller is only able to emulate the frequency-selective antenna interface at a single cancellation frequency, resulting in a 20-dB SIC bandwidth of about 3 MHz in Figure 9.2(c). This correlates well with the result in (9.1).

To enhance the cancellation bandwidth, second-order reconfigurable BPFs with amplitude and phase control in each path are introduced in the RF canceller (see Figure 9.3(a)). An RF canceller with a reconfigurable second-order RF BPF consists of 4 degrees of freedom (DOF), namely the amplitude response, phase response, quality factor (or Q factor), and center frequency. This allows for the replication of not only the magnitude and phase of the antenna interface isolation at a certain frequency point, but also of the slope of both the magnitude and phase (i.e., group delay), thereby enhancing the 20-dB SIC bandwidth from 3 to 19 MHz, as shown in Figure 9.3(b).

Using a filter bank with independent BPF parameters allows such replication to occur at multiple points in different frequency sub-bands, further enhancing RF SIC bandwidth. In Figure 9.3(c), the use of two BPFs enhances the 20-dB SIC bandwidth to 32 MHz, representing a 10× improvement over a conventional RF canceller. This approach is, essentially, FDE in the RF domain.

Similar to traditional FDE in the digital domain, the RF SIC equalizer coefficients can be derived from the SI channel transfer function $H_{SI}(j\omega)$. Furthermore, an iterative successive approach is employed to configure the RF BPFs in the FDE-based RF canceller. In Figure 9.3(c), $\hat{H}_1(j\omega)$ first emulates the antenna interface $H_{SI}(j\omega)$ at $\omega_{SIC,1} = 2 \times 1.370$ G rad/sec, then $\hat{H}_2(j\omega)$ replicates the residual response $H_{SI}(j\omega) - \hat{H}_1(j\omega)$ at $\omega_{SIC,2} = 2 \times 1.350$ G rad/sec. Because the introduction of $\hat{H}_2(j\omega)$ will adversely affect the cancellation originally achieved at $\omega_{SIC,1}$, an iteration of $\hat{H}_1$ and $\hat{H}_2$ is required.

9.2.2 Integrated RF Bandpass Filters

Reconfigurable RF BPFs with high-quality factors (e.g., as high as 81.2 in the example shown in Figure 9.3) are required to channelize the SI signal bandwidth for FDE in the RF domain. However, integrated RF BPFs have been notoriously difficult to implement. In this section, we first review two popular approaches to the implementations of integrated RF BPFs and then introduce our proposed two-port N-path G_m-C filter with embedded variable attenuation and phase shifting.

RF BPFs can be realized by using inductor-capacitor (LC) circuitry. However, the qualify factor of integrated RF BPFs based on LC topology is limited by the inductor qualify factor. To compensate the loss in the LC tank, the Q-enhanced LC-based technique has been proposed [7]. In an RF BPF design using the Q-enhanced technique, a negative resistor that can be implemented as a transconductor with positive feedback to compensate the LC tank loss [7, 8]. In theory, the achievable quality factor for Q-enhanced RF BPFs can be as high as desired. At large quality factor enhancements, the negative resistance must be carefully controlled to avoid oscillation, limiting the achievable quality factor. While a Q-factor of 400 at 2.4 GHz has been reported in [8], Q-enhanced LC RF BPFs have several disadvantages such as the need for inductors (which are not friendly with process scaling), limited frequency-tuning range and poor dynamic range performance. In addition, phase shifting capability that is required by the RF equalizer BPF cannot be easily embedded within a Q-enhanced BPF.

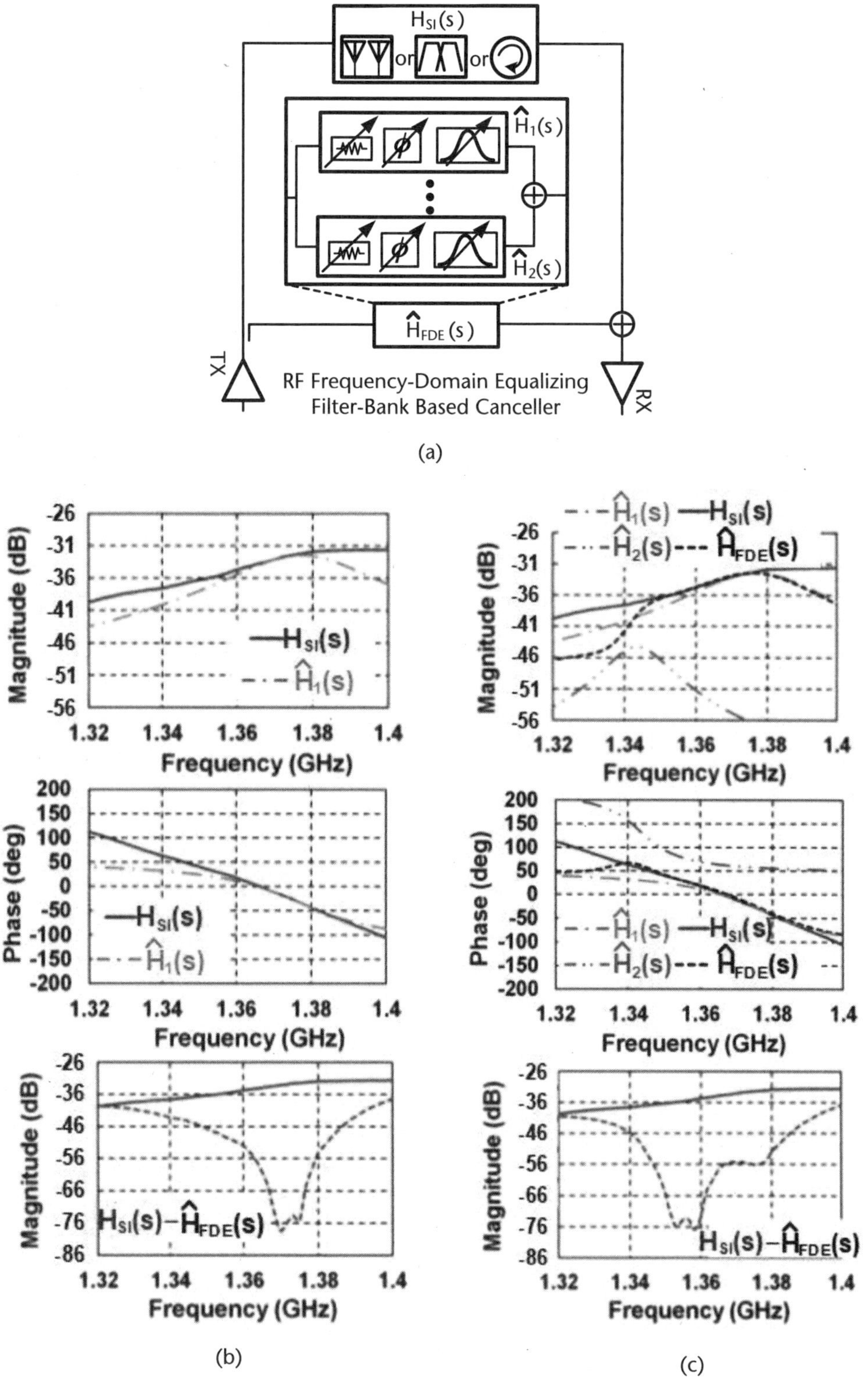

Figure 9.3 Concept of frequency-domain equalization. (a) Block diagram; (b) frequency responses of the antenna interface and RF canceller with one filter and the resulting SIC; and (c) frequency responses of the antenna interface and RF canceller with two filters and the resulting SIC. (© 2015 IEEE. Reprinted, with permission, from [3].)

The idea of N-path filters dates back to the 1940s and was demonstrated by using an electromechanical circuit [9]. Thanks to the advancement in modern ultra-scaled CMOS technology, transistors are able to efficiently switch within the gigahertz frequency range, making N-path filters a promising solution for the implementation of integrated widely tunable high-Q RF filters, as shown in Figure 9.4 [4]. When compared to the Q-enhanced LC-based RF BPFs, N-path filters in CMOS have more compact size, wider-frequency tuning range, and better dynamic range performance that is decoupled with Q-factors [4]. Furthermore, as we will discuss shortly, phase shifting can be embedded within a two-port N-path filter, making N-path filters a viable solution for the RF FDE technique.

The operation of N-path filters can be understood intuitively, as shown in Figure 9.4(b). An input signal is first downconverted to a baseband frequency, then processed by a resistor-capacitor (RC) lowpass filter, and finally upconverted back to the RF. This process essentially transfers an RC lowpass response to a bandpass response centered around the mixer switching frequency and its harmonics. Also, due to the transparency of the transistor switches, the downconversion and upconversion mixers can be merged into a set of switching transistors, as in Figure 9.4(a). For exact modeling of N-path filters, linear periodically time-varying (LPTV) analyses can be applied [4].

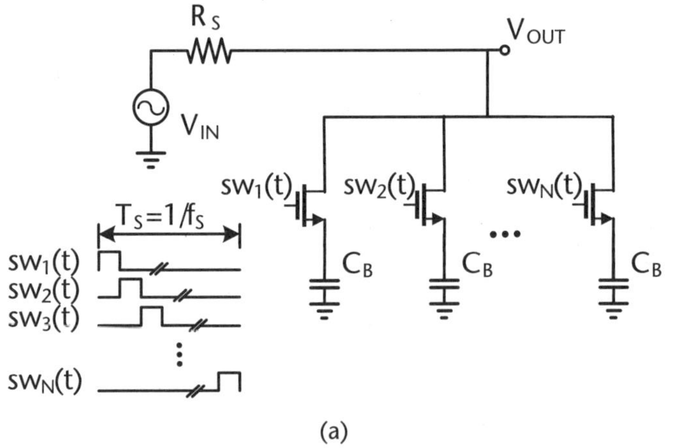

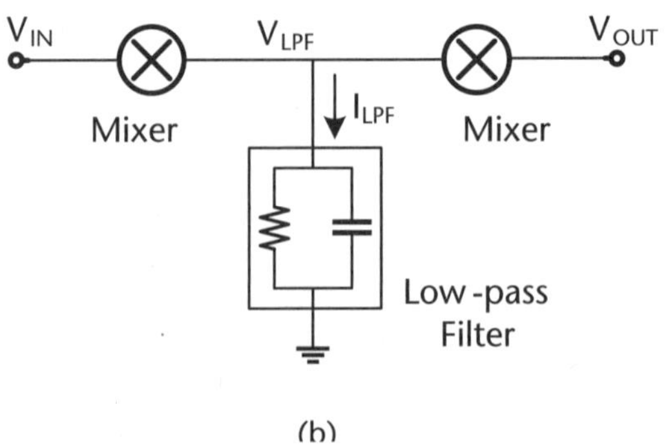

Figure 9.4 Integrated RF bandpass N-path filter: (a) circuit schematic and (b) functional diagram.

9.3 A 0.8–1.4-GHz Rx with FDE-Based SIC

9.3.1 Integrated Two-Port N-Path Canceller Filter with Embedded Phase Shifting

Our proposed two-port N-path filter for FDE is depicted in Figure 9.5(a), where R_S and R_L are, respectively, the resistive loads at the Tx and Rx sides. C_C weakly couples the cancellation signal to the receiver input for SIC. The quality factor of

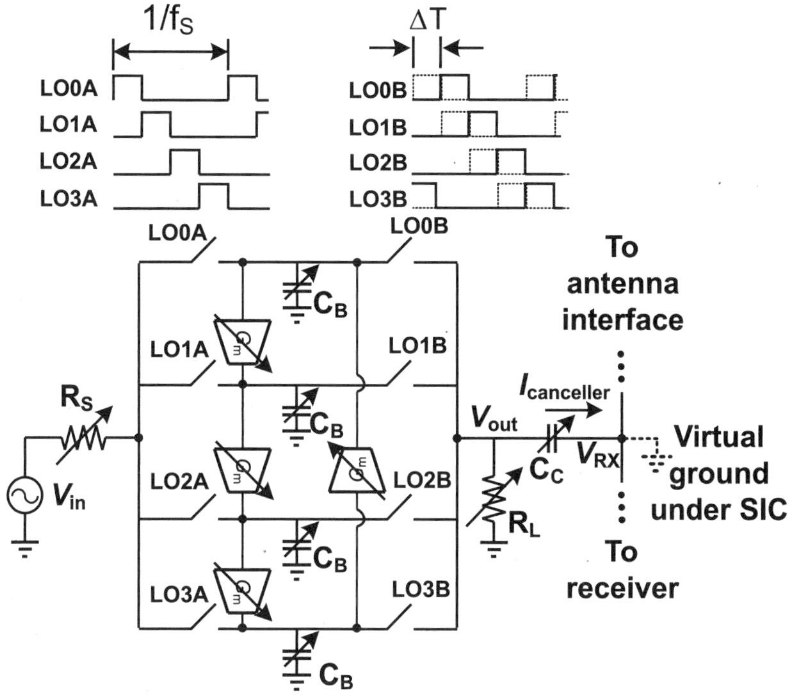

(a)

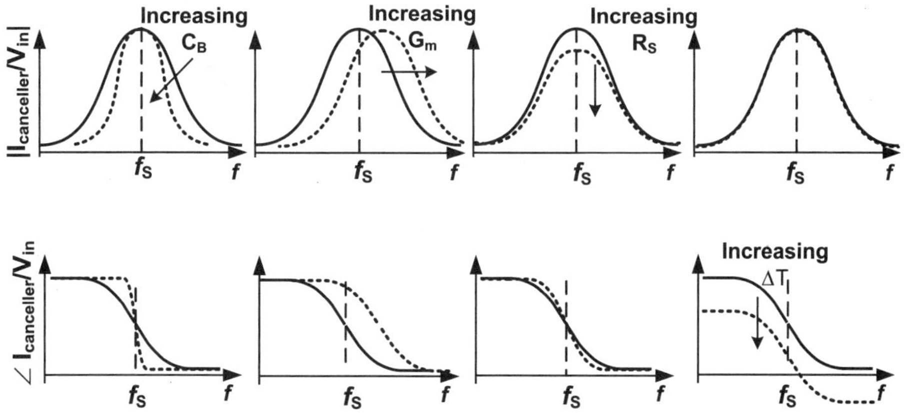

(b)

Figure 9.5 Two-port G_m-C N-path filter implementation with embedded variable attenuation and phase shift. (a) Block diagram; and (b) illustration of variable quality factor (group delay), frequency shift, attenuation, and phase shift. (© 2015 IEEE. Reprinted, with permission, from [3].)

an N-path filter may be reconfigured via the baseband capacitor C_B, given fixed R_S and R_L. LPTV analysis yields [4]:

$$Q = 4\pi f_S\left(\left(R_S + R_{on}\right)\|\left(R_L + R_{on}\right)\right)C_B, \tag{9.2}$$

where R_{on} is the on-resistance of the N-path switches, f_S is the switching frequency, and we have ignored the loading effect of C_C. Through clockwise/counterclockwise (only the counterclockwise connection is shown in Figure 9.5(a) for simplicity) connected baseband reconfigurable transconductors (G_m), an upwards/downwards frequency offset with respect to the switching frequency can be obtained [10]. The frequency offset of the center frequency to the switching frequency is indicated by $\Delta\omega = G_m/C_B$ [10]. Reconfiguring R_S and R_L relative to each other can introduce variable attenuation. The magnitude response at the center frequency is represented as:

$$H(j\omega_S) \approx \frac{8}{\pi^2}\frac{\left(R_L + R_{on}\right)\omega_S C_C R_0}{R_S + R_L + 2R_{on}}, \tag{9.3}$$

where we ignored the loading effect of C_C and R_0 is the system reference resistance, which is typically 50Ω. It merits noting that once cancellation is performed, V_{RX} node in Figure 9.5(a) is a virtual ground. The canceller transfer function can be computed by locating the $I_{Canceller}$ that flows into the virtual ground and then multiplying it by the reference resistance R_0. Also, in (9.3), the center frequency is assumed to be ω_S (i.e., $G_m = 0$) for simplicity.

Phase shifting, interestingly, can be embedded within a two-port N-path filter by phase shifting the local oscillators (LOs) driving the switches on the output side, as in Figure 9.5(b). A complete LPTV analysis of a two-port N-path filter with embedded phase shifting in [5] reveals that phase shifting the LOs driving the output-side switches imparts constant phase shifts to the two-port N-path filter frequency response with no other impact on the close-in response.

9.3.2 Implementation and Measurement Results

A canceller bank of two reconfigurable second-order G_m-C N-path filters together with a 0.8 to 1.4-GHz reconfigurable current-mode Rx is implemented within a 65-nm standard CMOS process, as is shown in the block diagram of Figure 9.6. These canceller filters have their own separate LO and Tx replica signal inputs, offering flexibility in their use (e.g., cancellation of two separate Tx signals for MIMO applications, as discussed in Chapter 14). The receiver uses a noise-cancelling common gate (CG), common source (CS) low-noise transconductance amplifier (LNTA) followed by 4-phase current-driven passive mixers and baseband transimpedance amplifier [11]. Programmable baseband recombination circuits combine the receiver outputs from the CG and CS paths for noise cancellation [11].

For each canceller filter, the coupling capacitor C_C bank is designed to enable a –10-dB coupling across the operating frequency range. Additionally, a fairly weak

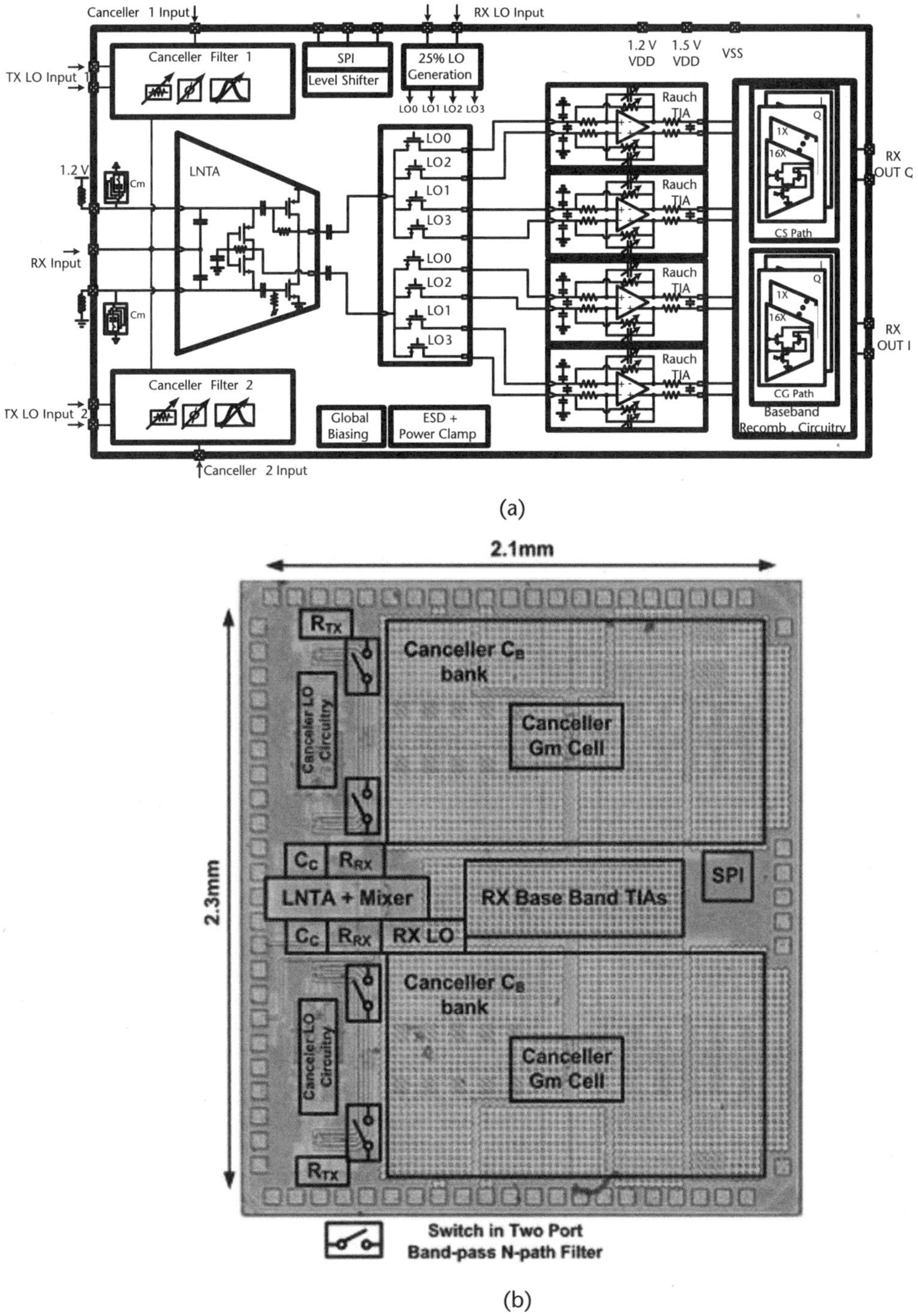

Figure 9.6 A 0.8 to 1.4-GHz, 65-nm CMOS Rx with FDE-based SIC in the RF domain, featuring a bank of two filters: (a) block diagram; and (b) chip photograph. (© 2015 IEEE. Reprinted, with permission, from [3].)

coupling is important for SIC at the Rx input to avoid degrading the Rx input matching and also prevent the N-path filters from interacting with each other and with the receiver. Next, R_L (nominal value 50Ω) is sized to be smaller than $1/|\omega C_C|$, so the capacitive loading effect of C_C on the N-path filter is weak. Based on the required attenuation range and (9.3), the resistance range of R_S at the Tx side can be obtained. Once R_S and R_L are set, the N-path filter baseband capacitance range can be determined by the required group delay or quality-factor range. Lastly, the transconductance range of the baseband G_m is designed using $\Delta\omega = G_m/C_B$ based on both the required frequency shift range and the C_B capacitance range. Based on the design principles mentioned earlier, both canceller filters have been designed to have a digitally controlled peak group delay that ranges from 1 to 28 ns, a frequency shift that ranges from –10 MHz to +10 MHz under the peak group delay setting, full 360° phase shift range, and an attenuation range from 20 to 40 dB, including the 10-dB Rx-side capacitive coupling. The capacitive load present at the receiver input as a result of the capacitive coupling is resonated outwards using a combination of wirebond and off-chip inductance. Programmable capacitor banks are additionally included to tune the input match to the desired frequency.

To ensure high canceller linearity, the G_m cells of the canceller filters are source-degenerated and operate under a slightly higher supply voltage (1.5V) than the other parts of the chip (1.2V). In addition, the on-resistance (R_{on}) of the N-path switches is designed to be 3.3Ω (much smaller than R_S and R_L), so that its impact on overall canceller linearity is minimized. The bulk nodes of the switches are connected to the source nodes to allow optimum DC biasing conditions. The LO path of each filter is comprised of a divide-by-2 quadrature divider, LO slew-rate-control filters, I/Q vector-interpolation phase shifters, and 25% duty-cycle generators. The slew-rate-control RC filters attenuate the harmonics, which ensures the linearity of the subsequent vector interpolators.

The chip microphotograph is shown in Figure 9.6 and has an active area of 4.8 mm^2. The chip is wire-bonded and packaged in a quad-flat no-leads (QFN) package and mounted on a printed circuit board (PCB) for all measurements.

Basic receiver measurements are performed with the canceller inactive and without SI. The receiver has a measured programmable conversion gain that ranges from 27 to 42 dB. The measured receiver noise figure ranges from 4.8 to 5.8 dB over the operating frequency range. The receiver has a measured out-of-band and in-band IIP3 of +17 dBm and –20 dBm, respectively. The measured receiver out-of-band and in-band blocker-induced 1-dB gain compression points (blocker P1dB) are +4 dBm and –30 dBm, respectively.

We use a 1.4-GHz narrowband antenna-pair interface with a peak isolation group delay of 9 ns, peak isolation magnitude of 32 dB, and 3 dB of isolation magnitude variation over 1.36 to 1.38 GHz, which is the same antenna interface used in Figures 9.2 and 9.3. The SI canceller is able to achieve a 20-dB cancellation bandwidth of 15/25 MHz (one/two filters) in Figure 9.7. When a conventional frequency-flat amplitude-and-phase-based canceller is used, the SIC bandwidth is about 3 MHz. When measured, the two canceller filters share the same LO.

The associated noise figure increase is 0.9 to 1.1 dB/1.1 to 1.3 dB. SIC of up to –8 dBm of peak in-band SI at the Rx input results in a negligible gain compression of a desired signal (as compared to nearly 22 dB of compression when SIC is absent). The SIC is also shown to improve the impact of receiver nonlinearity on the SI itself, improving effective in-band IIP3 from –20 dBm to +2 dBm and effective in-band input second-order intercept point (IIP2) from +10 dBm to +68 dBm.

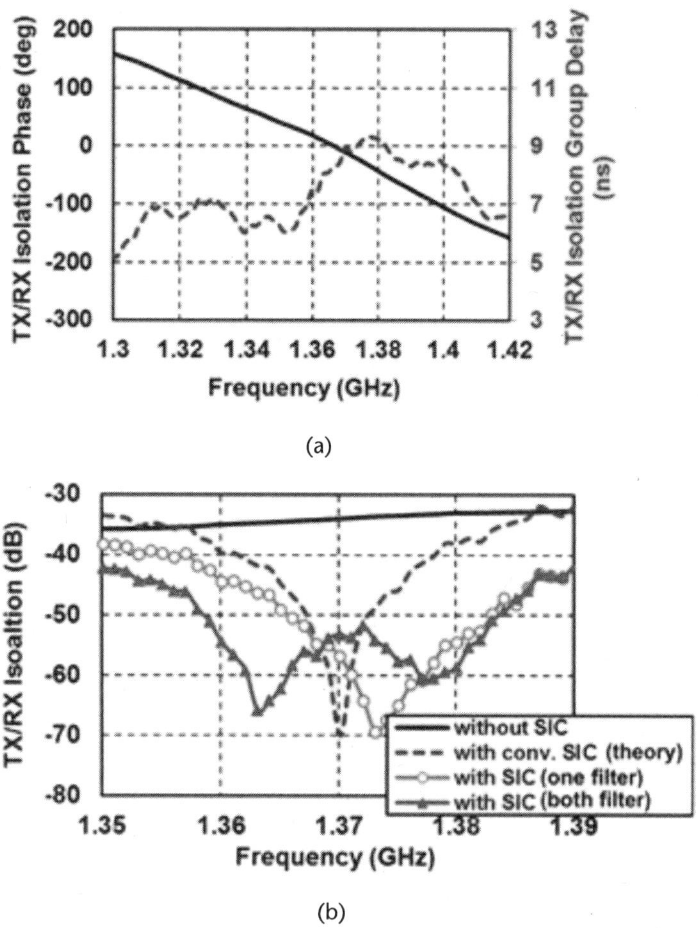

Figure 9.7 Measured Tx/Rx isolation at 1.4 GHz for a full-duplex antenna pair with and without the proposed SIC. (a) Tx/Rx isolation phase and group delay; and (b) Tx/Rx isolation magnitude without SIC, with conventional SIC (theoretical), and with the proposed SIC. Our proposed SI canceller achieves a 20-dB cancellation bandwidth of 15/25 MHz for one/two filters enabled, whereas a conventional frequency-flat amplitude-based and phase-based canceller has a 20-dB SIC bandwidth of only 3 MHz. (© 2015 IEEE. Reprinted, with permission, from [3].)

9.4 A 0.5–2.5-GHz Rx with N-Path-Filter-Based SIC

9.4.1 Integrated Self-Adaptive Wideband Interference Cancellation

As another analog/RF SIC example using N-path-filter-based delay generation, we discuss an integrated IBFD Rx that repurposes N-path delay generators for SIC adaptation circuitry [6]. Maintaining 20 to 30-dB RF/analog SIC in dynamic wireless environments is challenging. A time-varying wireless channel requires SIC to be reconfigured every 60 ms in moderately mobile environments (e.g., Wi-Fi hotspots) [12]. During the SIC adaptation, desired signals cannot be received, and hence the speed at which SIC can tune itself is an important metric.

A blind source algorithm and exhaustive search are used in [13, 14], respectively, to configure the RF cancellers, but result in long tuning time. Microsecond-scale adaptive RF SIC has been reported for IBFD wireless using off-the-shelf-component-based cancellers [12]. In [12], 24-μs overall adaptation overhead is shown based on SI channel estimation using DSP. However, the algorithm in [12] relies on accurate knowledge of canceller S-parameters; it is therefore questionable whether it can maintain microsecond-scale adaptation when applied to mass production using CMOS, which is subject to process, voltage, and temperature variations, as canceller calibration is likely required in the field.

Self-adaptive RF SIC using the least mean squares (LMS) algorithm does not require canceller calibration and has shown fast adaptation in IBFD radios [15]. However, this LMS algorithm uses the information at each tap, and the works in [15] use many off-the-shelf downconverters to access the baseband data at each RF canceller tap. Each LMS downconverter in [15] consumes 1.5-W DC power, which is too high for battery-powered applications. In addition, these LMS downconverters add significant complexity to IBFD MIMO systems. A CMOS LMS adaptive RF SIC is reported in [16] with 25-μs tuning time, but has a narrow SIC bandwidth due to the lack of on-chip delay generation and operates at a fixed frequency.

The work in [6] integrated LMS adaptive wideband RF SIC demonstrated for IBFD wireless on a CMOS chip. In contrast to [15], the LMS adaptive circuitry in [6] is codesigned with and partially embedded in a wideband RF/analog SIC and a gain boosted mixer-first Rx, as illustrated in Figure 9.8(a). This RF integrated circuit (RFIC) algorithm codesign method significantly reduces the adaptation overhead by minimizing the number of canceller taps and by fusing multiple functions into one circuit block.

To reduce the power and complexity associated with self-adaptive RF SIC, minimizing the number of canceller taps is essential, as the LMS algorithm needs to access each canceller tap. A wideband SIC based on RF finite impulse response (FIR) filters usually requires ≥5 taps [1, 13] and hence needs many downconverters, multipliers, and integrators. In this work, a Hilbert-transform-based wideband RF SIC was utilized for fewer canceller taps, given a fixed cancellation bandwidth, thanks to the introduction of complex (I/Q) canceller weights [14].

As mentioned in Sections 9.2 and 9.3, nanosecond-scale RF delay for wideband SIC has been demonstrated on chips using N-path filters [3]. In [6], the downconversion in delay-generating N-path filters is repurposed to translate RF signals to analog baseband for LMS self-adaptive operations. In addition, the I/Q downconverter in the N-path filter is essentially a widely tunable Hilbert transformer. Interestingly, a simple switched-capacitor N-path circuit, as in Figure 9.8(a), simultaneously acts as a delay generator, a Hilbert transformer, and an LMS Tx-side downconverter.

In contrast to [3, 14], where an upconversion mixer is used in the canceller for SIC at RF, the baseband cancellation signals in this work are directly injected at the Rx baseband, but the cancellation is transparent to RF by using a gain-boosted mixer-first Rx, such as in Figure 9.8(a). Furthermore, the mixer-first Rx is repurposed as the LMS downconverter at the Rx side. In essence, one passive mixer wrapping around an inverter-based LNTA replaces two mixers, one for the canceller upconversion and one for Rx downconversion, in [3, 14] while concurrently serving as the LMS Rx-side downconverter.

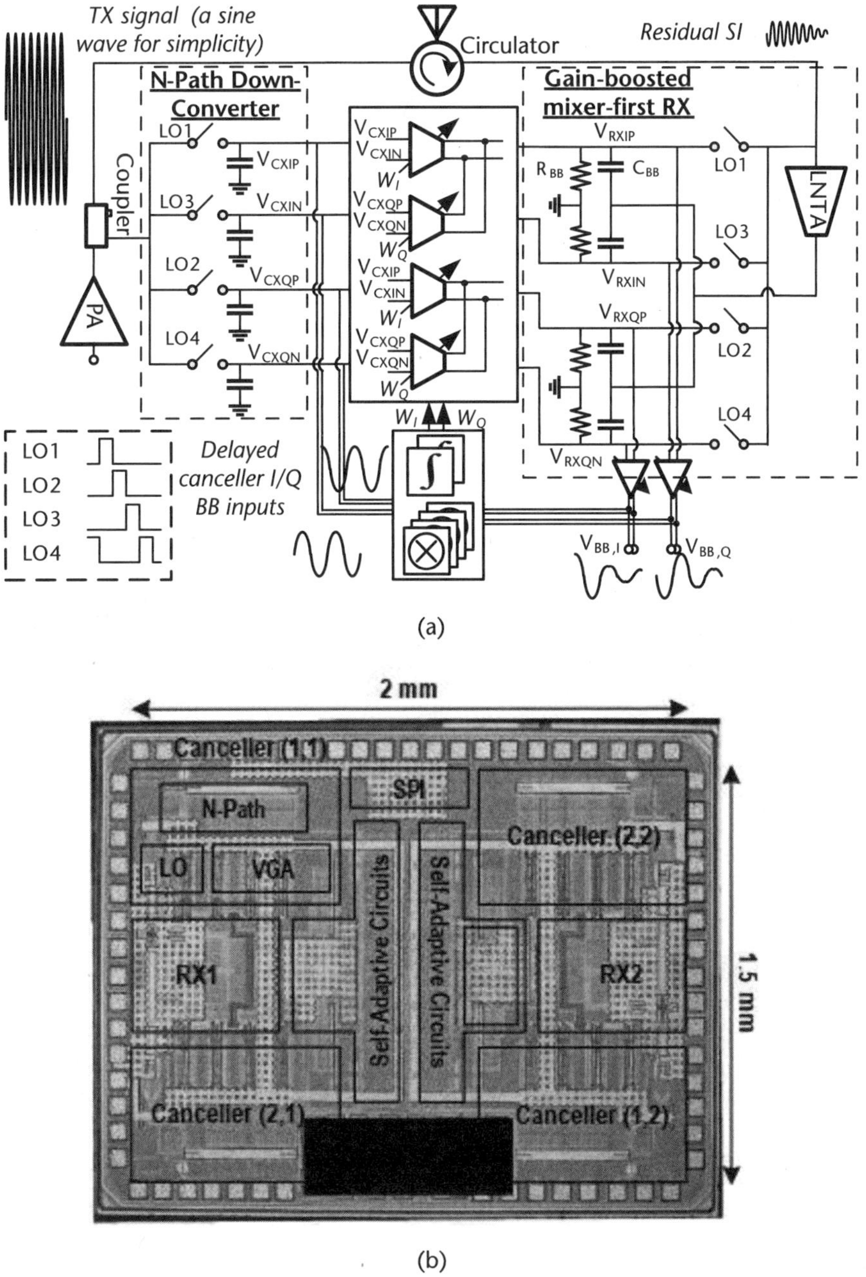

Figure 9.8 A 0.5–2.5-GHz IBFD MIMO Rx with self-adaptive N-path-filter-based interference cancellation: (a) block diagram; and (b) chip photograph. (© 2019 IEEE. Reprinted, with permission, from [6].)

9.4.2 Implementation and Measurement Results

In [6], a 2 × 2 IBFD MIMO Rx with integrated self-adaptive wideband analog/RF interference cancellation is implemented in 65-nm CMOS, as shown in Figure 9.8(b). There are 4 cancellers to suppress SI and crosstalk-interference (XI). Each canceller consists of an N-path filter, baseband variable-gain transconductance amplifiers (VGAs) and circuitry for the self-adaptive VGA weights. The baseband VGAs are split into 8 digitally controlled unit cells for MIMO power/noise figure scalability [6]. The self-adaptive circuitry includes analog multipliers implemented as choppers, analog integrators, and VGA buffers for LMS loop gain control. The canceller outputs are injected at the baseband of a 4-path gain-boosted mixer-first Rx and the cancellation is frequency-translated to RF. Finally, each Rx baseband resistor is split into two digitally controlled resistor banks with the cancellation point in between for MIMO power/noise figure scalability.

The prototype 65-nm CMOS IBFD MIMO Rx operates over 0.5 to 2.5 GHz in measurement. Each Rx element has a measured gain of 26 dB, noise figure of 3.1 dB, and IIP3 of –10.6 dBm with cancellation circuitry disabled.

The canceller self-adaptive operation is first measured with a single canceller enabled, as shown in Figure 9.9(a). Using a single-tone Tx signal at +18 dBm with antenna interface isolation of 24 dB, the canceller adapts itself to 35-dB SIC in 0.6 μs. The self-adaptive response time increases with weaker Tx power levels, and is measured at 6.7 μs with Tx power of –17 dBm. The LMS multiplier, integrator, and buffers consume 14 mW in total.

A test of simultaneous cancellation of SI and XI is depicted in Figure 9.9(b). The SI channel exhibits a group delay of 2.5 ns and 30 dB isolation from a circulator, while the XI channel is mimicked by 24-dB fixed attention with a 5.5-ns delay. The XI cancellation is enabled first and provides 6-dB suppression, and the SIC further suppresses the total interference by another 18 dB.

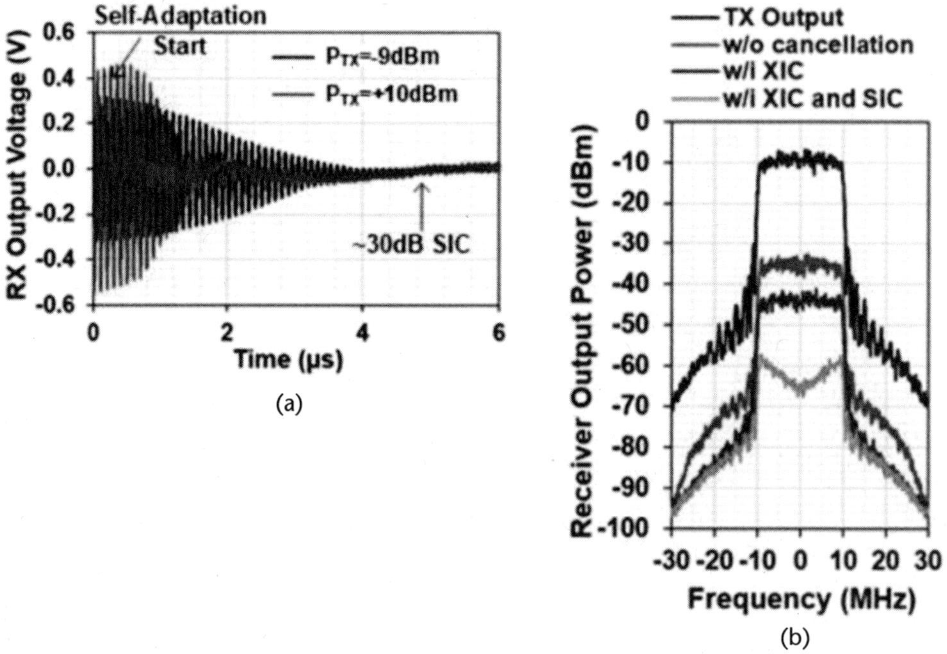

Figure 9.9 Measured self-adaptive analog/RF interference cancellation: (a) SIC using a single-tone Tx signal; and (b) SIC and crosstalk-interference cancellation (XIC) using a 20 MHz quadrature phase-shift keying (QPSK) modulated Tx signal. (© 2019 IEEE. Reprinted, with permission, from [6].)

9.5 Design and Experimentation Using Frequency-Domain RF Cancellers

In an effort to evaluate the performance of the FDE-based RF canceller described in Sections 9.2 and 9.3 at the system level, we prototyped FDE-based RF radios using a software-defined radio (SDR) platform composed of NI's universal software radio peripherals (USRPs). Figure 9.10(a) depicts the block diagram of an IBFD radio that incorporates SI suppression at the antenna interface and SIC in the analog/RF and digital domains.

Since interfacing an RFIC canceller (such as the one shown in Figure 9.6) with an SDR presents numerous technical challenges, we designed and implemented an FDE-based RF canceller with two FDE taps using discrete components on a PCB. This PCB canceller emulates its RFIC counterpart shown in Figure 9.6. As depicted in Figure 9.10(b), a reference signal is tapped from the SDR Tx output using a coupler and is split into two FDE taps through a power divider. Then the signals after each FDE tap are combined and RF SIC is performed at the SDR Rx input.

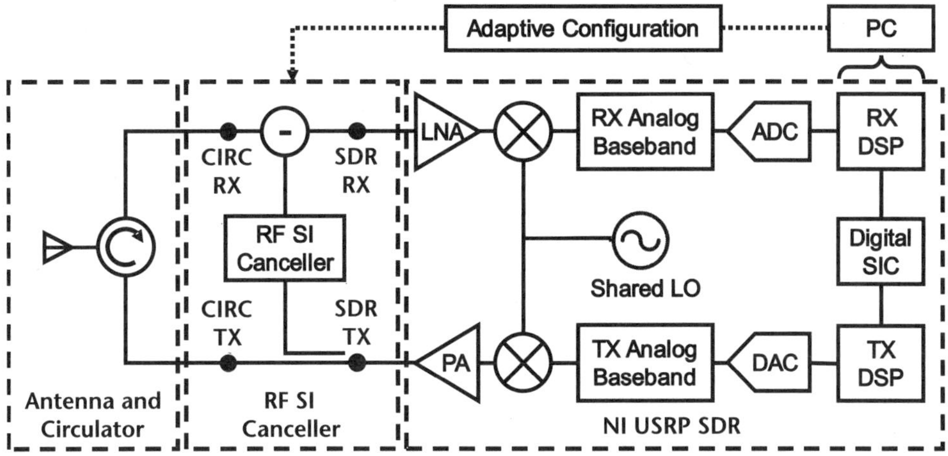

(a)

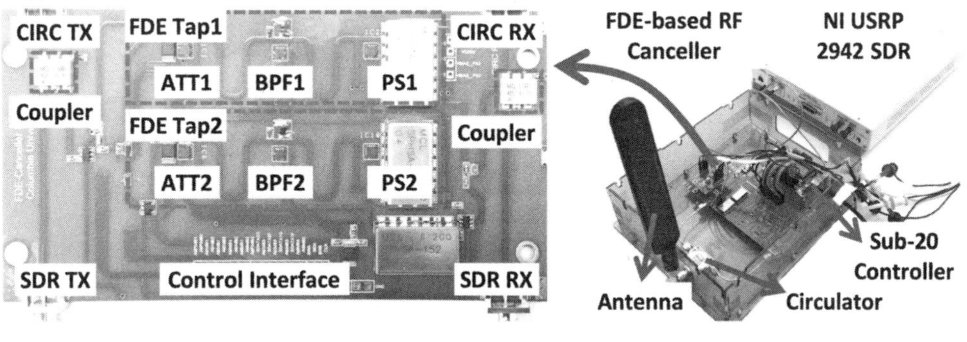

(b)

Figure 9.10 (a) Block diagram of an IBFD radio featuring SI suppression/cancellation at the antenna interface, and in the RF and digital domains; and (b) the implemented frequency-domain-based RF canceller (left) and wideband IBFD radio (right).

In particular, each FDE tap consists of a reconfigurable second-order BPF, as well as a variable gain attenuator and a phase shifter for amplitude and phase control, respectively. The BPF is implemented as a resistor-inductor-capacitor (RLC) filter with impedance transformation networks on both sides and is optimized around the 900-MHz operating frequency.

Figure 9.10(b) shows our prototyped FDE-based IBFD radio operating at a 900-MHz carrier frequency, which consists of an antenna, a coaxial circulator, an FDE-based PCB canceller, and the NI's USRP. We developed an efficient RF canceller configuration scheme so that the RF canceller adapts to environmental changes and achieves the best RF SIC performance. The residual SI after isolation and cancellation at the antenna interface and RF domain is further suppressed in the digital domain. The digital SIC is based on a truncated Volterra series with nonlinear-tapped delay lines and is implemented on the host computer.

Figure 9.11(a) shows the measurement results of the FDE-based IBFD radio, which achieves 52/43-dB SIC over a 20-MHz bandwidth in the RF/digital domain (an overall SIC of 95 dB). Because the USRP has a measured noise floor of –85 dBm in a realistic environment, the IBFD radio can support a maximum average Tx power of +10 dBm (a peak Tx power of +20 dBm).

To evaluate the performance of the FDE-based IBFD radio at the link level, we conducted experiments using two IBFD radios and measured the link packet

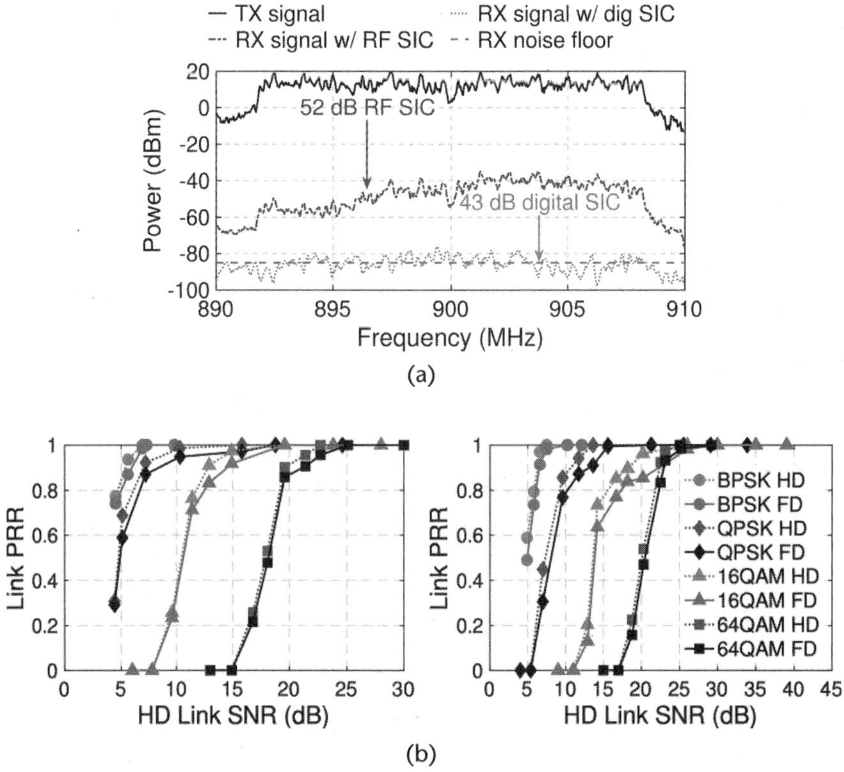

Figure 9.11 (a) Power spectrum of the received signal after SIC in the RF and digital domains with +10-dBm average Tx power, 20-MHz bandwidth, and –85-dBm Rx noise floor; and (b) half-duplex (HD) and full-duplex (FD) packet reception ratio (PRR) with varying HD link SNR and modulation and coding schemes (MCSs): (left) coding rate 1/2; and (right) coding rate 3/4. (© 2019 ACM. Reprinted, with permission, from [17].)

reception ratio (PRR) as a function of the half-duplex link SNR. In particular, OFDM data packets with varying modulation and coding schemes (MCSs) are sent over the wireless link simultaneously in the full-duplex mode or in alternating directions in the half-duplex mode (i.e., the two radios take turns and transmit to each other).

Figure 9.11(b) shows the relationship between the measured link PRR and the half-duplex link SNR with varying MCSs from BPSK-1/2 to 64QAM-3/4. The experimental results show that with sufficient link SNR values (e.g., 8 dB for BPSK-1/2 and 28 dB for 64QAM-3/4), the FDE-based IBFD radio achieves a link PRR of 100%. With insufficient link SNR values, the average full-duplex link PRR is 6.5% lower than that of the half-duplex link PRR across varying MCSs. Our extensive experiments in different scenarios show that our FDE-based IBFD radio achieves an average full-duplex link throughput gain of 1.85 to 1.91 times. More experimental evaluations of the FDE-based IBFD radios at the link level and the network level can be found in [18].

Moreover, to allow the broader community to experiment with IBFD wireless, we integrated an improved version of the FDE-based IBFD radio in the National Science Foundation (NSF) Platforms for Advanced Wireless Research (PAWR) Cloud Enhanced Open Software Defined Mobile Wireless Testbed for City-Scale Deployment (COSMOS) testbed, which is a programmable city-scale advanced wireless testbed being deployed in West Harlem, New York City. We open-sourced both the PCB canceller hardware design and example software programs [19] to support experimental evaluation of full-duplex-related algorithms from the physical layer to higher layers of the networking stack.

9.6 Practical Considerations

When designing a SIC Rx microchip, it is pivotal to take the antenna interface isolation and chip package into account early on. The amount of antenna interface isolation determines the strength of the SI. This, together with the Tx side coupling strength, dictates the SIC Rx linearity and noise performance. Also, RF connection cables used in measurements can have a significant delay. If the measurement setup was not carefully designed, the group delay in the measurement setup, rather than the group delay from the antenna interface, could limit the ultimate SIC bandwidth. Finally, similar to all microchip development, the type of package could significantly affect the chip performance due to its parasitics. A codesign of the package and the chip itself at an early stage is important.

9.7 Conclusion

9.7.1 Summary

Broadband SIC in the RF domain using FDE has been proposed where reconfigurable BPFs are introduced within the canceller. The reconfigurable BPFs allow for replication of not only the magnitude and phase of the frequency-selective antenna interface isolation but also its slopes in different sub-bands, which greatly enhances

SIC bandwidth. A two-port G_m-C N-path filter with embedded variable attenuation and phase shift is introduced to implement the reconfigurable canceller BPFs. A proof-of-concept 0.8–1.4-GHz SIC Rx prototype in 65-nm CMOS achieves >20-MHz cancellation bandwidth across a 1.4-GHz antenna pair. Another N-path-filter-based wideband analog/RF SIC prototype supporting IBFD MIMO and autonomous SIC operation is also briefly discussed. It operates from 0.5 to 2.5 GHz with ≥24-dB RF/analog interference cancellation across 20-MHz bandwidth with microsecond-scale adaptation. Finally, the system-level performance of full-duplex links is characterized using our custom-designed IBFD radios with FDE-based RF SIC.

9.7.2 Future Directions

Future efforts will include the investigation of the noise performance of the N-path-filter-based canceller in the presence of clock-path phase noise and unknown blockers. LO leakage suppression inside canceller N-path filters will also be investigated. Additional research will also be conducted on the fundamental limit of the SIC bandwidth, especially with large SI channel delay spread.

References

[1] Chu, K., et al., "A Broadband and Deep-TX Self-Interference Cancellation Technique for Full-Duplex and Frequency-Domain-Duplex Transceiver Applications," *Proc. IEEE ISSCC'18*, 2018.

[2] Kolodziej, K. E., J. G. McMichael, and B. T. Perry, "Multitap RF Canceller for In-Band Full-Duplex Wireless Communications," *IEEE Transactions on Wireless Communications*, Vol. 15, No. 6, 2016, pp. 4321–4334.

[3] Zhou, J., et al., "Integrated Wideband Self-Interference Cancellation in the RF Domain for FDD and Full-Duplex Wireless," *IEEE Journal of Solid-State Circuits*, Vol. 50, No. 12, December 2015, pp. 3015–3031.

[4] Ghaffari, A., et al., "Tunable High-Q N-Path Band-Pass Filters: Modeling and Verification," *IEEE Journal of Solid State Circuits*, Vol. 46, No. 5, 2011, pp. 998–1010.

[5] Reiskarimian, N., et al., "Analysis and Design of Two-Port N-Path Bandpass Filters with Embedded Phase Shifting," *IEEE Transactions on Circuits and Systems II: Express Briefs*, Vol. 63, No. 8, 2016, pp. 728–732.

[6] Cao, Y., and J. Zhou, "A CMOS 0.5-2.5GHz Full-Duplex MIMO Receiver with Self-Adaptive and Power-Scalable RF/Analog Wideband Interference Cancellation," *2019 IEEE Radio Frequency Integrated Circuits Symposium (RFIC)*, Boston, MA, 2019, pp. 147–150.

[7] Kuhn, W. B., F. W. Stephenson, and A. Elshabini-Riad, "A 200 MHz CMOS Q-Enhanced LC Bandpass Filter," *IEEE Journal of Solid-State Circuits*, Vol. 31, No. 8, 1996, pp. 1112–1122.

[8] Testi, N., et al., "A 2.4GHz 72dB-Variable-Gain 100dB-DR 7.8mW 4th-Order Tunable Q-Enhanced LC Band-Pass Filter," *Proc. IEEE Radio Frequency Integrated Circuits Symposium (RFIC)*, 2015.

[9] Busignies, H., and M. Dishal, "Some Relations Between Speed of Indication, Bandwidth, and Signal-to-Random-Noise Ratio in Radio Navigation and Direction Finding," *Proceedings of the IRE*, Vol. 37, No. 5, 1949, pp. 478–488.

[10] Darvishi, M., et al., "Widely Tunable 4th-Order Switched Gm-C Band-Pass Filter Based on N-Path Filters," *IEEE Journal of Solid-State Circuits*, Vol. 47, No. 12, 2012, pp. 3105–3119.

[11] Zhou, J., et al., "Low-Noise Active Cancellation of Transmitter Leakage and Transmitter Noise in Broadband Wireless Receivers for FDD/Co-Existence," *IEEE Journal of Solid-State Circuits*, Vol. 49, No. 12, 2014, pp. 3046–3062.

[12] Bharadia, D., and S. Katti, "Full Duplex MIMO Radios," *Proc. USENIX NSDI*, 2014.

[13] Zhang, T., et al., "Wideband Dual-Injection Path Self-Interference Cancellation Architecture for Full-Duplex Transceivers," *IEEE J. Solid-State Circuits*, Vol. 53, No. 6, 2018, pp. 1563–1576.

[14] El Sayed, A., et al., "A Hilbert Transform Equalizer Enabling 80 MHz RF Self-Interference Cancellation for Full-Duplex Receivers," *IEEE Transactions on Circuits and Systems I: Regular Papers*, Vol. 66, No. 3, 2018, pp. 1153–1165.

[15] Korpi, D., et al., "Full-Duplex Mobile Device: Pushing the Limits," *IEEE Communications Magazine*, Vol. 54, No. 9, 2016, pp. 80–87.

[16] Aparin, V., et al., "An Integrated LMS Adaptive Filter of TX Leakage for CDMA Receiver Front Ends," *IEEE Journal of Solid-State Circuits*, Vol. 41, No. 5, 2006, pp. 1171–1182.

[17] Chen, T., et al., "Wideband Full-Duplex Wireless Via Frequency-Domain Equalization: Design and Experimentation," *Proc. ACM MobiCom'19*, 2019.

[18] Raychaudhuri, D., et al., "Challenge: COSMOS: A City-Scale Programmable Testbed for Experimentation with Advanced Wireless," *Proc. ACM MobiCom'20*, 2020.

[19] "Tutorial: Open-Access Full-Duplex Wireless in the COSMOS Testbed," 2020, https://wiki.cosmos-lab.org/wiki/Tutorials/Wireless/FullDuplex.

CHAPTER 10

Digitally-Assisted RF Cancellation

Melissa Duarte

10.1 Introduction

In digitally-assisted RF cancellation, the filter taps for generating the SI estimate are implemented all or partly in the digital domain, as highlighted within the IBFD transceiver diagram of Figure 10.1(a) and SIC techniques tree of Figure 10.1(b). This is in contrast to time-domain and frequency-domain cancellation approaches (discussed in Chapters 8 and 9, respectively), where the filter taps reside completely in the analog domain. The benefits of digitally-assisted RF cancellation are that the digital domain implementation of taps: (1) provides the ability to model more taps of the SI channel, allowing SI cancellation for larger bandwidths; and (2) offers more flexibility on the implementation of each tap, since varying the amplitude, delay, and phase per tap can be done at finer resolution and span larger ranges than when done in the analog domain.

One drawback of digitally-assisted RF cancellation is that additional hardware such as DACs and mixers is required. Another drawback of this technique is the challenging task of generating, in the digital domain, a cancellation signal that captures the nonlinear behavior of the SI signal caused by the analog Tx chain. Recall that, in time-domain and frequency-domain cancellation approaches, the reference signal that is fed to the filter taps can be taken from the output of the Tx chain; thus, the reference signal already includes the nonlinearities of the SI signal, and, hence, there is no need to model them since they are naturally included in the cancellation signal. In contrast, because digitally-assisted RF cancellation has some of the filter taps in the digital domain, the input to these digital taps then has to be a digital signal, and the challenge is to either provide a digital input that includes the effects of Tx chain nonlinearities or to model and estimate these nonlinearities in the digital domain. Thus, computationally, the most demanding task of digitally-assisted RF cancellation is the modeling and estimation of the nonlinear components that are present in the SI signal due to Tx chain nonlinearities.

The digitally-assisted RF cancellation architectures presented in this chapter are divided into two main types of architectures: (1) those with all-digital input, as

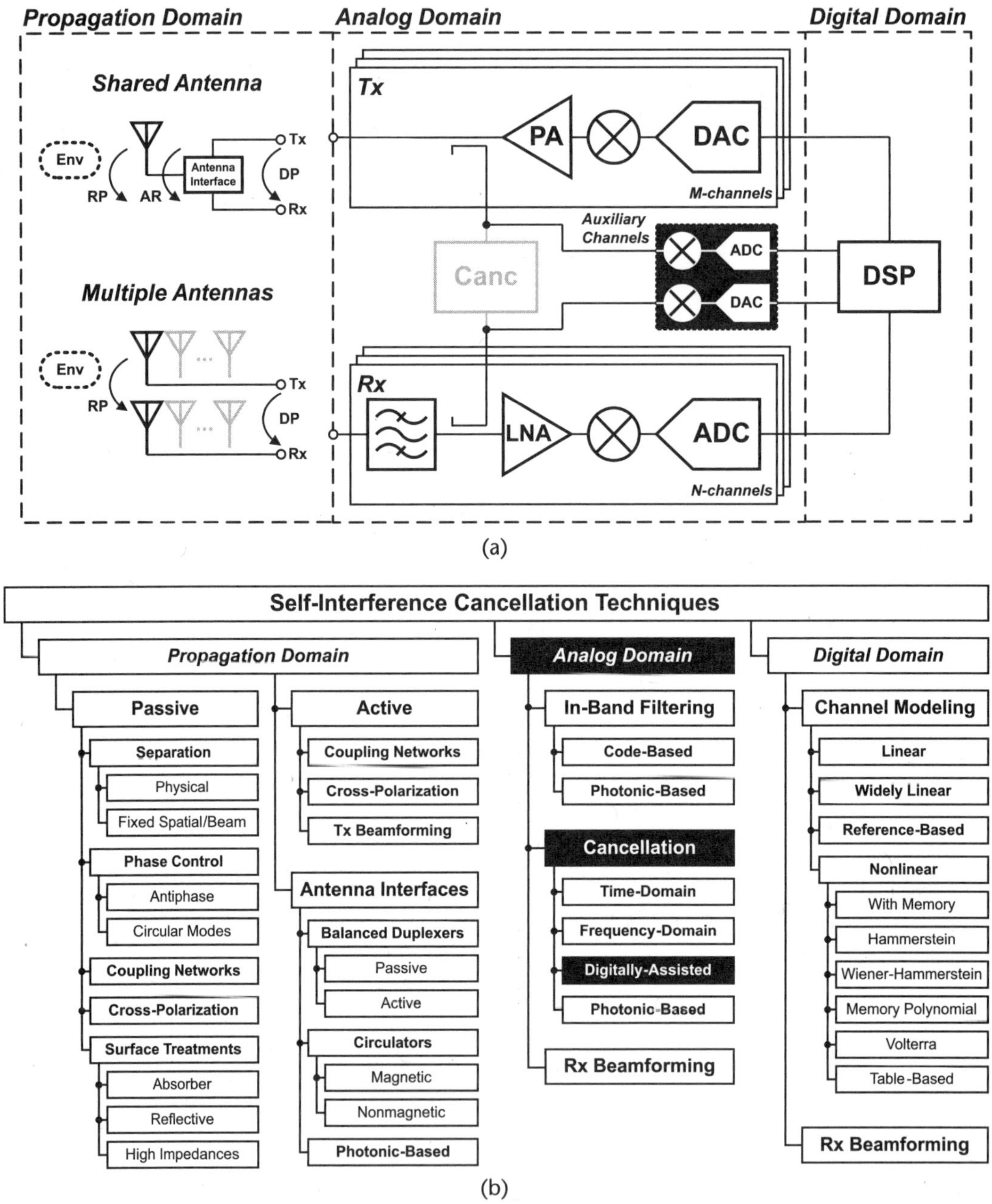

Figure 10.1 Digitally assisted RF cancellation and specific chapter focus highlighted within (a) a generic IBFD transceiver and (b) a SIC techniques tree diagram.

exemplified in Figure 10.2; and (2) those with combined analog and digital inputs, as shown in Figure 10.3, where the analog input is tapped from the output of the Tx chain. This chapter will explain how the difference between the various architectures is rooted in the way that they solve the problem of providing RF cancellation of the nonlinear components of the SI signal.

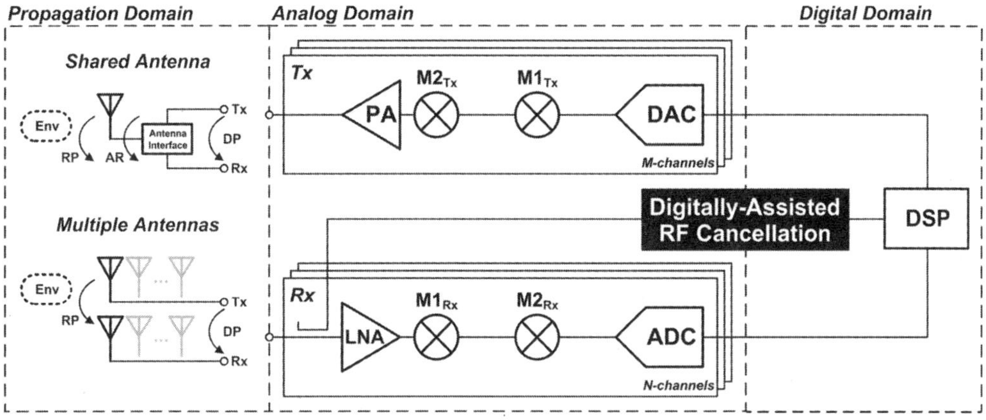

Figure 10.2 IBFD transceiver architecture with a digitally assisted RF cancellation using all-digital input.

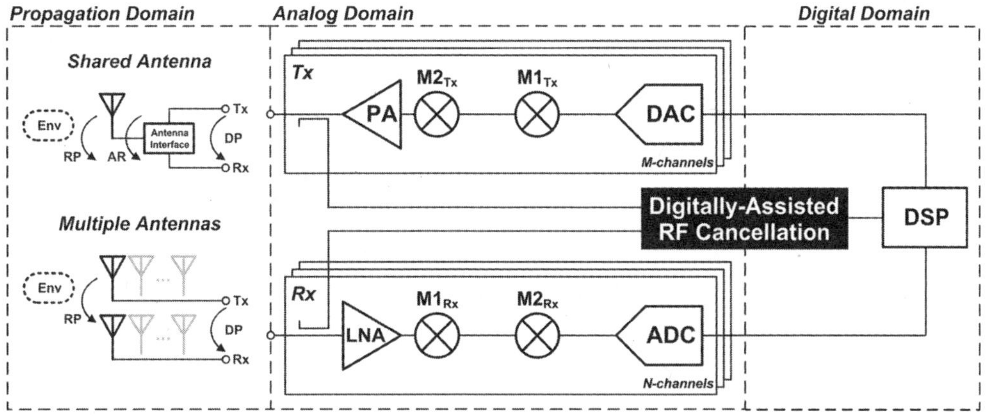

Figure 10.3 IBFD transceiver architecture with a digitally assisted RF cancellation using analog and digital inputs.

10.2 Digitally-Assisted RF Cancellation with All-Digital Inputs

In this architecture, the reference signal for SI estimation is tapped from the digital Tx signal path, as shown in Figure 10.4. An estimate of the negative of the SI signal is computed in the digital domain and converted to the analog domain via an auxiliary (AUX) Tx chain. The output of the AUX Tx chain is coupled to the Rx path at the input of the LNA, and the goal is that the signal provided by the AUX Tx cancels the SI component at this stage.

The architecture in Figure 10.4 was first demonstrated for IBFD communication in [1–3] and later extended to a MIMO scenario in [4]. The SI estimation proposed in these works only targeted cancellation of the linear components of the SI. Later works [5, 6] demonstrated that higher cancellation is achieved when both linear and nonlinear components of the SI signal are targeted for cancellation. This is particularly important as the Tx power increases, because operating at higher powers moves the operating point of analog components to the region of nonlinear behavior; for low Tx powers, cancellation of only the linear components may provide the required SI reduction [5, 7].

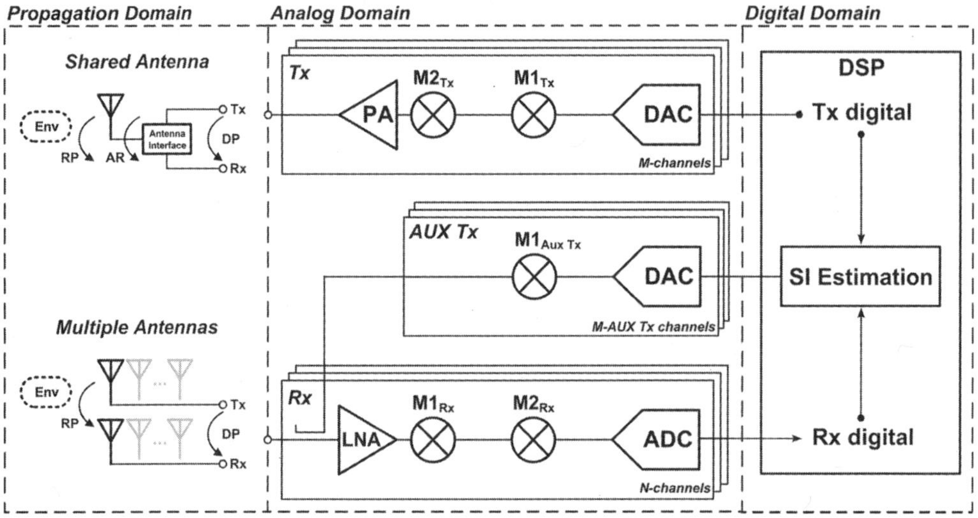

Figure 10.4 IBFD transceiver architecture with a digitally assisted RF cancellation using all-digital input. The SI is first estimated in the digital domain and a cancellation signal is fed to the AUX Tx, which generates the RF domain cancellation signal that is applied before the LNA.

The estimation of the nonlinearities of the SI is not an easy task, and different methods have been proposed, as discussed in Chapter 11. Initial work [6] considered the use of a Volterra series to model the nonlinearities and the implementation proposed in [6] showed that this improved the cancellation by about 3 dB with respect to the scenario where the digitally-assisted RF canceller estimates only the linear part of the SI.

More recent work in [5] considered a parallel Hammerstein model in order to account for the Tx PA nonlinearities and proposed a closed-loop parameter learning algorithm that tracks changes in the nonlinearities and accordingly updates the filter coefficients. The experiment results in [5] showed the importance of keeping track and updating the parameters of the nonlinear model. Specifically, the nonlinearities due to the Rx LNA will be stronger during initial operation of the algorithm because, at this stage, the input to the LNA is dominated by the strong SI. The SI Rx power will be eventually reduced with further iterations of the algorithm thanks to improved digitally-assisted RF cancellation; hence, the impact of the nonlinearities induced by the LNA will also be reduced with respect to its initial value. Consequently, an algorithm that keeps track of the changes in the dominant nonlinearities has better performance than an algorithm that does not. The experiment results in [5] showed that the closed-loop algorithm can improve the cancellation performance by 37 dB. Without closed-loop parameter learning, the performance of digitally-assisted RF cancellation was measured at 17 dB, and with closed-loop parameter learning, the performance was improved to 54 dB of cancellation over a 20-MHz bandwidth.

More recently, the work in [7] compared the use of the Wiener model versus the Hammerstein model for the nonlinearities showing different trade-offs between the number of training samples and the amount of cancellation achieved as a function of the Rx SI power.

10.3 Digitally-Assisted RF Cancellation with Analog and Digital Inputs

10.3.1 Analog and Digital Inputs for All-Digital Taps

This type of architecture, illustrated in Figure 10.5 and proposed in [8], uses an AUX Rx chain that taps the output of the Tx PA and generates a corresponding digital baseband signal that is used for estimation of the SI, which is similar to the reference-based digital cancellation schemes discussed in Chapters 12 and 13. Thus, the DSP block for SI estimation is provided with two versions of the SI signal. The first version is the Tx digital baseband before passing through the Tx chain, and the second version is the downconverted copy of the output of the Tx PA; hence, this second version does provide information about the Tx chain nonlinearities. Adaptive filtering at the digital domain uses these two input signals to generate the cancellation signal that is fed to the AUX Tx chain and then coupled to the input of the LNA for SI cancellation. The output of the AUX Tx chain is an estimate of the negative of the SI signal, which, thanks to the AUX Rx tapping, contains estimates of the linear and nonlinear components of the received SI.

Experimental results in [8] reported cancellation measurements based on their implementation of the architecture in Figure 10.5. As shown, the nonlinear components of the SI signal increase in power as the output Tx power increases due to the nonlinear behavior of the PA. The digitally-assisted analog cancellation measured in [8] was observed to provide at least 20 dB of cancellation of the nonlinear components of the SI signal over a 20-MHz bandwidth.

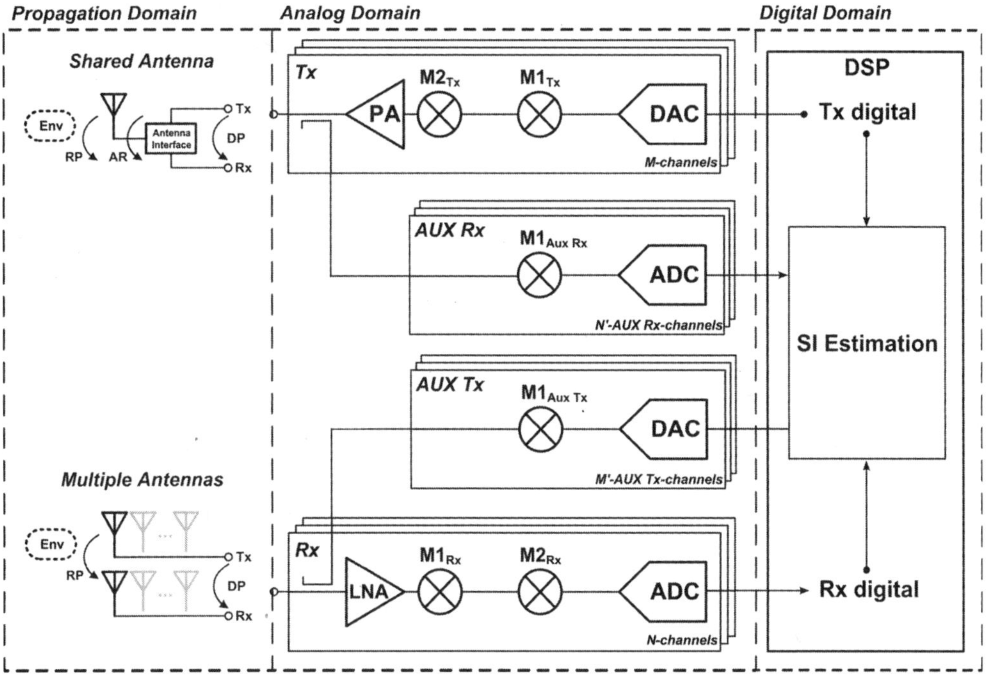

Figure 10.5 IBFD transceiver architecture with a digitally assisted RF cancellation using analog and digital inputs. The analog input is a reference signal tapped from the output of the Tx PA and converted to the digital domain via an AUX Rx. The SI is first estimated in the digital domain and a cancellation signal is fed to the AUX Tx, which generates the RF domain cancellation signal that is applied before the LNA.

10.3.2 Analog and Digital Inputs for Jointly Designed Analog and Digital Taps

Digitally-assisted RF cancellation can also be applied in conjunction with time-domain analog cancellation, as discussed in Chapter 8 and shown in Figure 10.6. Such types of architectures can be found in [9, 10]. As pointed out in [9], the advantage of such an architecture is that the time-domain cancellation and the cancellation via the AUX Tx can be jointly designed to reap the benefits of each cancellation stage by tailoring each stage to cancel the SI components that it can handle best. Specifically, time-domain cancellation is best-suited for mitigating the strongest, typically slow-varying, SI components that are mainly due to DP leakage, while the cancellation via the AUX Tx can target the mitigation of the multipath components or target, providing further cancellation of the residual after the time-domain analog canceller. This type of architecture is demanding in terms of the hardware required because there are both analog taps and AUX Tx chains. As argued in [9], some simplifications are possible. The time-domain canceller could be simplified because it does not have to handle multipath components thanks to the presence of the cancellation via the AUX Tx. The time-domain canceller total cancellation requirements could also be alleviated because there is still one more cancellation stage via the AUX Tx before the LNA. At the same time, as the cancellation via the AUX Tx is only targeting the residual after time-domain cancellation, this could simplify the requirements in terms of dynamic range of the AUX Tx.

Laboratory experiments reported in [9] demonstrated a total SI cancellation of up to 66 dB for an architecture as in Figure 10.6 for the cases where the SI is a 20-MHz Long Term Evolution (LTE) signal and a 100-MHz LTE advanced signal. The implementation of the AUX Tx cancellation in [9] is considered a digital domain processing that provided an estimate of the nonlinear components of the SI. As demonstrated in [9], accounting for these nonlinearities significantly improves the cancellation via the AUX Tx compared to the case of a linear model, especially as the Tx power of the SI increases, which results in stronger nonlinear components.

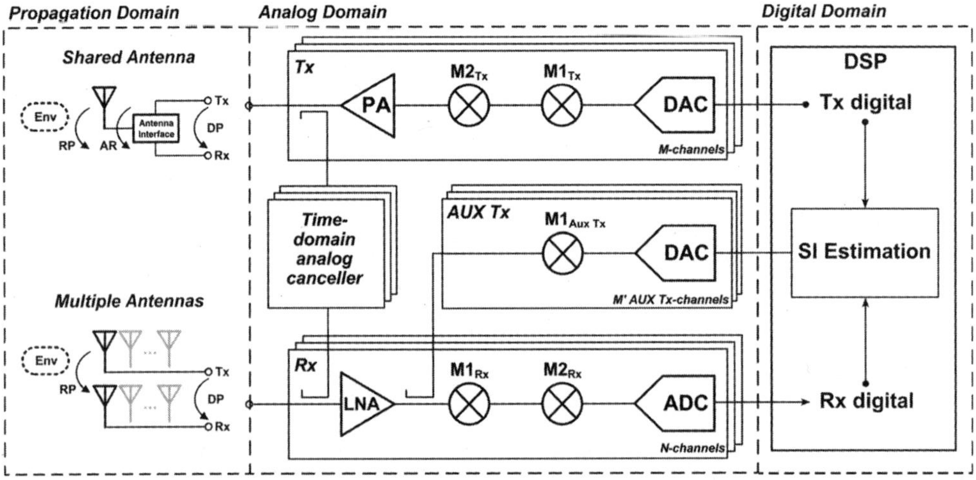

Figure 10.6 IBFD transceiver architecture with a digitally assisted RF cancellation using analog and digital inputs. The analog input is used for time-domain analog cancellation. The digital input is used to compute a second cancellation signal that is converted to RF via an AUX Tx and can be applied after the LNA (as shown) or before the LNA. The cancellation signal via the AUX Tx is designed taking into account the previously applied time-domain analog canceller.

10.3.3 Analog and Digital Inputs for RF Vector Modulators

Digitally-assisted RF cancellation has also been considered in the form of digital processing for control of a cancellation signal that is fed to the LNA input via an RF vector modulator, or RF vector multiplier (called RF VM here because there are vector multipliers that can be used as vector modulators [11]). The RF VM is used for phase and amplitude modification of the analog tapped signal and can effectively create a time-domain RF canceller architecture (similar to those discussed in Chapter 8) [12]. Work in [13, 14] considered an architecture as shown in Figure 10.7, where analog signals are tapped via AUX Rx chains (AUX downconverters) to have digital domain signals that are used to compute the control for the RF VM. For this architecture, experiment results in [13] reported RF cancellation of 40 dB for a SI bandwidth of 80 MHz. Researchers in [15, 16] have proposed an architecture as in Figure 10.8, where control of the RF VM is computed based on the available digital Tx and Rx signals and experiments reported in [15] demonstrate 50-dB cancellation for a SI with a bandwidth of 20 MHz. In both types of architectures shown in Figures 10.7 and 10.8, there is part of the SI estimation happening in the digital domain because the control signals that set the weights for the RF VM are computed in the digital domain. The output of the RF VM is an estimate of the negative of the SI such that it cancels the SI when coupled to the Rx path before the LNA.

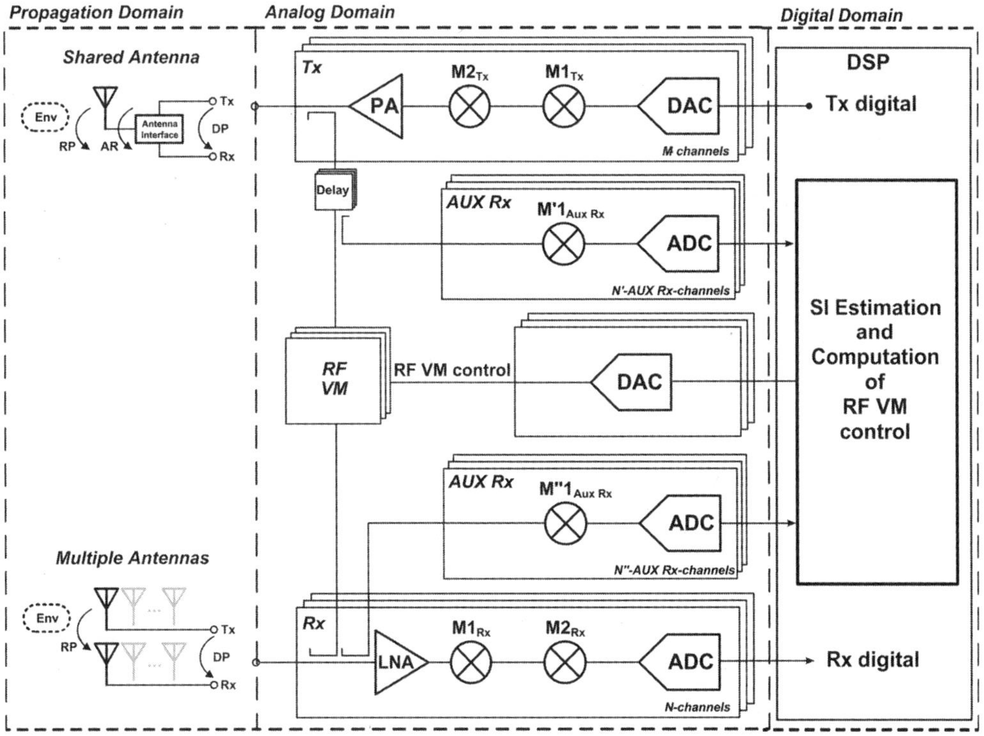

Figure 10.7 IBFD transceiver architecture with a digitally assisted RF cancellation using analog and digital inputs. The analog reference is tapped to the RF VM and also converted to the digital domain via an AUX Rx. A further reference from the Rx path is also digitized via supplementary AUX Rx. The digital information from the tapped signals is used to compute the weights that control the behavior of the RF VM, which outputs an estimate of the negative of the SI signal used for RF cancellation.

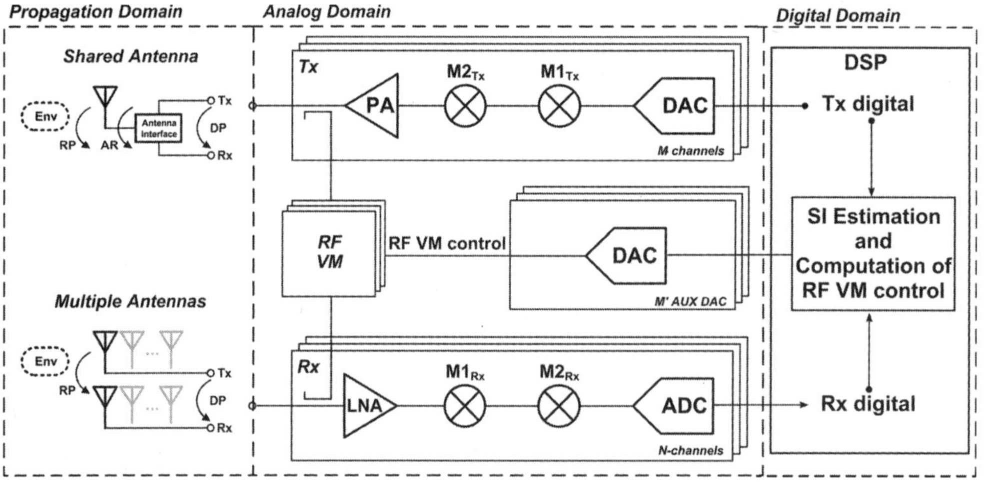

Figure 10.8 IBFD transceiver architecture with a digitally assisted RF cancellation using analog and digital inputs. The analog reference is tapped to the RF VM. The SI is estimated in the digital domain from the digital Tx and Rx signals and is used to compute the weights that control the behavior of the RF VM, which outputs an estimate of the negative of the SI signal used for RF cancellation.

It is worth noting that the earlier implementation of SI cancellation via the QHx220 noise cancellation chip described in [17] could also be considered as a digitally-assisted RF cancellation, where the QHx220 provides phase and gain adaptation of the RF signal based on the digital control obtained from digital processing and thus resembles the architectures in Figures 10.7 and 10.8.

The advantage of techniques like in Figures 10.7 and 10.8 is that, because the analog signal is tapped from the output of the Tx chain, it includes the nonlinear effects of the Tx chain. The RF VM has its own hardware imbalances that must be considered in the computation of the digital control signal as mentioned in [16]; hence, the nonlinearities are partly being handled in the digital domain.

10.4 Practical Considerations

One common denominator across all different architectures for digitally-assisted RF cancellation is the presence of an AUX Tx chain, an AUX Rx chain, or both. These AUX chains place a burden on the hardware requirements, but some simplifications may be possible. For example, as first pointed out in [2], the AUX Tx chain may not need a PA, which simplifies its implementation and also limits the unwanted nonlinear behaviors. Another possible simplification of the AUX paths could be the number of bits used for the ADCs and DACs; however, reducing them for simplified implementation may limit the amount of cancellation achieved.

Taking into account the effects of the AUX chains is important in the design of the digitally-assisted RF cancellation. For example, the phase and amplitude change that the AUX Tx chain applies to the cancelling signal must be considered, as in [2, 5]. Furthermore, these magnitude and phase shifts may change with the output power of the AUX Tx chain and should be carefully tracked or monitored and understood. This is important for correct design of the cancelling signal that is fed to the AUX Tx so that it takes into account any modifications to the signal when passing through this chain.

As pointed out throughout the chapter, a main challenge in the design of digitally-assisted RF cancellers is the modeling and estimation of the nonlinear components of the SI. While we have not discussed much about the nonlinearities of the AUX chains, these nonlinear signals are not expected to be dominant, so it may be possible to achieve the required levels of cancellation without taking this into account in the design. Referencing [18] as an example, SI cancellation using an AUX Tx may be limited by phase noise differences between the main Tx chain and the AUX Tx chain. Thus, AUX Tx chain nonlinearities may have to be considered to further improve the achieved cancellation.

The techniques for digitally-assisted RF cancellation that we have discussed in this chapter leave the Tx signal intact and are focused on the task of estimating and cancelling, at the Rx, the linear and nonlinear components of the SI signal due to the Tx signal. However, the task of estimating and cancelling the nonlinear components present in the SI can be simplified if we reduce the nonlinearity of the Tx signal. Indeed, as the experiments reported in [5] showed, when the Tx PA is operated in the linear region, there may be no need to estimate or cancel the nonlinear components of the SI signal because they are very weak. There is a vast literature on reducing Tx PA-induced nonlinearities by means of signal processing of the Tx signal. Digital predistortion (DPD) is one of these techniques and has been widely studied [19]. Applying such DPD techniques will improve the quality of the Rx signal in the remote receiver and at the same time reduce the undesired nonlinear effects, which complicate the task of digitally-assisted RF cancellation.

10.5 Conclusion

10.5.1 Summary

Different architectures for digitally-assisted RF cancellation were presented throughout the chapter. The advantages and constraints were discussed. We conclude that digitally-assisted RF cancellation can be either an alternative to time-domain and frequency-domain cancellation or complementary to them. The progress in this area of research is greatly due to more accurate modeling of the Tx chain nonlinearities. The work in [5] discussed the most recent implementations achieving 54 dB of cancellation over a 20-MHz bandwidth.

10.5.2 Future Directions

The discussions in this chapter are mainly focused on digitally-assisted RF cancellation for single-input single-output (SISO) antenna transceivers. When considering multi-antenna MIMO IBFD transceivers, a main advantage of digitally-assisted RF cancellation (over time-domain and frequency-domain approaches) is that, for architectures that do not tap the analog signal like the ones shown in Figures 10.2 and 10.4, the number of AUX chains scales only with the number of Rx antennas [2, 5]. This was demonstrated via experiments for two Tx antennas and one Rx antenna in [4].

In a MIMO scenario, a joint design of the Tx MIMO signal and the digitally-assisted analog canceller can provide RF cancellation while reducing the hardware requirements. An example of such a design is presented in [20], but this design has only been analyzed in simulation. As a result, future work would need to validate via experiments the use of digitally-assisted RF cancellation for MIMO systems.

References

[1] Duarte, M., and A. Sabharwal, "Full-Duplex Wireless Communications Using Off-the-Shelf Radios: Feasibility and First Results," *2010 Conference Record of the Forty Fourth Asilomar Conference on Signals, Systems and Computers*, Pacific Grove, CA, 2010, pp. 1558–1562.

[2] Duarte, M., C. Dick, and A. Sabharwal, "Experiment-Driven Characterization of Full-Duplex Wireless Systems," *IEEE Transactions on Wireless Communications*, Vol. 11, No. 12, December 2012, pp. 4296–4307.

[3] Krier, J. R., and I. F. Akyildiz, "Active Self-Interference Cancellation of Passband Signals Using Gradient Descent," *2013 IEEE 24th Annual International Symposium on Personal, Indoor, and Mobile Radio Communications (PIMRC)*, London, 2013, pp. 1212–1216.

[4] Duarte, M., et al., "Design and Characterization of a Full-Duplex Multiantenna System for WiFi Networks," *IEEE Transactions on Vehicular Technology*, Vol. 63, No. 3, March 2014, pp. 1160–1177.

[5] Kiayani, A., et al., "Adaptive Nonlinear RF Cancellation for Improved Isolation in Simultaneous Transmit–Receive Systems," *IEEE Transactions on Microwave Theory and Techniques*, Vol. 66, No. 5, May 2018, pp. 2299–2312.

[6] Askar, R., et al., "Active Self-Interference Cancellation Mechanism for Full-Duplex Wireless Transceivers," *2014 9th International Conference on Cognitive Radio Oriented Wireless Networks and Communications (CROWNCOM)*, Oulu, 2014, pp. 539–544.

[7] Ge, S., et al., "A Digital-Domain Controlled Nonlinear RF Interference Cancellation Scheme for Co-Site Wideband Radios," *IEEE Transactions on Electromagnetic Compatibility*, Vol. 61, No. 5, October 2019, pp. 1647–1654.

[8] Liu, Y., et al., "Digitally Assisted Analog Interference Cancellation for In-Band Full-Duplex Radios," *IEEE Communications Letters*, Vol. 21, No. 5, May 2017, pp. 1079–1082.

[9] Liu, Y., et al., "A Full-Duplex Transceiver with Two-Stage Analog Cancellations for Multipath Self-Interference," *IEEE Transactions on Microwave Theory and Techniques*, Vol. 65, No. 12, December 2017, pp. 5263–5273.

[10] Brodsky, I., J. Brand and M. Jain, "Freedom of Frequency: How the Quest for In-Band Full-Duplex Led to a Breakthrough in Filter Design," *IEEE Microwave Magazine*, Vol. 20, No. 2, February 2019, pp. 36–43.

[11] https://www.analog.com/media/en/technical-documentation/data-sheets/ADL5390.

[12] Kolodziej, K. E., and B. T. Perry, "Wideband Vector Modulator for RF Cancellers in STAR Systems," *2018 IEEE Radio and Wireless Symposium (RWS)*, Anaheim, CA, 2018, pp. 64–67.

[13] Korpi, D., et al., "Full-Duplex Mobile Device: Pushing the Limits," *IEEE Communications Magazine*, Vol. 54, No. 9, September 2016, pp. 80–87.

[14] Choi, Y., and H. Shirani-Mehr, "Simultaneous Transmission and Reception: Algorithm, Design and System Level Performance," *IEEE Transactions on Wireless Communications*, Vol. 12, No. 12, December 2013, pp. 5992–6010.

[15] King, B., J. Xia, and S. Boumaiza, "Digitally Assisted RF-Analog Self Interference Cancellation for Wideband Full-Duplex Radios," *IEEE Transactions on Circuits and Systems II: Express Briefs*, Vol. 65, No. 3, March 2018, pp. 336–340.

[16] Leem, S. K., and S. H. Cho, "RF Vector-Multiplier Compensation for Self-Interference Cancellation in Full-Duplex Radios," *2019 IEEE International Conference on Signal Processing, Communications and Computing (ICSPCC)*, Dalian, China, 2019.

[17] Il Choi, J., et al., "Achieving Single Channel, Full Duplex Wireless Communication," *Proceedings of 16th Annual International Conference on Mobile Computing and Networking, Association for Computing Machinery*, New York, pp. 1–12.

[18] Sahai, A., et al., "On the Impact of Phase Noise on Active Cancelation in Wireless Full-Duplex," *IEEE Transactions on Vehicular Technology*, Vol. 62, No. 9, November 2013, pp. 4494–4510.

[19] Dallinger, R., et al., "Adaptive Pre-Distortion Techniques Based on Orthogonal Polynomials," *2010 Conference Record of the 44th Asilomar Conference on Signals, Systems and Computers*, Pacific Grove, CA, 2010, pp. 1945–1950.

[20] Duarte, M., and G. C. Alexandropoulos, "Full Duplex MIMO Digital Beamforming with Reduced Complexity AUXTX Analog Cancellation," *2020 IEEE International Conference on Communications (ICC)*, 2020.

CHAPTER 11

Spline-Based Nonlinear Digital Cancellation

Dani Korpi, Mikko Valkama, Lauri Anttila, and Pablo Pascual Campo

11.1 Introduction

In order to achieve the benefits of STAR on the same frequency band, nearly complete suppression of SI is required. As discussed in Chapter 1, this calls for several cancellation stages occurring in different domains, starting at the antenna level. Regardless of the cancellation stages occurring before the ADC, the final cancellation stage must always be performed in the digital domain, as highlighted within the IBFD transceiver diagram of Figure 11.1(a) and SIC techniques tree of Figure 11.1(b). The objective of the digital canceller is no less than suppressing all the SI still remaining after the other cancellation stages.

In principle, digital SI cancellation follows the same simple principle as all the other cancellation stages covered in Chapters 8 through 10: as the transceiver obviously knows its own transmit signal, it can simply subtract it from the received signal. Assuming that the possible channel effects up to the point of subtraction are known, the SI could be perfectly cancelled with this simple principle. What makes digital SI cancellation challenging in reality is obtaining sufficiently accurate knowledge about the overall coupling channel (i.e., knowing exactly how the SI signal is distorted while propagating from the transmitter to the receiver). In particular, while the effects of the wireless coupling channel between the transmitter and the receiver can be compensated for in a relatively straightforward matter, the SI signal is often distorted also by the Tx and Rx circuitry. Such distortion, resulting from various analog impairments within the transceiver, cannot usually be captured by the same models that apply to wireless propagation, thereby making accurate digital SI cancellation rather cumbersome (a reference-based approach to this problem is presented in Chapter 12). Moreover, even a relatively insignificant level of unknown distortion in the SI waveform can be a very powerful source of interference for a weak received signal of interest.

Hence, one of the central research challenges in making wireless IBFD communications a reality is the modeling and canceling of the SI in the digital domain,

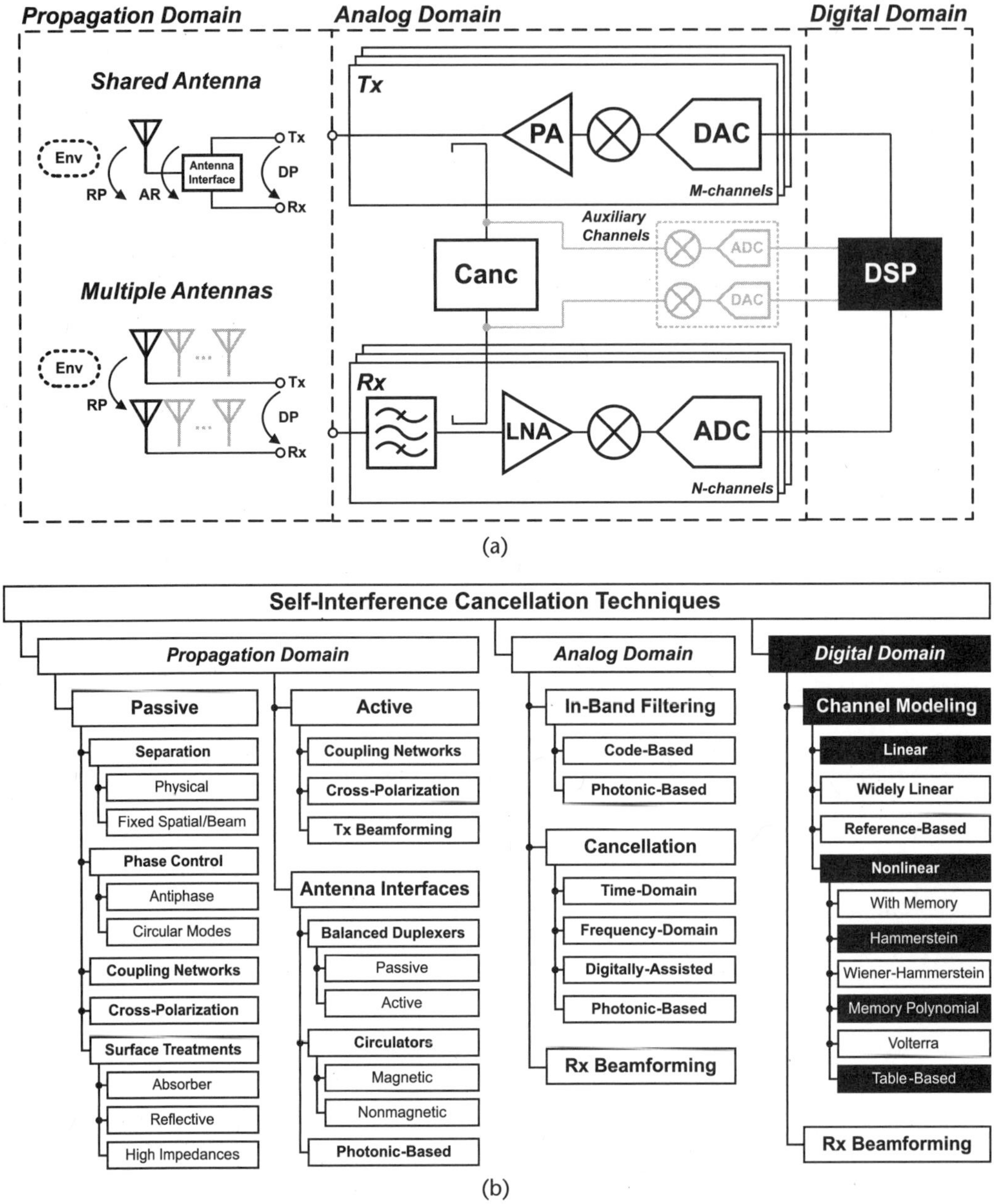

Figure 11.1 Spline-based nonlinear digital cancellation and specific chapter focus highlighted within (a) a generic IBFD transceiver and (b) a SIC techniques tree diagram.

both of which require extremely high accuracy. In the simplest scenario, the SI model consists only of the physical propagation channel between the transmitter and the receiver. As discussed above, facilitating IBFD communications in a practical transceiver requires more comprehensive models for the SI signal that also incorporate the analog impairments. This will facilitate the cancellation of SI beyond the limit established by the error vector magnitude (EVM), potentially down to the level of the Rx noise floor if all the significant impairments are taken into account [1–3].

Therefore, we will provide some insights and guidelines into developing a practical digital cancellation solution for IBFD radios here, taking into account the nonlinear distortion affecting the SI waveform. We begin by describing the most notable types of impairments occurring in a radio device, after which we will propose and discuss different modeling approaches. The parameter learning of the unknown coefficients within the signal models is also addressed. Lastly, we will show that combining such nonlinear digital cancellation with analog SI suppression can cancel the SI nearly perfectly. Therefore, our findings make it evident that IBFD communications is indeed possible also in reality. Other literature has also reported similar findings [2, 4–6].

11.2 Related Works

While several works have adopted a fully linear SI signal model [6–8], many of the recently reported digital cancellation solutions also incorporate a model for the nonlinear Tx sPA [2, 3, 5, 9]. Considering that the PA-induced nonlinearities are indeed the dominant source of distortion in most systems, such a nonlinear digital canceller is typically capable of highly efficient SI cancellation [2, 3, 5]. Recently, it was also shown that machine learning can be used to aid the modeling of the SI signal in the digital domain [10, 11].

Various IBFD transceiver prototypes have also been reported. While the key specifications and performance figures of the most notable prototype implementations are collected and discussed in [12], Table 11.1 provides an overview of the most prominent prototypes, including the one that utilizes the digital cancellers presented in this chapter.

Table 11.1 Key Specifications and Performance Figures of Some Prominent IBFD Prototype Implementations

Prototype	*Year*	*Frequency*	*Bandwidth*	*Structure*	*Analog Cancellation*	*Digital Cancellation*	*Total Isolation*
Rice I [8]	2010	2.4 GHz	0.625 MHz	Two antennas	Yes	Yes	80 dB
Stanford I [7]	2010	2.48 GHz	5 MHz	Three antennas	Yes	Yes	100 dB
Stanford II [5]	2013	2.45 GHz	80 MHz	Shared Tx/Rx antenna	Yes	Yes	110 dB
Rice II [13]	2014	2.4 GHz	20 MHz	Directional antennas	Yes	Yes	95 dB
Chengdu [14]	2014	2.535 GHz	20 MHz	2 × 2 MIMO	Yes	Yes	114 dB
Bristol [15]	2015	1.89 GHz	20 MHz	Shared Tx/Rx antenna	Yes	No	83 dB
Yonsei [6]	2015	2.52 GHz	20 MHz	Dual-polar antenna	Yes	Yes	103 dB
MIT [16]	2016	2.45 GHz	20 MHz	Shared Tx/Rx antenna	Yes	No	78 dB
NRL [17]	2017	4.5 GHz	700 MHz	Two antennas	No	Yes	120 dB
TUT and Aalto [3, 4]	2017 and 2018	2.56 GHz	80 MHz	Compact relay antenna	No	Yes	100 dB

11.3 Analog Impairments

11.3.1 IQ Imbalance

IQ imbalance is a prevalent issue in direct-conversion transceivers, stemming from the inherent phase and amplitude mismatches between the I and Q branches [12]. As a result of these mismatches, an image component is generated on top of the original signal, with Figure 11.2 illustrating this phenomenon in the frequency domain. The image component produced within a direct-conversion transceiver is the original signal, whose spectrum has been inverted with respect to the frequency axis. In the time domain, the image component is correspondingly the complex conjugate of the original signal. The magnitude of the image component in relation to the original signal is dictated by the severity of the IQ imbalance, although typically it is clearly weaker than the original signal [18].

To provide some insight into the magnitude of IQ imbalance, the 5G NR User Equipment (UE) specifications state the lowest acceptable image rejection ratio (IRR) as 25 dB [18], which means that the image component is 25 dB weaker than the actual signal. Although such IQ imbalance levels would be extremely detrimental for an IBFD radio [19], the IRR is typically considerably higher in reality, thanks to advanced calibration mechanisms. It is therefore assumed in this chapter that the IQ imbalance is well below the Rx noise floor, the validity of this assumption being confirmed by the measurement results in Section 11.5.

11.3.2 Nonlinear Distortion

As already discussed above, another fundamental issue, especially in low-cost communications devices, is the nonlinear distortion. It is primarily produced by the different active components, especially the transmitter PA, and can heavily distort the signal. In principle, it stems from the highest signal peaks being compressed; that is, when driven sufficiently close to saturation, the gain of the PA is smaller when the amplitude of the input signal is higher, and thereby the relationship between the input and output signals is nonlinear. In the frequency domain, nonlinear

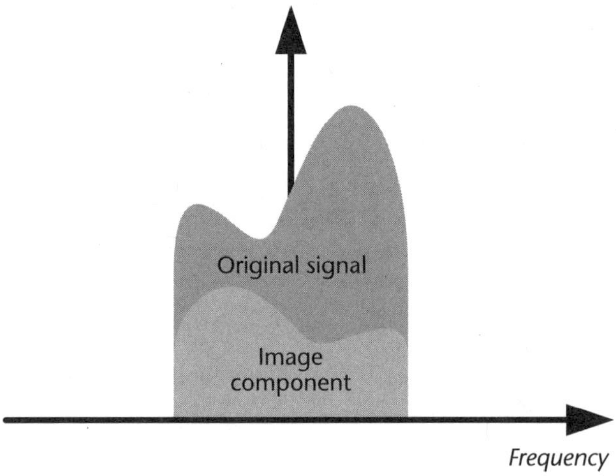

Figure 11.2 IQ imbalance in the frequency domain.

distortion can be illustrated as shown in Figure 11.3, where it exhibits itself as spectral regrowth.

There are various methods for modeling the PA-induced nonlinear distortion, such as the Volterra series or the Wiener model [20]. To limit the model's complexity and ensure efficient parameter estimation, the memory polynomial (MP) model has been widely adopted [21]. Denoting the baseband-equivalent PA input signal by $x_{PA}^{in}(t)$, its output signal can thereby be expressed with the adopted discrete-time MP model as follows [12]:

$$x_{PA}(n) = \sum_{\substack{p=1 \\ p \text{ odd}}}^{P} \sum_{m=0}^{M_{MP}} h_p(m) \left| x_{PA}^{in}(t)(n-m) \right|^{p-1} x_{PA}^{in}(t)(n-m), \tag{11.1}$$

where P is the nonlinearity order of the model, M_{MP} is the memory length of the model, and $h_p(m)$ contains the coefficients of the pth-order nonlinearity. This type of a model has been shown to be accurate for modeling a wide variety of practical PAs [22].

A traditional alternative to modeling memoryless nonlinear systems is to use look-up tables (LUTs). A particularly useful class of LUTs is spline-interpolated LUTs, where splines are used to interpolate between the entries of a small-sized LUT [23, 24]. Interpolated LUTs can be used as building blocks for memory models such as Wiener or Hammerstein models [25, 26]. In [25, 26], the concept of spline adaptive filters was established, where Wiener and Hammerstein models incorporating spline-interpolated LUTs and finite impulse response (FIR) filters were proposed, along with simple gradient descent-based learning rules. This concept was applied to nonlinear SI cancellation in [4], where it was found to be substantially less complex than classical approaches based on MP, while still achieving comparable modeling accuracy. The spline-based Hammerstein canceller is described in more detail in Section 11.4.3.

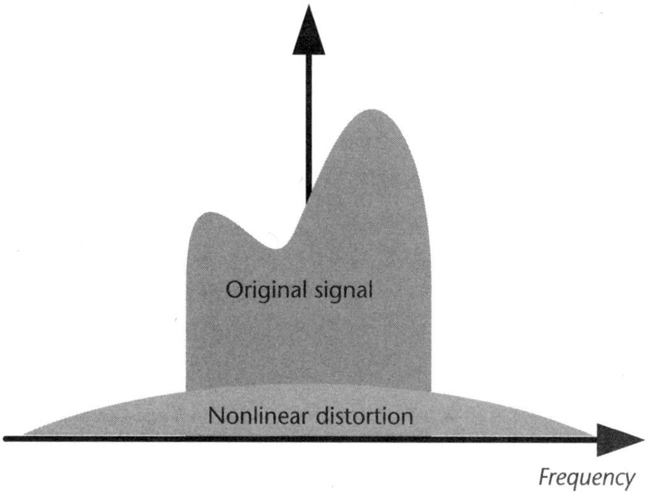

Figure 11.3 Nonlinear distortion in the frequency domain.

11.3.3 Quantization Noise

As opposed to the traditional half-duplex systems, the receiver ADC also plays a significant role in IBFD transceivers [27]. Namely, in a half-duplex receiver, the accuracy of the quantization upon analog-to-digital conversion is rarely a bottleneck, assuming proper automatic gain control (AGC) that amplifies the overall signal to match the dynamic range of the ADC [28]. The situation changes drastically when an IBFD transceiver is considered, because then the ADC input signal also contains some residual SI. Even after a high amount of passive isolation and RF cancellation, the power of the residual SI can still be significantly higher than the power of the received signal of interest. Hence, upon the analog-to-digital conversion, the dynamic range of the signal of interest is well below the dynamic range of the ADC, meaning that it is quantized with less bits than in a corresponding half-duplex device.

This phenomenon is illustrated in Figure 11.4, where two scenarios are shown. In the left side of the figure, there is no residual SI at the ADC input, and consequently the whole dynamic range of the ADC can be used to quantize the signal of interest. As can be observed, the quantization effects are only minor and the signal quality remains good. However, on the right side of the figure, the ADC input signal contains also some residual SI and, as a result, the AGC must use less gain in the receiver to avoid clipping in the ADC. Therefore, even after eliminating the SI with digital cancellation, the signal of interest is very noisy due to the quantization effects. These examples show that, in the context of IBFD transceivers, the dynamic range of the ADC and the analog SI cancellation performance must be carefully considered to ensure sufficient signal-to-interference-plus-noise ratio (SINR) for the signal of interest in the digital domain.

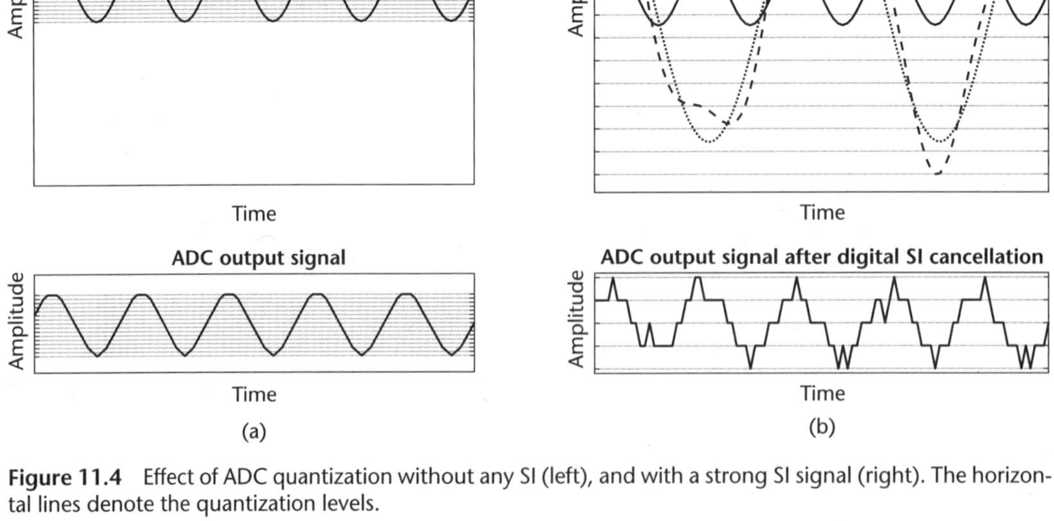

Figure 11.4 Effect of ADC quantization without any SI (left), and with a strong SI signal (right). The horizontal lines denote the quantization levels.

11.3.4 Phase Noise

The effect of phase noise has also been widely studied in the context of IBFD transceivers [29, 30]. Phase noise is caused by the varying phase of the local oscillator (LO) signal during upconversion and downconversion, which results in a multiplicative distortion component. Assuming that the transmitter and receiver rely on independent LOs for upconversion and downconversion, it has been shown that phase noise is a serious issue in IBFD transceivers [29]. Using independent LO signals in an IBFD transceiver is a rather pessimistic assumption as the transmitter and receiver operate on the same center frequency. Hence, in an IBFD radio, a more sensible option is to use the same LO signal for both the upconversion and downconversion. In such a case, assuming any reasonable delay between Tx and Rx chains, the effective phase noise of the main SI component is negligibly weak due to the self-cancellation effect [12]. Therefore, the effect of phase noise can typically be omitted in the analysis of SI cancellation solutions.

11.4 Digital Self-interference Modeling and Cancellation

11.4.1 Basic Transceiver Model

Referring to Chapter 1 and the primary block diagram in Figure 11.1(a), the digital cancellers presented in this chapter operate fully in the digital domain. Moreover, the cancellation solutions do not require any assumptions as to how the cancellation processing is performed in the analog/RF domain. In the forthcoming discussion, the original digital baseband transmit signal is denoted by $x(n)$, and the received digital-domain signal (with SI) is denoted by $y_{ADC}(n)$. The purpose of the digital canceller is to reconstruct the SI using $x(n)$ as accurately as possible, such that, when it is subtracted from $y_{ADC}(n)$, no SI remains.

11.4.2 Polynomial-Based Signal Models

A widely used approach to modeling the residual SI signal in the digital domain is to assume that the overall distortion is dominated by the nonlinearities produced by the Tx PA, as is done, for instance, in [2, 3, 5, 9]. Then the essential canceller architecture is as shown in Figure 11.5. In particular, the SI signal model consists of a parallel connection of static nonlinearities, followed by linear filters [2, 3], which is referred to as MP nonlinearity.

The digital-domain SI signal model can then be written as follows:

$$y_{ADC}(n) = \sum_{\substack{p=1 \\ p \text{ odd}}}^{P} \sum_{m=-M_{\text{pre}}}^{M_{\text{post}}} h_p^{NL}(m)\psi_p\big(x(n-m)\big) + e(n), \tag{11.2}$$

where $\psi_p(x(n)) = |x(n)|^{p-1} x(n)$ is the static pth-order nonlinear basis function, M_{pre} is the number of precursor taps, M_{post} is the number of postcursor filter taps, and $h_p^{NL}(m)$ is its total effective response between the transmitter and the receiver.

In order to actually cancel the SI in the digital domain, the above signal model in (11.2) should be used to regenerate the observed residual SI signal, which

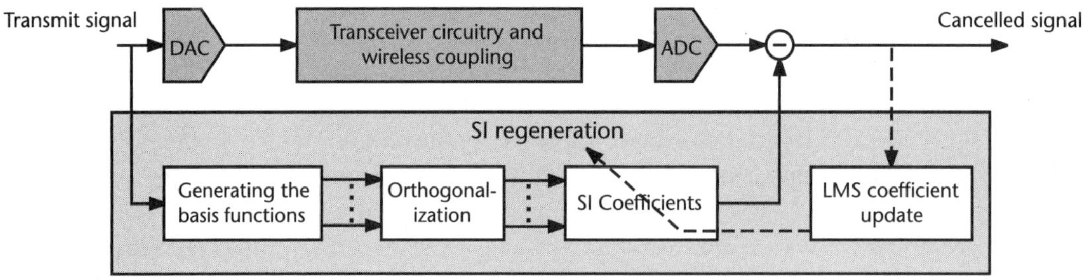

Figure 11.5 Illustration of an adaptive polynomial-based digital canceller.

obviously requires estimating the parameters of the corresponding signal model. A widely used solution for this type of an estimation problem is the least mean squares (LMS) parameter learning algorithm, which has been both described and evaluated in [2, 3, 9]. The basic operating principle of the LMS-based digital canceller is illustrated in Figure 11.5. Essentially, the LMS canceller aims at minimizing the power of its output signal [31], which it does by utilizing a predetermined signal model for the residual SI.

To this end, the static basis functions are collected into a vector, referred to as an instantaneous basis function vector, which is defined as

$$\varphi(n) = \left[\psi_1(x(n)) \;\; \psi_3(x(n)) \;\; \cdots \;\; \psi_P(x(n))\right], \tag{11.3}$$

where $\psi_p(x(n))$ is the static pth-order nonlinear basis function. Having obtained the vector in (11.3), the different static basis functions must be orthogonalized, as also shown in Figure 11.5. The reason for this is the poor convergence performance of the LMS algorithm if the elements of the input vector are highly correlated, stemming from the large eigenvalue spread of the input signal covariance matrix [9, 32, 33]. As the static basis functions can, in this case, certainly be expected to be correlated because they all rely on the original transmit signal, orthogonalizing them is necessary to ensure efficient parameter learning by the LMS algorithm. This can be done with the following orthogonalization matrix [9, 34]:

$$\mathbf{S}_\psi = \mathbf{U}_\psi \Lambda_\psi^{-\frac{1}{2}}, \tag{11.4}$$

where Λ_ψ is a diagonal matrix containing the eigenvalues of the covariance matrix $\mathbf{R}_\psi = E[\varphi^H(n)\varphi(n)]$, and $\mathbf{U}_\psi$ is a matrix containing the corresponding eigenvectors (for further details, please refer to [12, p. 58]).

The static basis functions can consequently be orthogonalized simply by

$$\tilde{\varphi}(\mathrm{n}) = \varphi(n)\mathbf{S}_\psi. \tag{11.5}$$

Note that other orthogonalization solutions can also be utilized to similar effect or, alternatively, one can use basis functions that have been drawn from a basis of orthogonal polynomials [35].

Having orthogonalized the basis functions, they can then be used to learn the SI channel coefficients using the LMS algorithm. The input vector of the LMS filter, now containing all the orthogonalized basis functions, is defined as follows:

$$\tilde{\Psi}(n) = \left[\tilde{\varphi}(n+M_{pre}) \quad \tilde{\varphi}(n+M_{pre}-1) \quad \cdots \quad \tilde{\varphi}(n-M_{post})\right]. \tag{11.6}$$

Then, denoting the LMS SI channel estimate after n iterations by $\hat{\mathbf{h}}(n)$, the cancelled signal is given by:

$$y_{DC}(n) = y_{ADC}(n) - \tilde{\Psi}(n)\hat{\mathbf{h}}(n), \tag{11.7}$$

after which the LMS algorithm updates the SI channel estimate using the following rule [9, 12]:

$$\hat{\mathbf{h}}(n+1) = \hat{\mathbf{h}}(n) + \mathbf{M}y_{DC}(n)\tilde{\Psi}^{H}(n), \tag{11.8}$$

where $\mathbf{M}$ is a diagonal matrix that contains the step sizes for the different orthogonalized basis functions on its diagonal. If no further side information is available, the channel estimate is initialized as $\hat{\mathbf{h}}(0) = 0$.

11.4.3 Hammerstein Spline-Based Signal Model

Another alternative solution to model the nonlinear SI signal is to use a cascaded Hammerstein structure, which builds upon a cascade of a static nonlinear function and a linear filter. These blocks model the nonlinear PA distortion and the SI coupling channel, respectively.

In this approach, the instantaneous nonlinearity can be implemented with various different nonlinear functions. In this work, a LUT with uniform spline interpolation is used to obtain a low-complexity solution [4, 26]. This approach is referred to as the spline-based Hammerstein (SPH) model and is illustrated in Figure 11.6. Spline interpolation builds on piecewise polynomials, which interpolate a set of control points under continuity and smoothness conditions [23, 24]. The set of control points is stored in a LUT, which is continuously estimated and updated.

With this approach, lower-order polynomials can be fitted in each region, as opposed to traditional approaches where one high-order polynomial models the nonlinearity over the whole input range. The regions for the spline interpolation

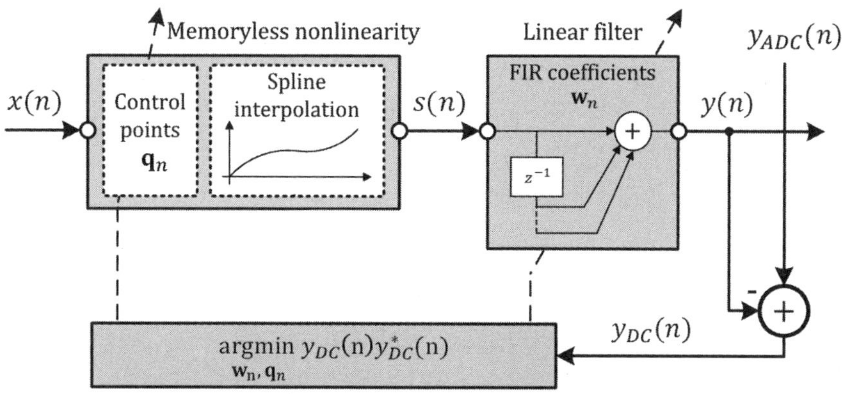

Figure 11.6 Architecture of the spline-interpolated Hammerstein scheme.

scheme are defined and accessed through the index value, abscissa value, and abscissa vector, defined as

$$i_n = \left\lfloor \frac{|x(n)|}{\Delta_x} \right\rfloor + 1 \tag{11.9}$$

$$u_n = \frac{|x(n)|}{\Delta_x} - (i_n - 1) \tag{11.10}$$

$$\mathbf{u}_n = \left[u_n^P \ u_n^{P-1} \cdots 1\right]^T, \tag{11.11}$$

where Δ_x is the region width and P is the spline interpolation order. In this context, i_n gives the region number, and u_n is the normalized value inside each region.

Denoting $s(n)$ as the instantaneous nonlinear output signal, the result of the spline interpolation can be written as

$$s(n) = x(n)\xi_n^T(\mathbf{1} + \mathbf{q}), \tag{11.12}$$

where $\mathbf{q}$ is the LUT containing the control points, and ξ^T is defined, in turn, as

$$\xi_n^T = \left[0 \ \cdots \ 0 \ \mathbf{u}_n^T\mathbf{C}_P \ 0 \ \cdots \ 0\right]^T, \tag{11.13}$$

where the first value of $\mathbf{u}_n^T\mathbf{C}_P$ is indexed in the i_nth element so that only the corresponding control points are multiplied by the spline basis functions. The term $\mathbf{C}_p$ is the spline basis matrix containing the spline interpolation coefficients, and it can be precomputed for given P and Δ_x [4, 25, 36].

In addition to the static nonlinearity, memory caused by the coupling channel is modeled with a linear FIR filter with M coefficients. Thus, the output signal of the complete model can be expressed as

$$y(n) = \mathbf{w}_n^T\mathbf{s}_n, \tag{11.14}$$

where $\mathbf{w}_n = [w_0 \ w_1 \ \ldots \ w_{M-1}]$ contains the filter coefficients, $\mathbf{s}_n = [s(n + M_{pre}) \ s(n + M_{pre} - 1) \ \ldots \ s(n - M_{post})]$ is the signal regression of $s(n)$, and $M = M_{pre} + M_{post} + 1$.

In this model, the control points and filter coefficients need to be estimated in order to accurately regenerate the SI signal. To that end, the gradient descent approach [33] is taken, where the gradient of the mean-squared error cost function is used in order to calculate the coefficient update at each step.

Following the notation of the previous section, the cost function can be expressed as a function of the residual SI signal (which is the error signal in this context), $y_{DC}(n) = y_{ADC}(n) - y(n)$, as

$$J(\mathbf{q}_n, \mathbf{w}_n) = y_{DC}(n)y_{DC}^*(n), \tag{11.15}$$

and, thus, the learning equations can be formulated as

$$\mathbf{q}_{n+1} = \mathbf{q}_n - \mu_q \nabla_{\mathbf{q}_n} J(\mathbf{q}_n, \mathbf{w}_n) \tag{11.16}$$

$$\mathbf{w}_{n+1} = \mathbf{w}_n - \mu_w \nabla_{\mathbf{w}_n} J(\mathbf{q}_n, \mathbf{w}_n), \tag{11.17}$$

where μ_q and μ_w are the learning rates for the LMS adaptation. Invoking elementary differentiation properties, the final learning equations for both parameters read [4]

$$\mathbf{q}_{n+1} = \mathbf{q}_n + \mu_q \, y_{DC}(n) \, \Sigma_n^T \mathbf{X}_n^* \mathbf{w}_n^* \tag{11.18}$$

$$\mathbf{w}_{n+1} = \mathbf{w}_n + \mu_w y_{DC}(n) \mathbf{s}_n^*, \tag{11.19}$$

where $\mathbf{X}_n$ contains the signal regression of $x(n)$ in its main diagonal, and $\Sigma_n = \left[\xi_{n+M_{pre}} \xi_{n+M_{pre}-1} \cdots \xi_{n-M_{post}} \right]^T$. If no further information is available, the control points and the filter coefficients are initialized as $\mathbf{q}_0 = \mathbf{w}_0 = 0$.

The learning update for $\mathbf{q}_n$ involves a relatively high computational complexity in the term $\Sigma_n^T \mathbf{X}_n^* \mathbf{w}_n^*$ if the memory order of the model is large. Therefore, an approximation is proposed to ease the computational load, while retaining high accuracy. The approximation consists on selecting only τ elements in the temporal dimension (rows) of Σ_n. Generally, the largest filter taps in $\mathbf{w}_n$ are the ones contributing the most to the update, and those are usually located in the samples nearest to the present moment. Therefore, by selecting the most relevant memory taps, the performance degradation remains negligible while the complexity is greatly reduced. This is

$$\Sigma_n = \left[\xi_{n+M_{pre}} \cdots \underbrace{\xi_{n+\tau_1} \cdots \xi_n \cdots \xi_{n-\tau_2}}_{\tau} \cdots \xi_{n-M_{post}} \right]^T, \tag{11.20}$$

where $\tau = \tau_1 + \tau_2 + 1$.

11.5 Measurement Results

In order to test and verify the two aforementioned digital cancellers, a set of RF measurements is carried out with a complete IBFD prototype working in the 2.4-GHz industrial, scientific, and medical (ISM) band, whose structure is discussed in more detail in [3]. The full-duplex prototype contains a PA in the transmit chain, which is producing some nonlinear distortion, and a back-to-back relay antenna to provide the transmitter/receiver analog isolation. Note that no RF canceller is employed in order to stress the digital canceller as much as possible. The transmit power of the device is +24 dBm.

The parametrization used to generate the results is for the MP-based canceller $P = 11$, $M_{pre} = 25$, $M_{post} = 50$, and canceller $P = 2$, $Q = 5$, $\tau = 5$, $M_{pre} = 25$, $M_{post} = 50$ for the SPH-based canceller. Note that, in the case of the MP model, the nonlinearity order P refers to the order of the whole polynomial, while in the case of the SPH-based model it represents the order of the individual splines. With this configuration, Figure 11.7 shows the spectra of the transmit, received, and residual signals after cancellation, for three different instantaneous bandwidths of 20, 40, and 80 MHz. With the narrowest bandwidth of 20 MHz, both MP and SPH models perform similarly, achieving similar levels of digital cancellation. In this specific case, the amount of cancellation achieved is 44.4 and 45.3 dB, respectively. This same conclusion can be drawn for a bandwidth of 40 MHz. In this case, the amount of cancellation achieved is 43 dB for both models. Finally, with the widest bandwidth of 80 MHz, the MP-based canceller outperforms the SPH model by 2 dB. The amounts of digital cancellation in this case are 40 and 38 dB,

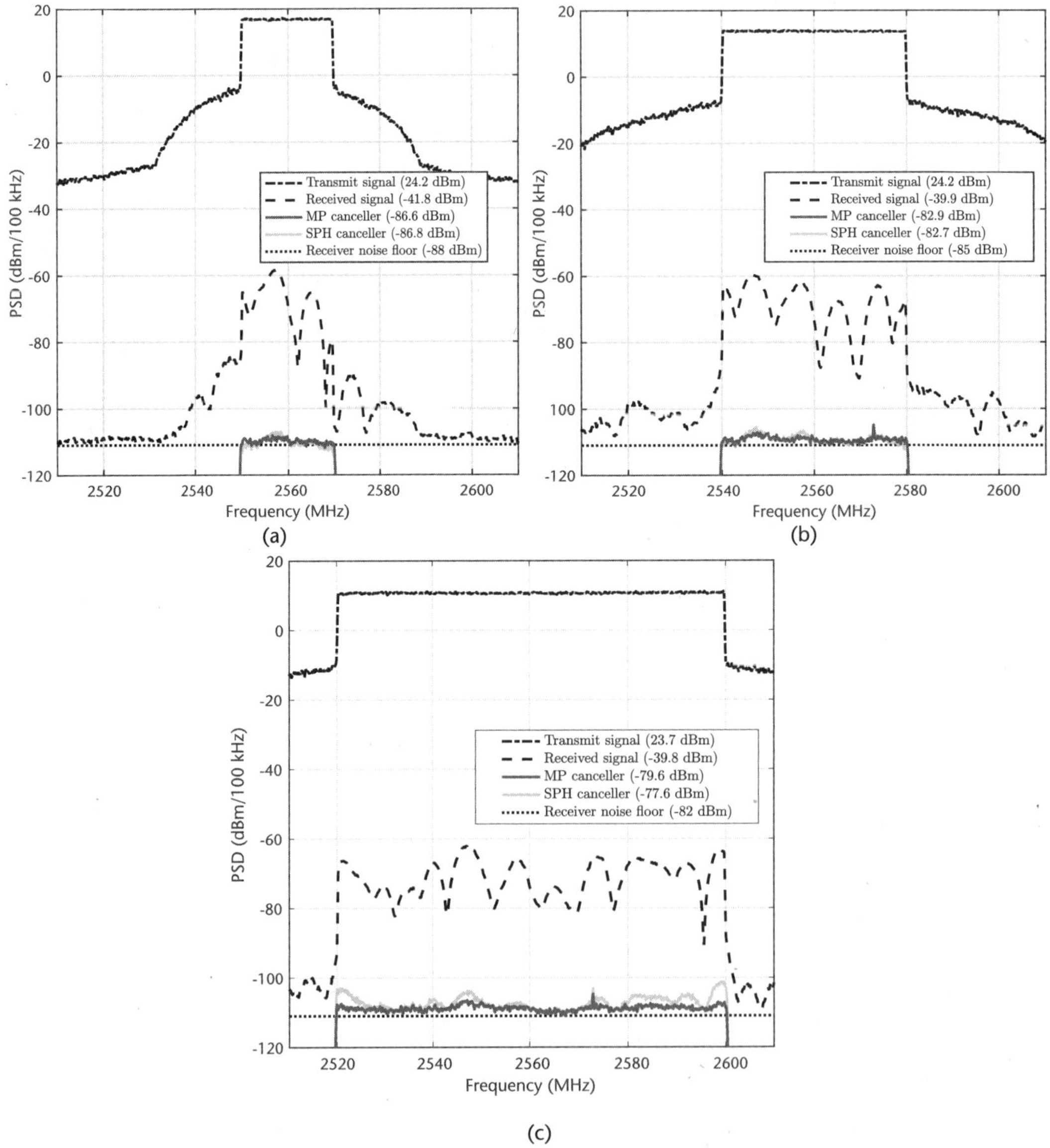

Figure 11.7 Power spectral densities (PSDs) for the transmit, received, and cancelled signals with an instantaneous bandwidth of 20, 40, and 80 MHz.

respectively. Despite the small performance difference with the widest bandwidths, in most cases, both of the proposed digital cancellers achieve a sufficient amount of digital cancellation, indicating that the algorithms have excellent nonlinear SI modeling accuracy.

In order to investigate the convergence of the models, Figure 11.8 shows the residual power after the two digital SI cancellers with respect to the number of iterations. When 20 MHz is considered, both approaches can suppress the SI interference near the receiver's noise floor, although the convergence of the SPH algorithm is faster, thus reaching the steady state earlier. With a bandwidth of 40 MHz, the performance and convergence of the models are similar, both reaching the steady state at approximately the same time. Finally, with 80 MHz, the overall behavior

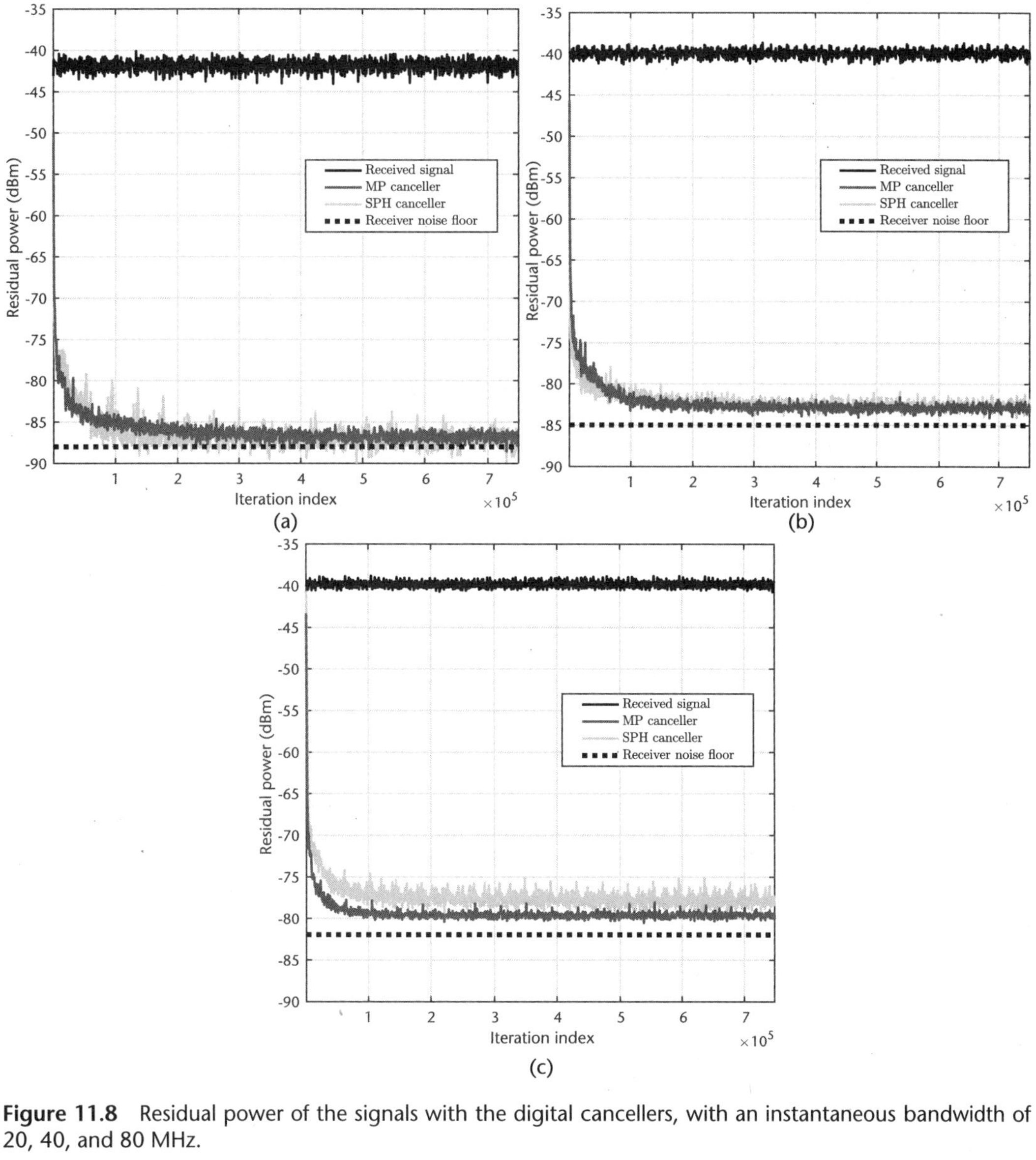

Figure 11.8 Residual power of the signals with the digital cancellers, with an instantaneous bandwidth of 20, 40, and 80 MHz.

of the SPH model seems somewhat degraded when compared to the MP-based, as observed before.

The processing complexity of the described digital cancellers is also an important metric in the full-duplex context, and it is thus analyzed within the next lines. The complexity analysis of the MP and SPH approaches is carried out by following the signal processing elements described within Sections 11.4.2 and 11.4.3, respectively. The analysis is presented in terms of real multiplications per input sample, as it is an important metric in hardware implementations.

The computational complexity expressions for both models are collected in Table 11.2, as a function of the modeling parameters of each model. In the MP canceller, the cancellation stage includes the basis function generation and SI regeneration, and the learning update stage includes the LMS learning update for the SI coupling channel, $\hat{\mathbf{h}}(n+1)$. With the SPH model, the cancellation stage also includes the SI regeneration, where the square root in (11.12) is done with the alpha max beta min algorithm [37]. The learning update stage is composed of updating the interpolation control points and the linear filter coefficients.

Considering the same parametrization as the one chosen to obtain the measurement results, specific complexity numbers can be obtained for both models. It has been assumed that one complex multiplication is performed with three real multiplications. In the cancellation stage, the MP and the SPH models require 1,470 and 240 real multiplications per input sample to regenerate the SI signal, respectively. In the learning update stage, the approaches require 1,352 and 298 real multiplications to learn and update the model coefficients. Therefore, the SPH model poses 83% and 78% complexity reduction in the cancellation and learning update stages, respectively.

In conclusion, the MP model is a robust and reliable technique to estimate and cancel the SI signal, but the SPH model achieves largely the same performance while greatly reducing computational complexity. However, it should be noted that the SPH model may suffer from slight performance degradation when very wide bandwidths are used, as was observed with the 80-MHz instantaneous bandwidth measurements.

11.6 Practical Considerations

This chapter provides measurement-based evaluations of fully adaptive nonlinear digital cancellers. An important aspect of such adaptive algorithms is to carefully adjust the various hyper parameters of the models, such as the learning rates,

Table 11.2 Computational Complexity of the Algorithms in Terms of Real Multiplications Per Input Sample

	MP	*SPH*
Cancellation	$3\left(\frac{P+1}{2}\right)^2+(3M+2)\left(\frac{P+1}{2}\right)$	$P^2+3P+3M+5$
Learning Update	$3M\left(\frac{P+1}{2}\right)+2$	$3P\tau+3P+3M+6\tau+7$

memory lengths, and nonlinearity orders. Proper selection of these hyper parameters can often be the difference between the state of the art and poor performance.

Another crucial aspect very particular to SI cancellation in IBFD radios is the interplay between the different isolation and cancellation mechanisms. The considered prototype provided the necessary analog isolation using a back-to-back relay antenna, meaning that no active RF cancellation was involved. While this simplified the hardware architecture, it also greatly increased the memory requirements in the digital canceller. This likely happens because the relay antenna has been designed to strongly attenuate the direct leakage and short coupling paths, while the reflections from the surrounding environment can freely propagate to the Rx side. This had to be considered in the digital canceller by using a rather large amount of memory taps.

11.7 Conclusion

11.7.1 Summary

In this chapter, we presented two alternative digital SI canceller algorithms, both of which are capable of modeling a nonlinearly distorted SI signal. The first one is based on a memory polynomial, which models the nonlinear response with a single high-order polynomial, while the second one is based on splines that model the nonlinear system in a piece-wise manner. Both of these cancellers are then evaluated as part of a real-world IBFD prototype, which incorporates a back-to-back relay antenna to provide the propagation domain isolation. It was shown that both of the proposed cancellers can suppress the SI to the level of the receiver noise floor, bringing the total amount of cancellation to a level beyond 100 dB. Of this, 40–45 dB was provided by the digital cancellers, the exact amount depending on the bandwidth. Remarkably, the spline-based digital canceller was able to match the high performance of the memory polynomial-based canceller while requiring 80% less multiplications.

11.7.2 Future Directions

Although the proposed digital cancellers can already achieve the required SI cancellation performance with the considered moderate transmit power levels, further work is needed in order to achieve sufficient modeling accuracy with very high transmit powers. Then the impairments that have been negligible so far might also have an impact on the SI waveform. Another important future research direction is to study the interplay between active analog and digital cancellation in greater detail. This might provide further insights into the overall SI cancellation architecture and requirements.

References

[1] Korpi, D., et al., "Full-Duplex Transceiver System Calculations: Analysis of ADC and Linearity Challenges," *IEEE Transactions on Wireless Communications,* Vol. 13, No. 7, July 2014, pp. 3821–3836.

[2] Korpi, D., et al., "Full-Duplex Mobile Device: Pushing the Limits," *IEEE Communications Magazine,* Vol. 54, September 2016, pp. 80–87.

[3] Korpi, D., et al., "Compact Inband Full-Duplex Relays with Beyond 100 dB Self-Interference Suppression: Enabling Techniques and Field Measurements," *IEEE Transactions on Antennas and Propagation,* Vol. 65, February 2017, pp. 960–965.

[4] Campo, P. P., et al., "Nonlinear Digital Cancellation in Full-Duplex Devices Using Spline-Based Hammerstein Model," *2018 IEEE Globecom Workshops,* Abu Dhabi, United Arab Emirates, 2018, pp. 1–7.

[5] Bharadia, D., E. McMilin, and S. Katti, "Full Duplex Radios," *Proc. SIGCOMM'13,* August 2013.

[6] Chung, M., et al., "Prototyping Real-Time Full Duplex Radios," IEEE Communications Magazine, Vol. 53, September 2015, pp. 56–63.

[7] Choi, J. I., et al., "Achieving Single Channel Full Duplex Wireless Communication," *Proc. 16th Annual International Conference on Mobile Computing and Networking (MobiCom),* September 2010, pp. 1–12.

[8] Duarte, M., and A. Sabharwal, "Full-Duplex Wireless Communications Using Off-the-Shelf Radios: Feasibility and First Results," *Proc. 44th Asilomar Conference on Signals, Systems, and Computers (ASILOMAR)*, November 2010, pp. 1558–1562.

[9] Korpi, D., et al., "Adaptive Nonlinear Digital Self-Interference Cancellation for Mobile Inband Full-Duplex Radio: Algorithms and RF Measurements," *Proc. IEEE Global Communications Conference (GLOBECOM)*, December 2015.

[10] Balatsoukas-Stimming, A., "Non-Linear Digital Self-Interference Cancellation for In-Band Full-Duplex Radios Using Neural Networks," *Proc. IEEE 19th International Workshop on Signal Processing Advances in Wireless Communications (SPAWC)*, 2018.

[11] Guo, H., et al., "Realtime Software Defined Self-Interference Cancellation Based on Machine Learning for In-Band Full Duplex Wireless Communications," *Proc. International Conference on Computing, Networking and Communications (ICNC)*, 2018, pp. 779–783.

[12] Korpi, D., "Full-Duplex Wireless: Self-Interference Modeling, Digital Cancellation, and System Studies," Doctoral dissertation, Tampere University of Technology, December 2017.

[13] Everett, E., A. Sahai and A. Sabharwal, "Passive Self-Interference Suppression for Full-Duplex Infrastructure Nodes," *IEEE Transactions on Wireless Communications,* Vol. 13, No. 2, 2014, pp. 680–694.

[14] Zhang, Z., et al., "Full Duplex 2x2 MIMO Radios," *Proc. Sixth International Conference on Wireless Communications and Signal Processing (WCSP)*, 2014.

[15] Laughlin, L., et al., "A Widely Tunable Full Duplex Transceiver Combining Electrical Balance Isolation and Active Analog Cancellation," *Proc. 81st IEEE Vehicular Technology Conference (VTC Spring)*, 2015.

[16] Kolodziej, K. E., J. G. McMichael, and B. T. Perry, "Multitap RF Canceller for In-Band Full-Duplex Wireless Communications," *IEEE Transactions on Wireless Communications,* Vol. 15, No. 6, 2016, pp. 4321–4334.

[17] Boglione, L., et al., "Wide Band, High Power, Same-Channel Full Duplex Transceiver System Demonstration," *Proc. IEEE MTT-S International Microwave Symposium (IMS)*, 2017.

[18] Third Generation Partnership Project (3GPP); Technical Specification Radio Access Network; NR, User Equipment (UE) Radio Transmission and Reception (Release 16), 2020.

[19] Korpi, D., et al., "Widely Linear Digital Self-Interference Cancellation in Direct-Conversion Full-Duplex Transceiver," *IEEE Journal on Selected Areas in Communications,* Vol. 32, September 2014, pp. 1674–1687.

[20] Abdelhafiz, A., et al., "Digital Predistortion of LTE-A Power Amplifiers Using Compressed-Sampling-Based Unstructured Pruning of Volterra Series," *IEEE Transactions on Microwave Theory and Techniques*, Vol. 62, November 2014, p. 2583–2593.

[21] Schoukens, M., R. Pintelon, and Y. Rolain, "Parametric Identification of Parallel Hammerstein Systems," *IEEE Transactions on Instrumentation and Measurement*, Vol. 60, December 2011, p. 3931–3938.

[22] Isaksson, M., D. Wisell, and A. D. Rönnow, "A Comparative Analysis of Behavioral Models for RF Power Amplifiers," *IEEE Transactions on Microwave Theory and Techniques*, Vol. 54, January 2006, pp. 348–359.

[23] De Boor, C., *A Practical Guide to Splines*, Vol. 27, New York: Springer-Verlag, 1978.

[24] Prautzsch, H., W. Boehm, and M. Paluszny, *Bézier and B-Spline Techniques*, New York: Springer Science & Business Media, 2013.

[25] Scarpiniti, M., et al., "Nonlinear Spline Adaptive Filtering," *Signal Processing,* Vol. 93, 2013, pp. 772–783.

[26] Scarpiniti, M., et al., "Hammerstein Uniform Cubic Spline Adaptive Filters: Learning and Convergence Properties," *Signal Processing,* Vol. 100, 2014, pp. 112–123.

[27] Sabharwal, A., et al., "In-Band Full-Duplex Wireless: Challenges and Opportunities," *IEEE Journal on Selected Areas in Communications*, Vol. 32, September 2014, pp. 1637–1652.

[28] Gu, Q., *RF System Design of Transceivers for Wireless Communications*, New York: Springer, 2006.

[29] Syrjälä, V., et al., "Analysis of Oscillator Phase-Noise Effects on Self-Interference Cancellation in Full-Duplex OFDM Radio Transceivers," *IEEE Transactions on Wireless Communications,* Vol. 13, June 2014, pp. 2977–2990.

[30] Sahai, A., et al., "On the Impact of Phase Noise on Active Cancelation in Wireless Full-Duplex," *IEEE Transactions on Vehicular Technology*, Vol. 62, November 2013, p. 4494–4510.

[31] Haykin, S., *Neural Networks: A Comprehensive Foundation*, 2nd ed., Upper Saddle River, NJ: Prentice Hall, 1999.

[32] Emara, M., et al., "Nonlinear Digital Self-Interference Cancellation with Reduced Complexity for Full Duplex Systems," *Proc. International ITG Workshop on Smart Antennas (WSA)*, March 2017.

[33] Haykin, S., *Adaptive Filter Theory*, Upper Saddle River, NJ: Prentice-Hall, 1996.

[34] Hyvärinen, A., J. Karhunen and E. Oja, *Independent Component Analysis*, New York: John Wiley & Sons, 2001.

[35] Dallinger, R., et al., "Adaptive Pre-Distortion Techniques Based on Orthogonal Polynomials," *Proc. 44th Asilomar Conference on Signals, Systems and Computers,* 2010, pp. 1945–1950.

[36] Scarpiniti, M., et al., "Novel Cascade Spline Architectures for the Identification of Nonlinear Systems," *IEEE Transactions on Circuits and Systems I: Regular Papers*, Vol. 62, 2015, pp. 1825–1835.

[37] Frerking, M., *Digital Signal Processing in Communications Systems*, New York: Springer Science & Business Media, 2013.

CHAPTER 12

Reference-Based Nonlinear Digital Cancellation

Joel Goodman, Luciano Boglione, and Clayton G. Davis

12.1 Introduction

Unlike its commercial counterpart, full-duplex operation for U.S. Department of Defense (DoD) systems in many cases requires employing both a high-power broadband transmitter and a wideband sensitive receiver. The U.S. Naval Research Laboratory (NRL) in Washington, D.C., for instance, has developed an IBFD system with a transmitter that has 60 dBm of effective radiated power (ERP) and a colocated receiver operating with roughly 700 MHz of instantaneous bandwidth, tunable anywhere from 2 to 7 GHz. The receiver is protected by 130-dB isolation relative to the ERP (120 dB relative to the 50-dBm transmit power): separate transmit and receive apertures provide 80-dB isolation relative to ERP (70 dB relative to transmit power), and digital cancellation provides an additional 50-dB isolation, as highlighted within the IBFD transceiver diagram of Figure 12.1(a) and SIC techniques tree of Figure 12.1(b).

An illustration of power levels at the input to such a receiver from various transmitter and environmental artifacts is shown in Figure 12.2. NRL used a separate transmit and receive aperture, discussed in Section 12.2, that was able to achieve, on average, 70 dB of RF isolation across 2 to 7 GHz with less than 25 cm of separation without any analog cancellation [1]. With a 50-dBm (100-W) transmitter and 70 dB of isolation, the receiver was subject to not only direct path (DP) transmitter leakage, but also external environmental multipath as well as transmit and receiver distortions, as discussed in Chapter 1. Because many DoD systems operate on platforms where the antenna system does not have an unobstructed spatial field of view, multipath is a significant concern. It is not uncommon for the multipath strength to be equal to the DP leakage from the transmitter.

Some DoD high-power transmitters are biased at or beyond the 1-dB compression point, employ class AB or even C amplifiers, and/or transmit multiple waveforms simultaneously that are separated in frequency [2]. Like their commercial counterparts, these systems have spectral mask requirements; in some cases, these

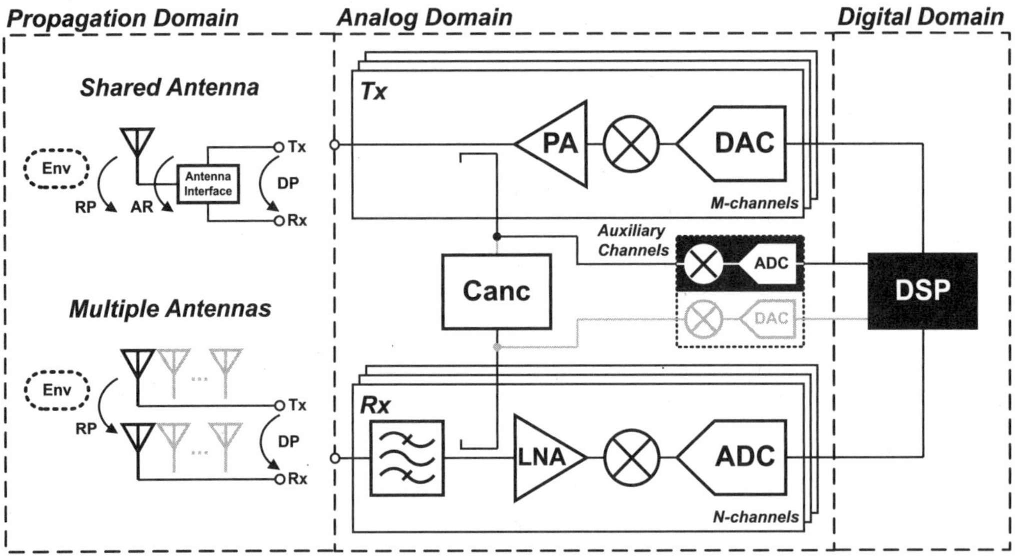

(a)

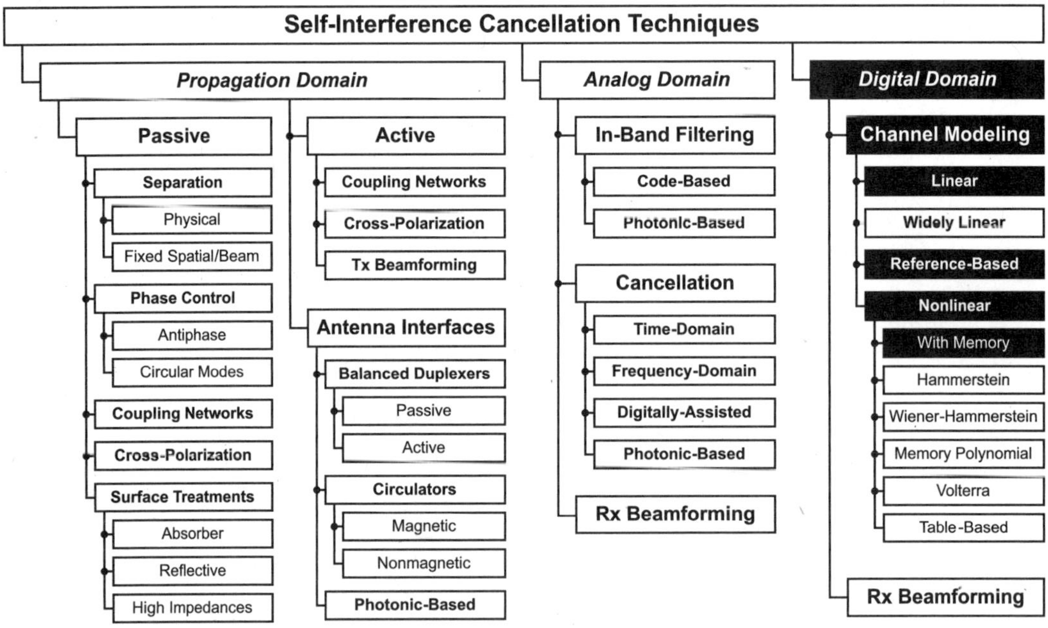

(b)

Figure 12.1 Reference-based nonlinear digital cancellation and specific chapter focus highlighted within (a) a generic IBFD transceiver and (b) a SIC techniques tree diagram.

requirements are significantly relaxed, especially for in-band transmission in applications such as radar. This means that the distortion and/or phase noise generated by these broadband amplifiers will be difficult or impossible to correct using predistortion, for example, a traveling-wave tube (TWT) amplifier biased at or beyond its 1-dB compression point with a peak-to-average power ratio (PAPR) of >10 dB is uninvertible [3].

The RF receiver itself presents an interesting challenge, given the high interference power at its input. In general, to achieve a relatively low cascaded noise figure,

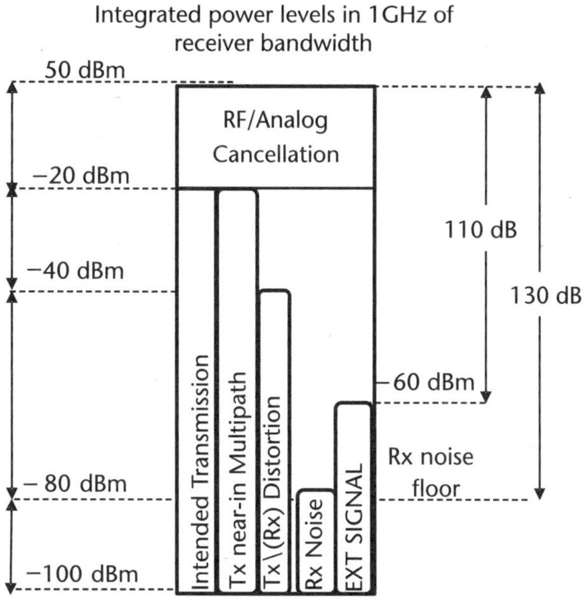

Figure 12.2 The received signal power of the various transmitter and environmental artifacts from a 50-dBm transmitter. This illustration represents the impact of Tx and Rx distortions, as well as multipath on the signal strength at the receiver input after 70 dB of RF isolation. Buried in this sea of interference is the desired external signal (EXT SIGNAL) that we wish to recover.

the first-stage LNA should have relatively high gain to suppress the noise contribution of the following stages. However, a high-gain, broadband LNA operating in the presence of strong interference can saturate the receiver, generating unwanted distortions. Further, most commercial ADCs reach full scale at or near 0 dBm, limiting the overall achievable gain and, by extension, cascaded noise figure. Nonlinear equalization on receive can be employed to subtract off distortions generated by the receiver [4], as distortions are an unavoidable artifact of high (input) power operation, extending the dynamic range in many cases by ≥20 dB.

Finally, as mentioned previously, noise and distortions generated by high-power (>1 kW) nonlinear broadband amplifiers driven into compression may be difficult to invert, significantly limiting both analog and digital cancellation performance. The following sections describe an architecture to overcome many of the aforementioned factors that can limit IBFD operations in practice.

12.2 Robust In-Band Nonlinear Digital Cancellation

12.2.1 Overview of the Approach

As discussed in Section 12.1, the objective of the NRL full-duplex system is to mitigate broadband noise and distortions generated by both the transmitter and receiver. To that end, NRL has developed a dual-channel receiver that is illustrated in Figure 12.3. This system employs only passive RF isolation and digital cancellation. However, the architecture is fully compatible with analog cancellation, which can further enhance the system performance.

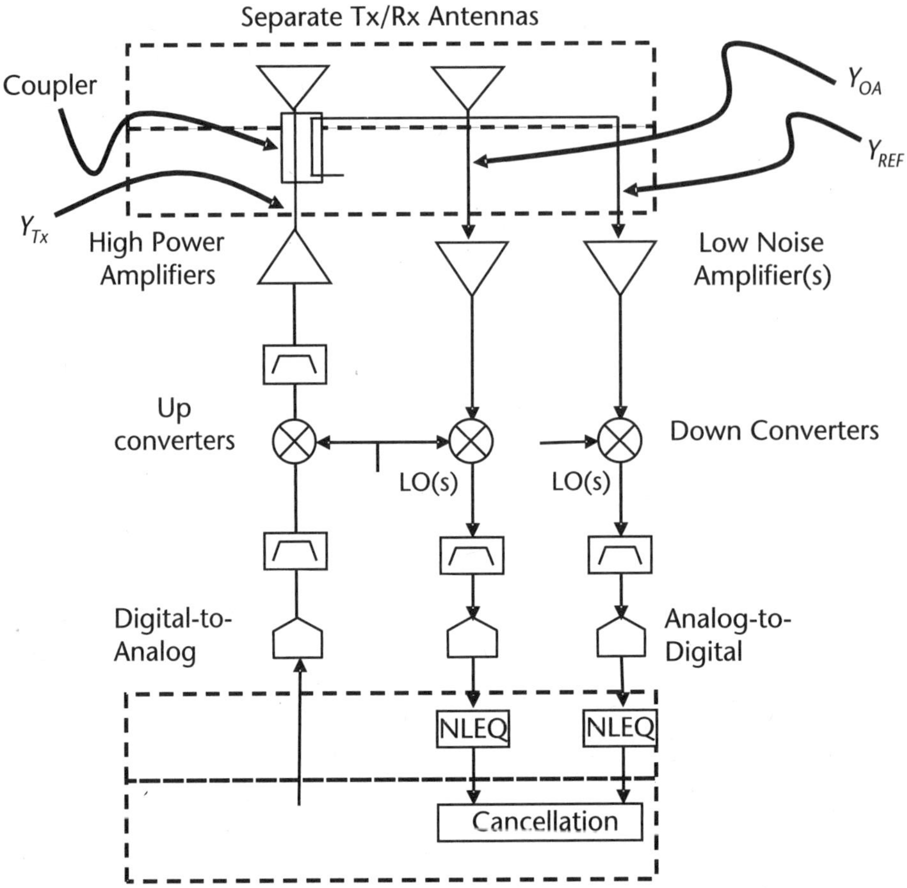

Figure 12.3 Full-duplex dual-channel receiver and transmitter.

The principle of operation is straightforward. Referring to Figure 12.3, a digital waveform is synthesized by the digital processor, converted to an analog waveform by a DAC, mixed up to a carrier frequency, and transmitted. Let the transmitted signal be represented by

$$y_{Tx}(t) = x_{Tx}(t) + g_{Tx}(x_{Tx}(t)) + v_{Tx}(t), \tag{12.1}$$

where $x_{Tx}(\cdot)$ is the intended transmission, $g_{Tx}(\cdot)$ is a function that adds nonlinear distortion generated by the active and mixed signal components of the transmitter, and $v_{Tx}(t)$ is a combination of broadband noise terms that the transmitter generates. The output from the high-power amplifier (HPA) of the transmitter is coupled into one port of the receiver, which we will call the reference (REF) port. The signal at the output of the digitizer at discrete time n from the REF port

$$y_{REF}(n) = \left(y_{Tx} * h_{REF} + g_{REF}\left(y_{Tx} * h_{REF}\right) + v_{REF}\right)(n), \tag{12.2}$$

includes the transmitted signal convolved with the impulse response $h_{REF}(t)$ of the analog components and voltage standing-wave ratio (VSWR) of the reference channel, nonlinear distortion $g_{REF}(\cdot)$ generated by the active and mixed signal components, and noise term $v_{REF}(\cdot)$. The digitized output of the second port of the receiver, which we will refer to as the over-the-air (OA) port, is directly connected to the receive antenna such that

$$y_{OA}(n) = \left(y_{Tx} * h_{OA} + g_{OA}\left(y_{Tx} * h_{OA}\right) + v_{OA} + s\right)(n), \tag{12.3}$$

which differs from (12.2) in that the OA impulse response h_{OA} not only includes the response of the analog components, but also multipath as well as the external signals s in the environment that we wish to recover. If we are able to project and subtract off the reference signal y_{REF} from the OA signal y_{OA}, we can suppress the common mode signal $y_{Tx}(n)$ and recover the signal of interest $s(n)$. This projection operation, which is an affine transformation as described in Section 12.2.3, is ill-equipped to suppress the nonlinear terms in (12.2) and (12.3) generated by the receiver. We consider the impact of these terms, $g_{OA}(\cdot)$ and $g_{REF}(\cdot)$, in the following section and approaches to mitigate these unwanted distortions.

12.2.2 A Practical View of Distortion

12.2.2.1 Receiver Nonlinearities

A source of signal degradation in a receiver is due to nonlinearities principally generated by the RF/analog and mixed signal components of the system. For example, amplifiers generate distortions when there is a nonlinear relationship between the input and output power, as shown in Figure 12.4.

Because many systems tend to have a narrow bandwidth with respect to the carrier frequency, it is common to consider only odd-order terms, as the even-order polynomial distortions are filtered out. To see this, consider a superheterodyne receiver with both the second-order distortions

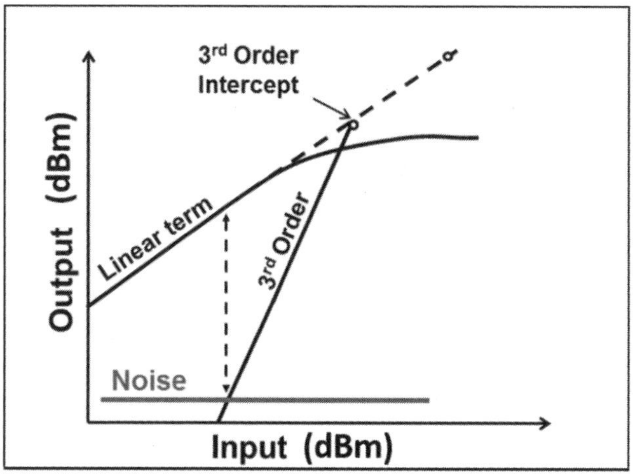

Figure 12.4 Input-output characteristics of amplifiers. Note that, in this case, a third-order nonlinearity with respect to the linear response is shown. The location where a purely linear system and third-order distortion have equal output power is the third-order intercept point.

$$\alpha_2\left(x(t)e^{j\omega_c t} \bullet x(t)e^{j\omega_c t}\right) = \alpha_2 x(t)^2 e^{j2\omega_c t} \tag{12.4}$$

$$\alpha_2'\left(x(t)e^{j\omega_c t} \bullet x^*(t)e^{-j\omega_c t}\right) = \alpha_2'\left|x(t)\right|^2, \tag{12.5}$$

and third-order order distortions:

$$\alpha_3\left(x(t)e^{j\omega_c t} \bullet x(t)e^{j\omega_c t} \bullet x(t)e^{j\omega_c t}\right) = \alpha_3 x(t)^3 e^{j3\omega_c t} \tag{12.6}$$

$$\alpha_3'\left(x(t)e^{j\omega_c t} \bullet x^*(t)e^{-j\omega_c t} \bullet x(t)e^{j\omega_c t}\right) = \alpha_3'\left|x(t)\right|^2 x(t)e^{j\omega_c t}, \tag{12.7}$$

of the narrowband signal $x(t)$, which is a complex representation of the analog signal that is input to a device that generates the nonlinearities. In (12.4) through (12.7), the α_is and α'_is correspond to polynomial coefficients of order i, and ω_c is the carrier frequency. Equations (12.4) and (12.5) represent second-order distortions, located at twice the carrier frequency and at DC (0-Hz frequency). Because it is common for most receivers to have filters that reject these distortion components, it is typical to ignore second-order distortion effects in amplifiers and mixers. One can show that this phenomenon is expressly manifest in all even-order distortions. However, odd-order distortions do have components that land at the carrier frequency, as quantified in (12.7). These distortion components are passed by the system and represent unwanted nonlinear artifacts generated by the analog devices in the RF system prior to ADC or after DAC.

Unlike its analog counterpart, the mixed signal components (ADC) can pass both even and odd distortions. ADCs have various architectures for converting between the analog and digital domains, including pipelined, interleaved, and flash [5]. Each of these architectures introduces a type of structured distortion that is inherently different from the compressing input-output curves associated with amplifiers. A detailed discussion of these architectures can be found in [6]. In this exposition, we will consider a common source of distortions in converters that is graphically illustrated in Figure 12.5, where the differential nonlinearity is a measure of the deviation from an ideal quantization level, and the integral nonlinearity

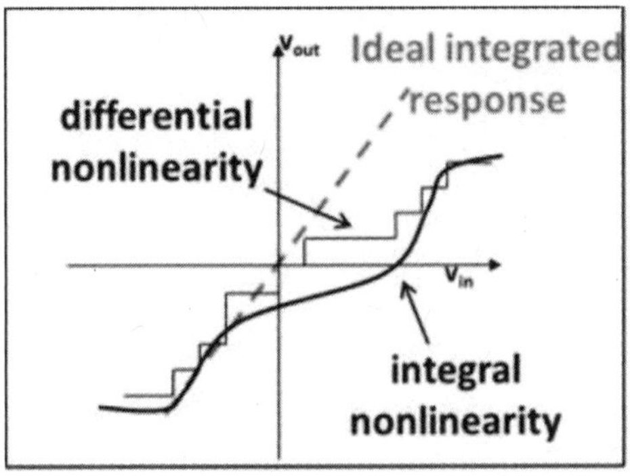

Figure 12.5 ADC nonlinearities can be identified as differential (stepwise deviations from ideal level) and integral (accumulated deviations from ideal linear response).

represents these deviations averaged across all quantization levels (i.e., the input-output characteristics of the converter).

Because there is no filter in place to remove these distortions, both even-order and odd-order distortions are present at the output of the converter. In fact, ADC out-of-band distortions end up aliasing back in-band. We consider removal of these types of distortions in Section 12.2.2.4.

The mathematical representation of the distortions in (12.4) through (12.7) is highly oversimplified and represents the memoryless case, that is, the output is a nonlinear function of the input at the current time instant. In practice, there are memory effects that reflect the fact that the output is not only a nonlinear function of the input at the current time instant, but also at previous time instances as well. As illustrated in Figure 12.6, the measured input-output relationship of an amplifier can be quite complicated, with one input mapping to many outputs depending on the previous trajectory of the input.

There have been a number of approaches to model distortions with memory, as discussed in Chapter 11. Two popular approaches are memory polynomials (MPs) and generalized memory polynomials (GMPs) [7]. MPs and GMPs are mathematically modeled at baseband as

$$y_{MP}(n) = \sum_{m} h(m)\left|x(n-m)\right|^{p-1} x(n-m) \tag{12.8}$$

$$y_{GMP}(n) = \sum_{m_1,m_2} h(m_1,m_2)\left|x(n-m_1-m_2)\right|^{p-1} x(n-m_1), \tag{12.9}$$

with (12.8) and (12.9) representing a pth-order nonlinear model where p is generally odd, with the coefficients $h(\cdot)$ replacing the αs in (12.7). Note that, in this case, the objective is to determine the coefficients $h(\cdot)$ that best model the measured

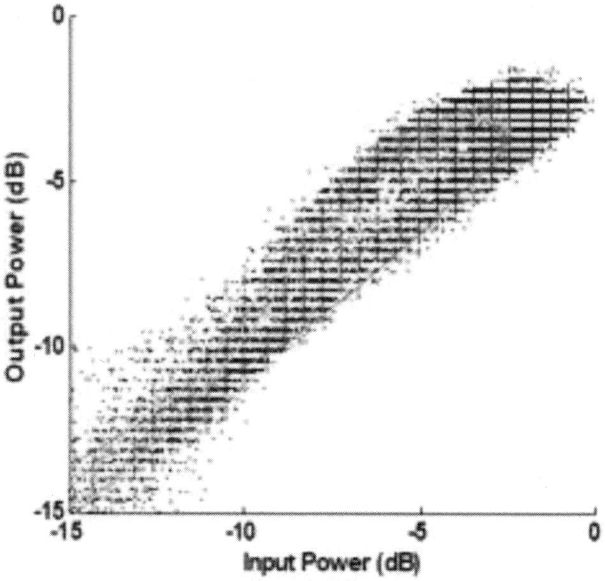

Figure 12.6 Input-output relationship of a realistic amplifier. Memory effects make modeling of the system quite complicated.

pth-order distortions $y_{(G)MP}(n)$ in a mean square error (MSE) sense based on a known input $x(n)$. Both MP and GMP capture memory effects, but only partially. This can best be visualized geometrically by considering the coefficient space for a third-order distortion, as illustrated in Figure 12.7. In this coefficient space, a MP can capture only those distortions lying along a diagonal line, while a GMP can capture distortions lying along a plane. However, most of the coefficient space remains unrepresented. The reason for the popularity of using MPs and GMPs to model distortions is twofold: first, there is a simple closed-form solution for the coefficients, and second, it was found in practice to do a reasonable job representing the bulk of the distortions generated by amplifiers, as mentioned in Chapter 11.

Similar to (12.7), MPs and GMPs as formulated in (12.8) and (12.9) capture only odd-order distortions and make them well-suited for modeling amplifiers and mixers in narrowband systems. An alternative to GMPs and MPs is cube coefficient subspaces (CCS), whose geometric representation is illustrated in Figure 12.8 [8].

The pth-order nonlinear d-dimensional CCS components are mathematically represented as:

$$y_{CCS}(n) = \sum_{\alpha \in A} \sum_{m_1=0}^{M} \cdots \sum_{m_d=0}^{M} h(m_1+\alpha_1,\cdots,m_d+\alpha_d,\alpha_{d+1},\cdots,\alpha_p) \prod_{k=1}^{\left\lceil \frac{d}{2} \right\rceil} x(n-m_k-\alpha_k) \prod_{l=\left\lceil \frac{d}{2} \right\rceil+1}^{d} x^*(n-m_l-\alpha_l) \prod_{q=d+1}^{\left\lfloor \frac{p+d+1}{2} \right\rfloor} x(n-\alpha_q) \prod_{r=\left\lfloor \frac{p+d+1}{2} \right\rfloor+1}^{p} x^*(n-\alpha_r), \quad (12.10)$$

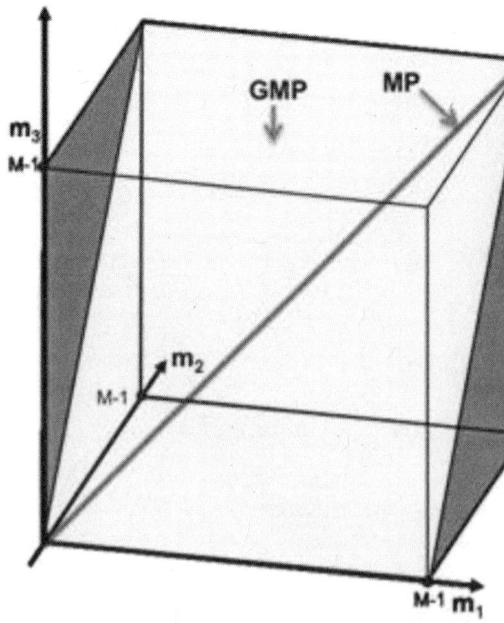

Figure 12.7 Geometrical characterization of MP and GMP coefficient space for nonlinear modeling. (© 2008 IEEE. Reprinted, with permission, from [8].)

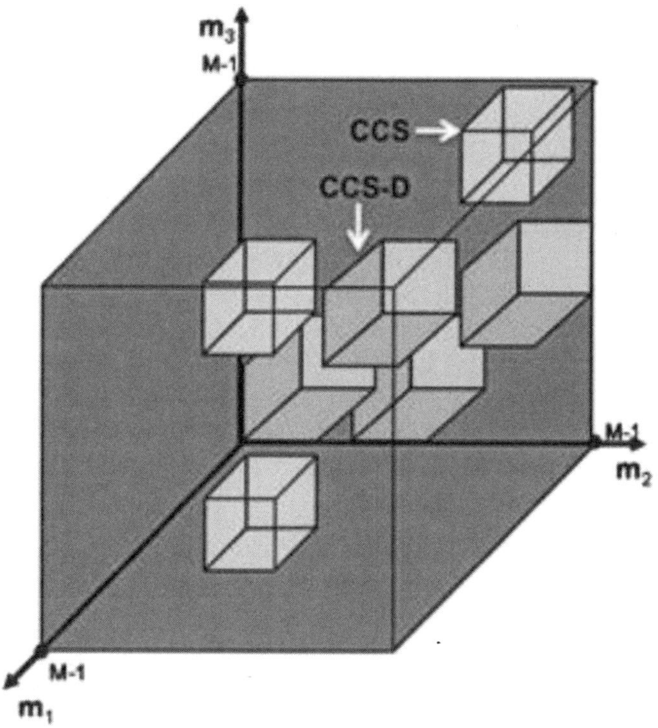

Figure 12.8 CCS for a 3-dimensional (3-D) coordinate system. CCS-diagonal (CCS-D) corresponds to parallel-pipeds, as described in [8]. (© 2008 IEEE. Reprinted, with permission, from [8].)

where the α_is are the fixed integer delays of the cube represented by vector α in the set A representing all cubes included in the model. In (12.10), the dimension d corresponds to the dimension of the coordinate system where the cubes are constructed [8], whereby definition $d \leq p$. Unlike GMP and MP, solving for the coefficients $h(\cdot)$ can be computationally burdensome, as the space can grow geometrically large with increasing delays. It was shown in [8], however that it is straightforward to apply a matching pursuit-like algorithm [4] to select the cubes to optimize the MSE that fit within the computational constraints of the system. We discuss this in more depth in Section 12.2.2.2. Note that, in practice, the cubes are selected once (e.g., during laboratory calibration) and then a least squares approach is subsequently used to update the coefficients based on device aging and/or changes in system temperature. As it has been demonstrated [8] that CCS can do a better job capturing nonlinearities with memory effects than either MP or GMP, we will use it going forward to both model and suppress distortions generated by the RF/analog and mixed signal components of the system.

12.2.2.2 Receiver Nonlinearity Mitigation

Receiver nonlinearities share many of the same characteristics as transmitter nonlinearities (i.e., the amplifier and mixer distortions exhibit the same phenomenology whether they are used in transmitters or receivers). Additionally, ADC distortions, which tend to be a significant source of distortions in the receive chain, are generated by on-chip sample-and-hold amplifiers, nonuniform quantization step sizes, and converter mismatches in time-multiplexed architectures and other architecture specific nonidealities [4]. To address these nonlinearities, we consider modeling and subtracting these distortions using CCS. As formulated, (12.10) is only able to model nonaliased odd-order distortions. To simultaneously address both even-order and aliased distortions as well as nonaliased odd-order distortion, we consider the following modifications. First, and for simplicity of exposition, consider the 1-d complex pth-order CCS formulation from (12.10):

$$X_i^{PE}(n) = \sum_m h(m+\alpha_1,\alpha_2,\cdots,\alpha_p)\underbrace{x(n-m-\alpha_1)\prod_{k=2}^{\lceil \frac{p}{2} \rceil} x(n-\alpha_k)\prod_{l=\lceil \frac{p}{2} \rceil+1}^{p} x^*(n-\alpha_l)}_{X_i^{PE}(n,m)}, \quad (12.11)$$

where $X_i^{PE}(\cdot,m)$ represents the ith processing element or PE at a delay m. A PE represents the delayed product of the input, which enables the multilinear decomposition of a higher-order nonlinearity and straightforward recovery of the coefficients. To include both even and odd aliased distortions, it is possible to write a real-valued analog of (12.11):

$$\chi_j^{PE}(n) = \sum_m g(m+\gamma_1,\gamma_2,\cdots,\gamma_p)\underbrace{x(n-m-\gamma_1)\prod_{k=2}^{p} x(n-\gamma_k)}_{\chi_j^{PE}(n,m)}, \quad (12.12)$$

To jointly solve for the coefficients $g(\cdot)$ and $h(\cdot)$, we consider the following mixed basis optimization. Let

$$X_j^{PE}(n, m = 0:M-1) = \downarrow 2(\mathcal{H}([\chi_j^{PE}(n,0),\cdots,\chi_j^{PE}(n,1)e^{\frac{j\pi}{2}(M-2)},\chi_j^{PE}(n,M-1)e^{\frac{j\pi}{2}(M-1)}])), \quad (12.13)$$

where in (12.13), $\mathcal{H}(\cdot)$ represents the Hilbert transform operator, $\downarrow 2$ represents frequency shifting by $-\pi/2$ and downsampling by 2, and the subscript $0{:}M-1$ corresponds to the values indexed by 0, 1, ..., $M-1$. In effect, (12.13) transforms a real sequence of data into in-phase and quadrature, or IQ, components. Equations (12.12) and (12.13) can be combined in matrix form as

$$\begin{bmatrix} y_{OA}(n) \\ y_{OA}(n+1) \\ \vdots \\ y_{OA}(n+K-2) \\ y_{OA}(n+K-1) \end{bmatrix} =$$

$$\begin{bmatrix} X_1^{PE}(n,-\alpha_1) & X_1^{PE}(n,1-\alpha_1) & \cdots & X_L^{PE}(n,M-1-\gamma_1) \\ X_1^{PE}(n+1,0-\alpha_1) & & & X_L^{PE}(n+1,M-1-\gamma_1) \\ \vdots & & \ddots & \vdots \\ X_1^{PE}(n+K-1,0) & & \cdots & X_L^{PE}(n+K-1,M-1-\gamma_1) \end{bmatrix}, \qquad (12.14)$$

$$\times \begin{bmatrix} h(0+\alpha_1,\cdots) \\ h(1+\alpha_1,\cdots) \\ \vdots \\ g(M-2+\gamma_1,\cdots) \\ g(M-1+\gamma_1,\cdots) \end{bmatrix}$$

where in (12.14) $y_{OA}(\cdot)$ represents the signal received on the OA channel, and the $X_i^{PE}(\cdot)$s are the PEs defined in (12.11) through (12.14). We make the following observations:

1. Because both aliased and nonaliased distortions are represented, both amplifier and mixer distortions whose even-order terms and filtered out-of-band odd-order products are modeled, as well as ADC distortions that alias back in-band.
2. Post-distortion to model or correct for nonlinearities that are generated and filtered in the analog domain and sampled near the critical sampling frequency can, in general, only be modeled correctly using a multidimensional filter, $d = p$ [9].
3. The coefficients $g(\cdot)$and $h(\cdot)$, which span the subspace of both aliased and nonaliased distortions, respectively, are complex.
4. Because the coefficient space grows geometrically with polynomial order (see Figure 12.9), (12.11) is solved using a basis pursuit-like algorithm to select processing elements. In this case, each processing element is tried one at a time and the one with smallest $\mathcal{L}_2$ (Euclidean norm) error is selected and placed in to a set of selected PEs. This greedy optimization continues until either the $\mathcal{L}_2$ error is sufficiently small or the number of PEs meets and/or exceeds an upper bound on the computational complexity of the solution.

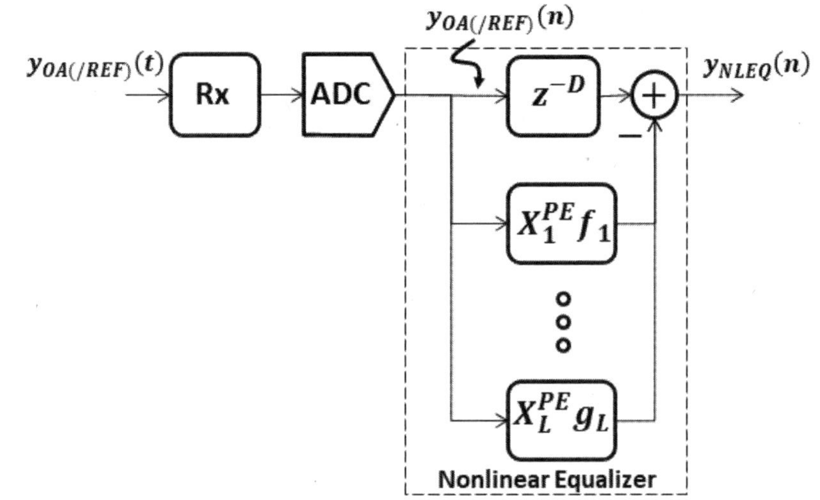

Figure 12.9 Nonlinear equalization architecture. Each of the PEs and its coefficients are estimated and used to subtract off the appropriately delayed incoming signal.

Another significant source of distortions in a receiver is due to interleaving onboard some ADCs. To see how interleaved distortions are manifested, consider the case where $N_I = 2$ (two-way interleaver), where each converter has a baseband impulse response $q_1(n)$ and $q_2(n)$, respectively. Because the output samples are time multiplexed, the upsampled outputs of the converters can be modeled as

$$x_1(n) = \frac{1}{2}\left(q_1(n) + q_1(n)e^{j\pi n}\right), \text{ and}$$
$$x_2(n) = \frac{1}{2}\left(q_2(n+\Delta) - q_2(n+\Delta)e^{j\pi n}\right), \quad (12.15)$$

where Δ in (12.15) corresponds to the sampling phase error between the two converters. It is straightforward to show [4] that for an interleaving factor N_I, the constituent output of each converter is given by

$$x_i(n) = q_i(n+\Delta_i)\sum_{k=0}^{N_I-1} e^{\frac{j2\pi k}{N_I}(n+i-1)}, \quad (12.16)$$

where Δ_is represents the relative sampling phase errors of the constituent converters. To remove the distortions landing at the image frequencies, we construct linear filters

$$X_k^{PE}(n) = \sum_{m=0}^{M-1} f_k(m)x(n-m)e^{\frac{j2\pi k}{N_I}n}, \quad (12.17)$$

for $k = 1,2, \ldots, N_I - 1$. Using (12.15) with the PEs in (12.17) included to address interleaver spurs, the architecture of the nonlinear equalizer consists of series of PEs that in parallel cancel the distortions generated by the receiver, as illustrated by the nonlinear equalizer architecture in Figure 12.9.

The nonlinear equalization architecture in Figure 12.9 uses the incoming signal generate nonlinear distortions using the PEs selected during basis pursuit, which relies on the weakly nonlinear assumption [4]. To test the efficacy of the approach, we constructed a training testbed used to extract the architecture of the nonlinear equalizer, as pictured in Figure 12.10.

Signal generators were used to inject tone sets into the RF receiver and the digitized samples were captured and stored on a custom FPGA-based card with 6 GB of memory. Each tone set consisted of one, two, or three tones, which, although ideal for capturing up to third-order nonlinearities, was a compromise for characterizing and subtracting off higher-order (e.g., more than third-order) distortions [4]. Once the data was captured, a basis pursuit-like algorithm selected PEs and solved for the coefficients, as described previously. There were two second-order PEs, three third-order PEs, a fourth-order and fifth-order PE, and an interleaved PE selected, which met the computational complexity constraints imposed by the FPGA hardware architecture chosen, in this case, a Xilinx Ultrascale+. Cross-validation was used in training and validation, with the measured validation results shown in Figure 12.11.

Distortion suppression exceeded 20 dB on average. We briefly note here that because a multitone excitation meets the persistence of excitation of nonlinear systems, the nonlinear equalizer was able to generalize to any waveform, not just tones, that were input into the receiver.

12.2.3 Digital Cancellation for Simultaneous Transmit and Receive

With the receiver's nonlinear terms suppressed using the techniques described in Section 12.2.2.2, the received signals on both channels

$$\begin{aligned} y_{OA}(n) &= (y_{Tx} * h_{OA} + v_{OA} + s)(n) \\ y_{REF}(n) &= (y_{Tx} * h_{REF} + v_{REF})(n) \end{aligned}, \tag{12.18}$$

are linear functions of the terms in (12.2) and (12.3), with the OA and REF receiver noise v_{OA} and v_{REF}, respectively. The signals are matched by means of an adaptive

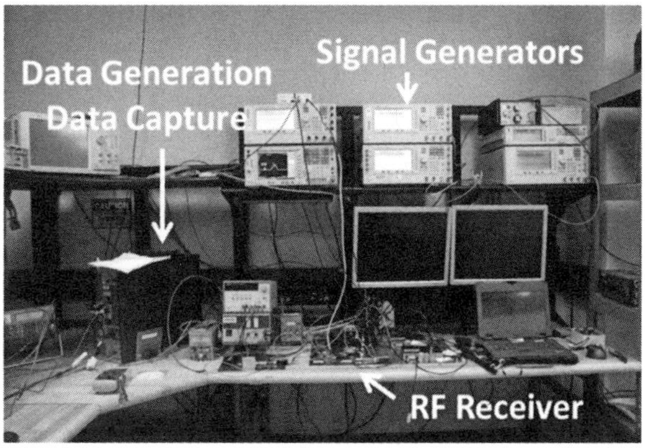

Figure 12.10 Laboratory setup for performing nonlinear equalization architecture identification.

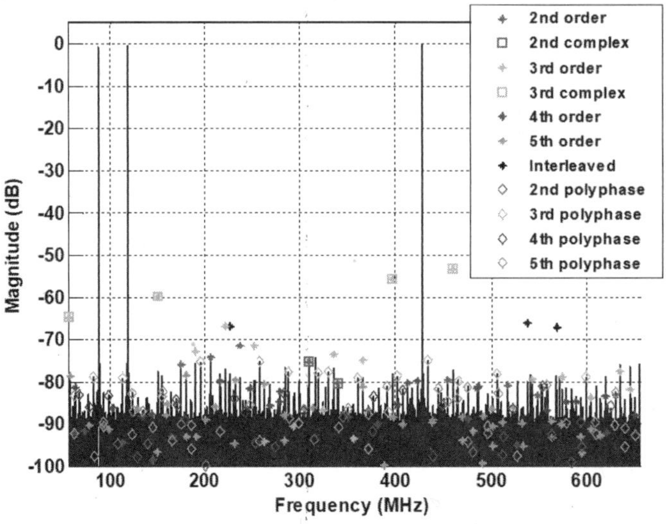

Figure 12.11 Measured nonlinear equalization performance. The icons and stems show the performance before and after equalization, respectively.

filter h_{EQ}, as shown in Figure 12.12. The error term $e = \hat{s} = y_{OA} - y_{REF} * h_{EQ}$ is an estimator for the signal of interest s.

By applying the adaptive filter approach, we make the usual assumptions that the signals y_{Tx}, v_{OA}, v_{REF}, and s are uncorrelated. The estimated equalizer h_{EQ} is a biased estimator of $h_{REF}^{-1} * h_{OA}$ for three reasons:

1. In general, $h_{REF}^{-1} * h_{OA}$ has infinite length, whereas h_{EQ} must be windowed.
2. The reference signal y_{REF} has measurement error v_{REF}.
3. The dual-channel receiver cannot fully suppress out-of-band energy from being aliased by the digitizer.

Regarding the first source of bias, we note that a system can be designed to ensure that h_{REF}^{-1} is as well-behaved as possible. The bandwidth of h_{REF} should be larger than that of h_{OA} and its stopband more gradual to discourage magnification of any frequency components of v_{REF} above the scale of v_{OA}. The response h_{REF} should also be as concentrated in time as possible. These design goals will reduce the windowing bias in h_{EQ}.

The second source of bias arises from reference channel noise v_{REF} (measurement error) and is commonly called regression dilution [10]. Common filter estimators will produce estimates h_{EQ}, which are biased towards zero compared with the ideal filter $h_{REF}^{-1} * h_{OA}$. However, the objective is not truly to estimate $h_{REF}^{-1} * h_{OA}$

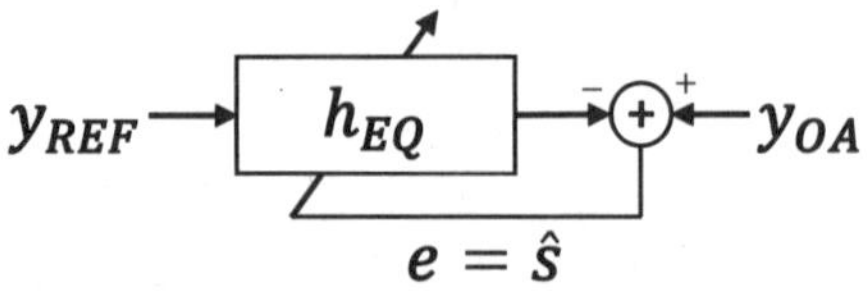

Figure 12.12 Adaptive filter block diagram for equalizer h_{EQ}.

but rather to find the least MSE estimator $\hat{s} = e = y_{OA} - h_{EQ} * y_{REF}$ for s. Regression dilution reduces the MSE of the estimate of s and is beneficial for STAR systems.

The third source of bias is aliasing by the ADCs. The process of frequency-folding from an infinite-bandwidth transfer function $H(f)$ to the discrete-time Fourier transform (DTFT) of a digital impulse response, $H_{2\pi}(\omega)$ with $\omega \in [0, \pi]$, is modeled by:

$$H_{2\pi}(\omega) = \sum_{n=0}^{\infty} H\left(\frac{\omega}{2\pi} f_s + n f_s\right) + H^{*}\left((n+1) f_s - \frac{\omega}{2\pi} f_s\right). \tag{12.19}$$

This means that, in the frequency domain, the digital equalizer will be biased towards a combination of the equalizers applicable to the many Nyquist regions that are aliased by the digitizer. In steady-state configurations, the effect is that digital cancellation is limited by the level of aliased SI, reinforcing the general guidance that an RF receiver should employ anti-aliasing filters. In cases where the SI power spectrum is nonstationary, more complicated bias effects may occur. Supposing that $h_{EQ} = h_{REF}^{-1} * h_{OA}$ holds, then the reconstruction error is $\hat{s} - s = v_{OA} - h_{EQ} * v_{REF}$.

A first approximation would be that both the REF and OA channels have equal SNR, and that the OA noise floor increases by 3 dB due to the combining of REF and OA noise. This approximation is useful but can be misleading for a few reasons. First, it is easier to control the gain of the REF channel than the OA channel because the OA propagation environment is continually changing. Second, the power available on the REF channel is typically much greater than that available at the input to the OA channel, which may simplify the design of a REF channel with high SNR. Third, in many applications, the PAPR of the transmitted waveform is optimized to reduce demands on the PA, which will improve achievable SNR on the REF channel. On the OA channel, such PAPR reduction will typically be degraded by the OA channel response (which, assuming good DP antenna isolation, will often approximate a Rayleigh channel model). For these reasons, in practical circumstances, the SNR of the OA channel may limit overall STAR performance.

12.2.4 Measured Performance

A full implementation of the NRL digital cancellation approach was tested in an anechoic chamber with static and dynamic multipath and two external signals of interest. The digital subsystem was combined with a custom-designed dual-channel receiver, high-power transmitter, and separate Tx/Rx antenna to provide a full STAR test system.

The dual-channel receiver was designed with COTS components, except for a custom absorptive filter designed by NRL, to demonstrate exceptionally low noise and spurious levels over a wide instantaneous bandwidth. The ADC is a 12-bit converter that operates at 1.8-GHz clock speed. The selected off-the-shelf ADC limits the performance of the system bandwidth to 700 MHz. Single-stage downconversion in wideband receivers generates mixer spurs that cannot be filtered [11]; two-stage conversion suppresses these spurs, so the receiver performs upconversion to

23.5 GHz followed by downconversion to the digitizer's second Nyquist frequency of 1.45 GHz.

Results shown in this section were generated using a 2 to 7-GHz cavity-embedded horn-loop bistatic STAR antenna designed at the University of Colorado Boulder, which relied on physical separation of 25 cm, high impedance, and absorber treatment. The antenna had 5 dBi of gain and achieved 70 to 80 dB of isolation. For additional details, the reader is referred to [1] and Chapter 2 of this book.

The transmitter generated SI driven by a high-power, MtronPTI gallium nitride (GaN) 100-W amplifier. To demonstrate tolerance to transmit nonlinearity, the amplifier was driven by a continuous broadband 700-MHz QPSK communications waveform without predistortion to its 3-dB compression point, where the power in-out curve is shown in Figure 12.13.

The full-duplex subsystems are pictured in Figure 12.14. The full-duplex antenna was seated in an anechoic chamber and was subject to both static and dynamic multipath, as well as the output of two transmitters on the other end of the chamber 10m away. The dynamic multipath generator was mounted on a motorized pedestal and operated by continuously rotating through 20° to 30°, resulting in a roughly 25-dB difference in amplitude depending on the position of the pedestal in azimuth, as shown in Figure 12.15.

The external signals were a 50-MHz binary phase-shift keying (BPSK) communications system and a 100-MHz linear frequency modulated (LFM) chirp radar signal. Before digital cancellation was applied, even the nominal 70 dB of RF isolation was insufficient to recover the communications signals, as shown in Figure 12.16(a). Digital cancellation further reduced SI by 50 dB across the full 700-MHz instantaneous bandwidth, and the communications signal was accurately received and demodulated (Figure 12.16(b)). These results have been previously disclosed in [11]. Note that RF distortions generated by the transmitter had no effect on digital cancellation performance.

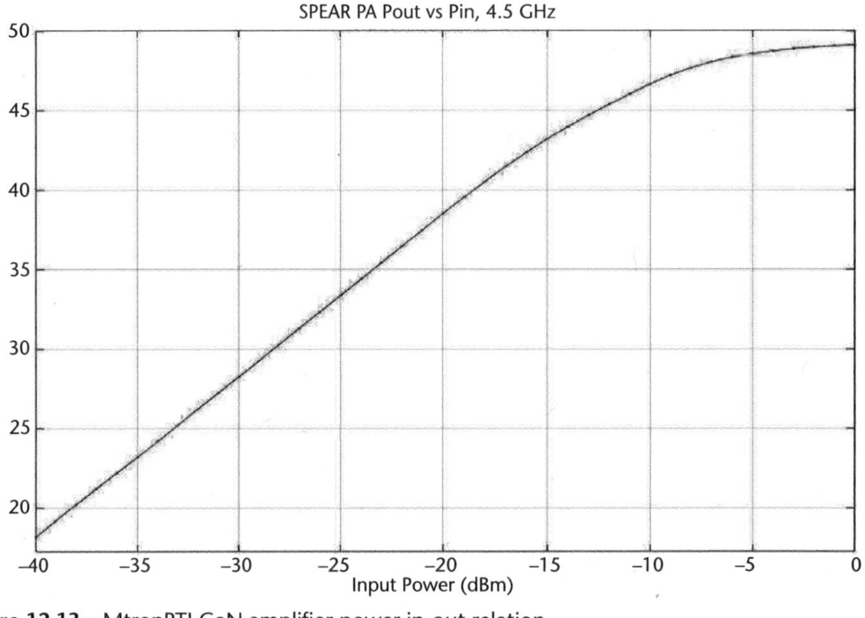

Figure 12.13 MtronPTI GaN amplifier power in-out relation.

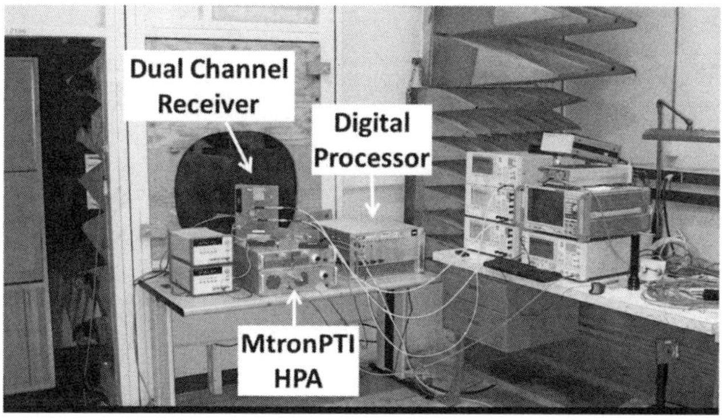

(a)

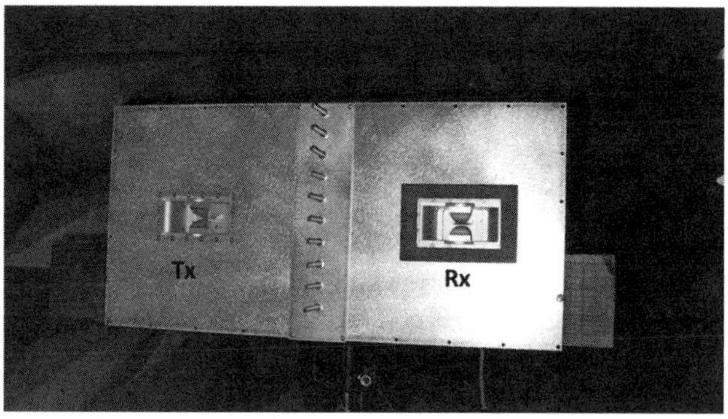

(b)

Figure 12.14 NRL full-duplex setup. (a) Transmitter and receiver subsystems; and (b) Tx and Rx antenna aperture.

12.3 Practical Considerations

Digital adaptive filter implementation has been well-studied, and the STAR system described here does not require any new adaptation method. Requirements typical of STAR applications may drive the choice of implementation. Specific requirements may include:

- Sample throughput may be quite high; Section 12.2.4 demonstrates cancellation of 700-MHz instantaneous bandwidth.
- The required filter may be both long (due to the wide bandwidths considered) and rapidly varying depending on the multipath environment under consideration.
- Some applications may be sensitive to latency.
- Traditional adaptive filter algorithms are feedback algorithms. For example, the least mean squares (LMS) algorithm requires one dot-product and one scalar-vector multiplication for each new input sample. Commercial FPGAs today cannot perform sample-by-sample filter adaptation at RF sample rates.

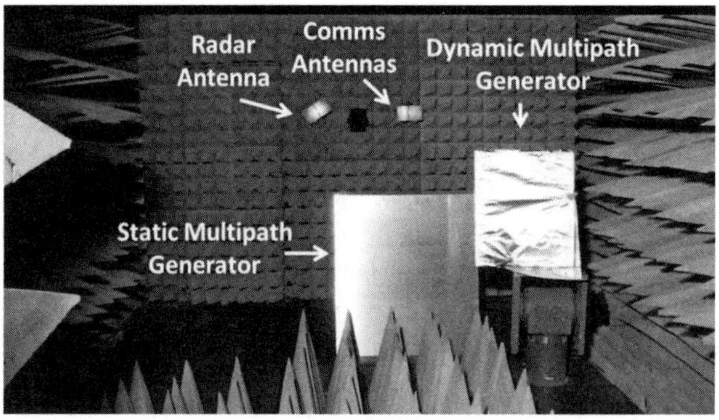

(a)

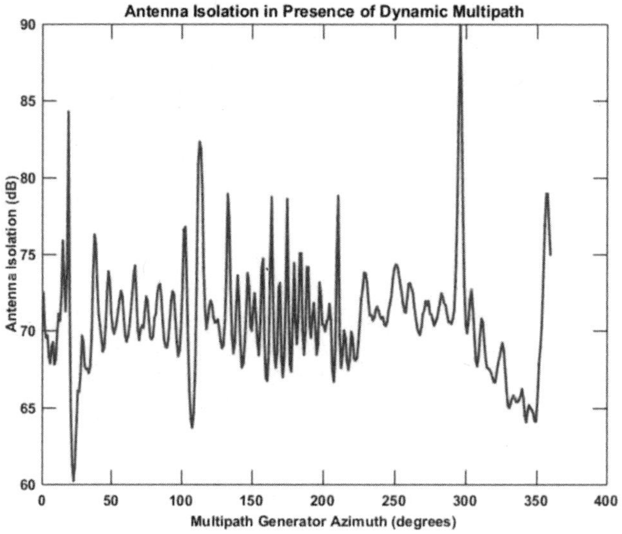

(b)

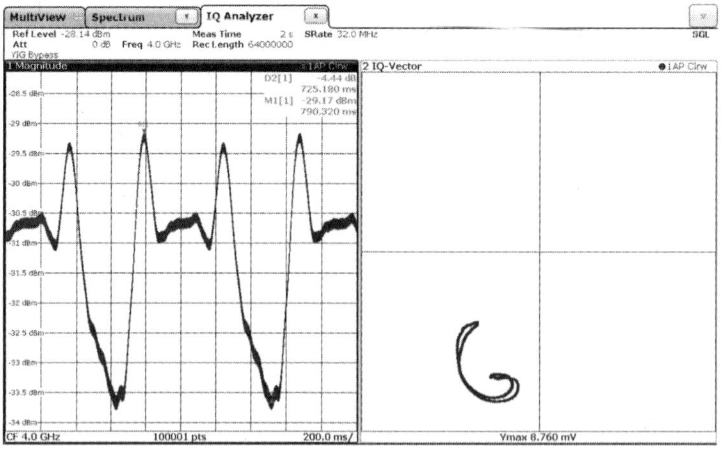

(c)

Figure 12.15 Static and dynamic multipath environment. (a) Multipath generators viewed from camera colocated with full-duplex antenna; (b) realized antenna isolation as dynamic multipath generator performs one full rotation; and (c) realized antenna isolation over time as dynamic multipath generator wobbles.

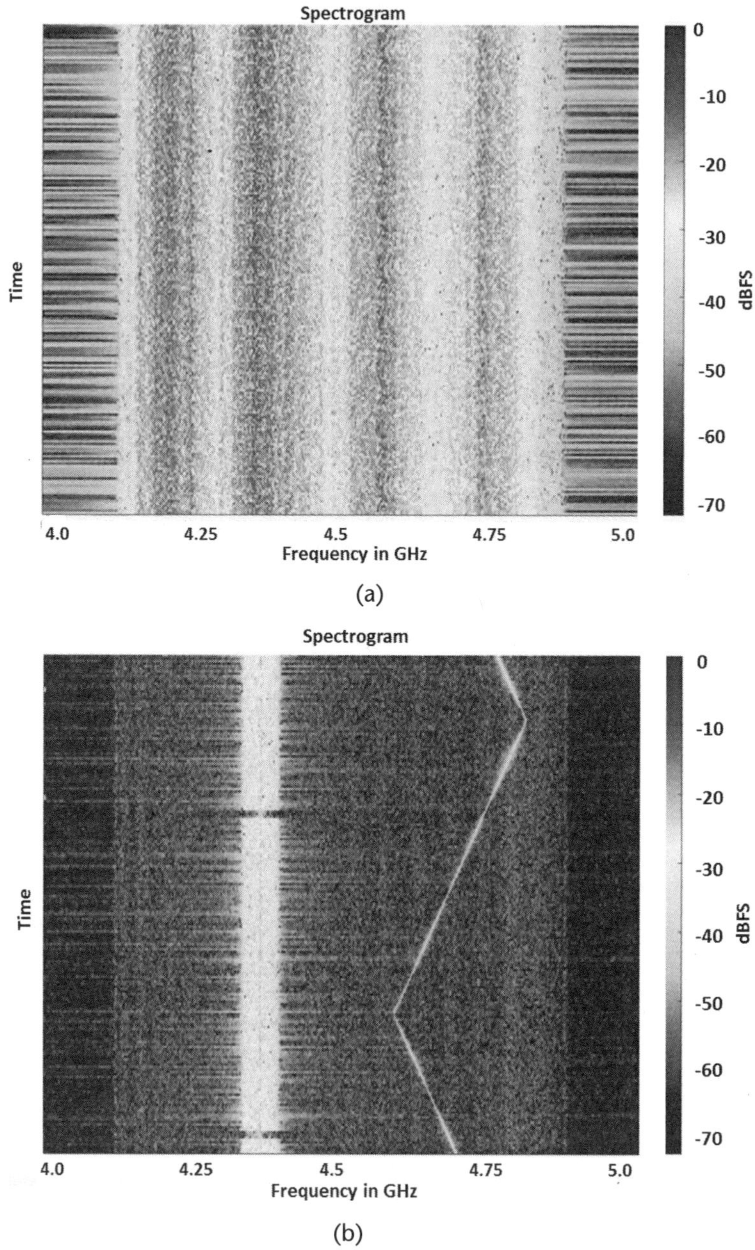

Figure 12.16 Received signal spectrogram: (a) before digital cancellation and (b) after digital cancellation.

For these reasons, block adaptive filter methods such as the fast block LMS (FBLMS) algorithm, as described in [12], are essential. FBLMS estimates the error gradient on blocks of data, which are sufficiently large so that convolution can be performed using fast Fourier transforms (FFTs). For block size L, an implementation must complete four FFTs in time less than L/fs. A low-latency FFT is a critical building block for high-throughput FBLMS implementations.

STAR processing uses the error signal from the LMS algorithm as the output of interference cancellation. However, this requires the reference signal to pass through two FFTs, adding latency. Latency-sensitive applications can apply a low-latency finite impulse response (FIR) filter separately, which may be either a direct time-domain implementation or low-latency fast-convolution implementation [13]. Latency from translating the filter coefficients into the time domain increases misadjustment error slightly, but the RF samples themselves will have interference canceled with minimal added delay.

Sub-band adaptive filtering is another possible approach, as described in [14]. In this approach, the input signal is decomposed into lower-rate frequency sub-bands (analysis), adaptive filtering is applied within each sub-band, and the full-rate signal is reconstituted (synthesis). This approach has the advantage that only feed-forward circuits operate at the full rate; the adaptation circuits operate at the reduced sub-band sample rate. A disadvantage of sub-band adaptive filters is latency: unlike FBLMS, there is no easy way to convert a filter from sub-band form into a low-latency direct form.

12.4 Conclusion

12.4.1 Summary

This chapter considered the process of digital cancellation for full-duplex operation in the presence of distortion generated by the RF and mixed-signal components of both the transmitter and receiver. Transmitter and receiver distortions can significantly limit cancellation performance and potentially prevent IBFD operation due to insufficient isolation. To ameliorate the effects of transmitter distortion, we used a two-channel receiver in which a sample of the transmitted signal was used as a template to cancel both the intended transmission and its distortion products at the receiver [15, 16]. As both the signal and distortion are sampled, the cancellation process is resistant to transmitter nonidealities. On receive, nonlinear equalization suppressed distortions generated by the RF and mixed signal components. In the OA testing, a combined 120 dB of cancellation was obtained, pushing interference to within 3 dB of the noise floor, which is the theoretical limit for a two-channel receiver.

12.4.2 Future Directions

One interesting direction of research is to apply STAR to an array system, as discussed in Chapters 13 and 14. The extension to an array-based approach is relatively straightforward but, if implemented naively, is computationally costly. Another active area of research involves the design of an analog canceller whose input is coupled to the transmitter and whose output is coupled to the receiver, as discussed in Chapters 8 and 9. Current implementations do not meet the challenging requirements for broadband high-power full-duplex operation in terms of both noise injection and/or nonlinear distortion.

References

[1] Elmansouri, M., et al., "0.5-45GHz Simultaneous Transmit and Receive (STAR) Antenna System for Electronic Attack," *Proceedings GomacTech Conference,* 2016.

[2] Goodman, J., et al., "Physical Layer Considerations for Wideband Cognitive Radio," *IEEE Military Communications Conference,* 2010, pp. 2113–2118.

[3] Qiu, J. X., et al., "Traveling-Wave Tube Amplifier Performance Evaluation and Design Optimization for Applications in Digital Communications with Multilevel Modulations," *IEEE Transactions on Microwave Theory and Techniques,* Vol. 51, No. 8, August 2003, pp. 1911–1919.

[4] Goodman, J., et al, "Polyphase Nonlinear Equalization of Time-Interleaved Analog-to-Digital Converters," *IEEE Journal of Selected Topics in Signal Processing*, Vol. 3, 2009.

[5] Walden, R. H., "Analog-to-Digital Converter Survey and Analysis," *IEEE Journal on Selected Areas in Communications,* Vol. 17, No. 4, April 1999, pp. 539–550.

[6] Razavi, B., *Principles of Data Conversion System Design*, New York: Wiley-IEEE Press, 1995.

[7] Morgan, D. R., et al., "A Generalized Memory Polynomial Model for Digital Predistortion of RF Power Amplifiers," *IEEE Transactions on Signal Processing,* Vol. 54, No. 10, October 2006, pp. 3852–3860.

[8] Herman, M., B. Miller, and J. Goodman, "The Cube Coefficient Subspace Architecture for Nonlinear Digital Predistortion," *The 42nd Asilomar Conference on Signals, Systems and Computers*, Pacific Grove, CA, 2008, pp. 1857–1861.

[9] Davis, C., et al., "Alias-Free Products of Signals Near Nyquist Rate," *IEEE Transactions on Signal Processing*, Vol. 66, No. 16, June 15, 2018, pp. 4151–4159.

[10] Frost, C., and S. G. Thompson. "Correcting for Regression Dilution Bias: Comparison of Methods for a Single Predictor Variable." *Journal of the Royal Statistical Society. Series A (Statistics in Society)*, Vol. 163, No. 2, 2000, pp. 173–189.

[11] Boglione, L., et al., "Wide Band, High Power, Same-Channel Full Duplex Transceiver System Demonstration," *2017 IEEE MTT-S International Microwave Symposium (IMS),* Honololu, HI, 2017, pp. 684–686.

[12] Farhang-Boroujeny, B., "Block Implementation of Adaptive Filters," in *Adaptive Filters: Theory and Applications*, 2nd ed., New York: Wiley, 2013.

[13] Gardner, W. G., "Efficient Convolution Without Input-Output Delay," *J. AES,* Vol. 43, No. 3, March 1995, pp. 127–136.

[14] Farhang-Boroujeny, B., "Subband Adaptive Filters," in *Adaptive Filters: Theory and Applications*, 2nd ed., New York: Wiley, 2013.

[15] Goodman, J., et al., "Extending the Dynamic Range of RF Receivers Using Nonlinear Equalization," *IEEE International Waveform Diversity and Design Conference,* Kissimmee, FL, 2009, pp. 224–228.

[16] Lorenz, K. S., et al., "Generalized Transmitter Compensation of Frequency Dependent I/Q Imbalance," *IEEE Transactions on Signal Processing,* Vol. 64, No. 9, May 1, 2016, pp. 2220–2231.

CHAPTER 13

Phased Array Systems

Jonathan Doane, Ian T. Cummings, Kenneth E. Kolodziej, and Bradley T. Perry

13.1 Introduction

While significant STAR work has been presented throughout this book, the primary focus has been on low-channel-count, fixed-beam antenna systems utilizing techniques that impose limitations on system performance when channel count increases. In this chapter, the focus is shifted to the extension of techniques for phased array antennas, as highlighted within the IBFD transceiver diagram of Figure 13.1(a) and the SIC techniques tree of Figure 13.1(b), including considerations for beam scanning and impacts to system performance and complexity. STAR phased arrays must provide high isolation between transmit and receive functions, while forming directional transmit and receive beams that can be electronically steered to desired locations within some scan volume. Phased arrays are generally utilized for applications with demanding link budgets that require very high effective radiated power and/or sensitive reception of very low-power incident signals. The total isolation required to realize a practical STAR phased array may therefore be substantially higher than that required for an omnidirectional system.

For the purpose of discussion, STAR and IBFD will both be assumed to mean reception of small signals in the presence of high-power transmissions at the same frequency and time with minimal degradation in system performance. Although this does not preclude the application of the techniques discussed herein to out-of-band emission control, this is not the focus of the discussion. STAR and IBFD systems use various techniques to mitigate SI and enable simultaneous operation of colocated transmit and receive systems in the same frequency band. Applicability of each SI mitigation approach must be carefully balanced with system trades to

DISTRIBUTION STATEMENT A. Approved for public release. Distribution is unlimited. This material is based upon work supported by the Under Secretary of Defense for Research and Engineering under Air Force Contract No. FA8702-15-D-0001. Any opinions, findings, conclusions or recommendations expressed in this material are those of the author(s) and do not necessarily reflect the views of the Under Secretary of Defense for Research and Engineering.

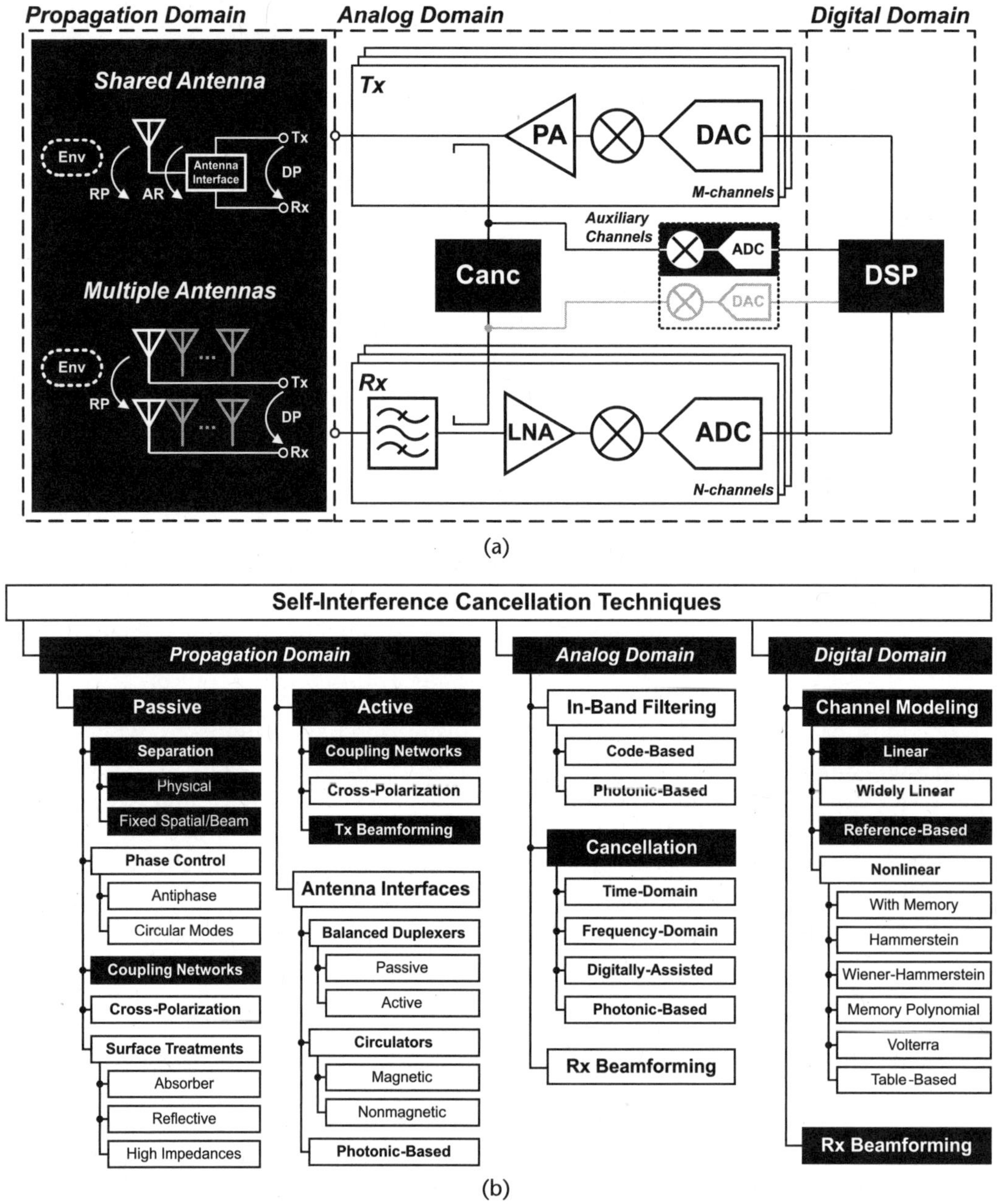

Figure 13.1 Phased array systems and specific chapter focus highlighted within (a) a generic IBFD transceiver and (b) a SIC techniques tree diagram.

create a STAR phased array system. Of great importance in the STAR array trade space is the class of phased array system being considered.

Throughout this chapter, techniques applicable to several classes of STAR phased arrays will be covered, including separate-aperture STAR, aperture-level STAR (ALSTAR), and element-level STAR (ELSTAR). Separate-aperture STAR refers to the use of separate transmit and receive arrays with physical spacing providing the primary SI mitigation, as utilized for the prototype discussed in Chapter 12. Passive surface treatments may also be used to reduce propagation of surface waves along a common ground plane. ALSTAR is a single aperture approach that allows for reconfiguration of transmit and receive portions of a common aperture during operation, and may utilize beamforming techniques to improve isolation between sections of the array. Both separate-aperture STAR and ALSTAR allow for grouping of elements with dedicated transmit and receive functions. In contrast, ELSTAR arrays can transmit and receive from the same set of elements and thus require SI mitigation at the individual element level. This generally necessitates specialized front-end components, including analog-based cancellers akin to those used in low-channel-count systems, whose performance parameters must be carefully balanced against system requirements, as discussed in Chapters 8 through 10. For these reasons, ELSTAR continues to be an active research area, motivating the need for further development of novel high-performance and highly integrated SIC technologies.

When considering the use of STAR in phased arrays, it is important to understand the associated challenges involved. An electronically scanned beam imposes fundamental constraints on element spacing and array geometry, and these parameters cannot generally be independently designed for optimal isolation while still providing desired scan performance. In other words, lattice spacing inherently constrains the size, weight, and power (SWaP) available for SI mitigation. Additionally, each channel in the array may represent an independent source of noise and distortion, requiring special techniques to mitigate multiple uncorrelated SI components. A phased array with numerous channels therefore requires a method for high-dimensional cancellation and consideration of the associated SWaP and computational complexity. All of these challenges require attention when implementing a STAR phased array for a particular application and should be carefully considered with the overall system performance goals.

13.2 Isolation Metrics for Phased Arrays

A critical metric for any STAR/IBFD system is the system isolation, which characterizes the residual SI due to undesired coupling of the transmitted signal into the receive path. The electronic isolation (I_e) of a system is defined as

$$I_e = \frac{P_t}{P_d}\mathrm{SINR}_o, \tag{13.1}$$

where P_t is the transmitted power, P_d is the received power of a desired external signal, and SINR_o is the resulting signal-to-interference-plus-noise ratio (SINR) after any SIC or other interference mitigation has been applied. Although this metric is commonly used for single-channel systems, it has several drawbacks and is not well-suited for characterizing phased arrays. First, it does not account for the antenna gain (or loss) experienced by the desired signals. Low gain or lossy antennas permit trivial improvement of I_e without any actual increase in STAR/IBFD performance. Conversely, high-gain antenna systems such as phased arrays can improve the desired link independently of the SI coupling, resulting in an improvement of effective isolation not reflected by (13.1).

A second limitation arises in extending (13.1) to multichannel systems. It is not immediately obvious how P_d should be defined when there are multiple receive channels. A consistent definition requires

$$P_d = P_d^{iso} G_r^a, \tag{13.2}$$

where G_r^a is the receive antenna gain in the desired direction and $P_d^{iso} = W_d \lambda^2/4\pi$ is the equivalent power that would be received by an ideal isotropic antenna, with W_d the incident power density and λ the wavelength. However, applying (13.2) to (13.1) is rather unsatisfying for multichannel systems because it implies that P_d no longer represents a physical power level at the input to the receiver(s). The received signal may undergo amplification, SIC, and/or digitization prior to the realization of coherent array gain, and it is awkward to define the input receiver power as if antenna gain had already been applied.

An alternate metric, effective isotropic isolation (EII), avoids these shortcomings by including the response of the antenna to the desired signals [1],

$$\text{EII} = \frac{\text{EIRP}}{P_d^{iso}} \text{SINR}_o. \tag{13.3}$$

As discussed above, P_d^{iso} represents the incident power density of the desired signal at the receive antenna. The transmitted power is similarly generalized as effective isotropic radiated power (EIRP), with $\text{EIRP} = P_t G_t^a$ and G_t^a the transmit antenna gain in the desired direction. In contrast to I_e, which characterizes the isolation at the interface between the antenna and electronics, EII characterizes the isolation relative to the externally propagating signals and thus the antenna (or array) is included as an intrinsic component of the system. The EII metric provides a consistent, meaningful, and useful measure of isolation for both single-channel and multichannel systems regardless of implementation. Note that

$$\text{EII} = G_t^a G_r^a I_e, \tag{13.4}$$

and that EII and I_e are equivalent for isotropic systems with unity transmit and receive antenna gain.

13.3 Signal Model for Phased Arrays

Active phased arrays involve complex multichannel coupling environments with many signal and noise components that must be properly accounted for. In this section, a simple but useful signal model is developed to characterize the signal and noise components in a STAR phased array. To minimize complexity, we assume linear, narrowband, additive white Gaussian noise (AWGN) channels, which are useful for first-order analysis and understanding of basic principles. This approach can be extended to wideband and/or multisignal models without altering the essential concepts and results. The reader is also referred to Chapters 11 and 12 for discussion of nonlinear channel modeling and compensation.

Consider a phased array with J transmitting elements and K receiving elements, as depicted in Figure 13.2. In general, the transmitting and receiving channels may share a common set of radiating elements (ELSTAR). Alternately, the transmitting and receiving elements may be partially overlapping or may be fully disjoint as in ALSTAR and separate-aperture STAR.

The transmitted signal vector $\boldsymbol{t}(n) \in \mathbb{C}^{J\times 1}$ at time n is

$$\boldsymbol{t}(n) = \boldsymbol{b}_t x(n) + \boldsymbol{n}_t(n), \tag{13.5}$$

where $x(n) \in \mathbb{C}$ is the desired signal to be transmitted with $E\left[\left|x(n)\right|^2\right] = 1$, and $\boldsymbol{b}_t \in \mathbb{C}^{J\times 1}$ is the vector of transmit beamforming coefficients with total transmitted power $P_t = \boldsymbol{b}_t^H \boldsymbol{b}_t$. The transmitter noise $\mathbf{n}_t(n)$ has covariance $\mathbf{N}_t \triangleq E\left[\mathbf{n}_t \mathbf{n}_t^H\right]$. The far-field EIRP of the transmitted beam in the direction (θ_t, ϕ_t) is

$$\text{EIRP}(\theta_t, \phi_t) = g(\theta_t, \phi_t)\boldsymbol{b}_t^H \boldsymbol{q}_t(\theta_t, \phi_t)\boldsymbol{q}_t^H(\theta_t, \phi_t)\boldsymbol{b}_t, \tag{13.6}$$

where the embedded antenna gain $g(\theta_t, \phi_t)$ is assumed to be common across all elements. $q_t(\theta_t, \phi_t) \in \mathbb{C}^{J\times 1}$ is the far-field steering vector for a beam at angle (θ_t, ϕ_t), which for a planar array is

$$\boldsymbol{q}_t(\theta_t, \phi_t) = \exp\left[-j\frac{2\pi}{\lambda}\left(\boldsymbol{x}_t \cos(\phi_t)\sin(\theta_t) + \boldsymbol{y}_t \sin(\phi_t)\sin(\theta_t)\right)\right], \tag{13.7}$$

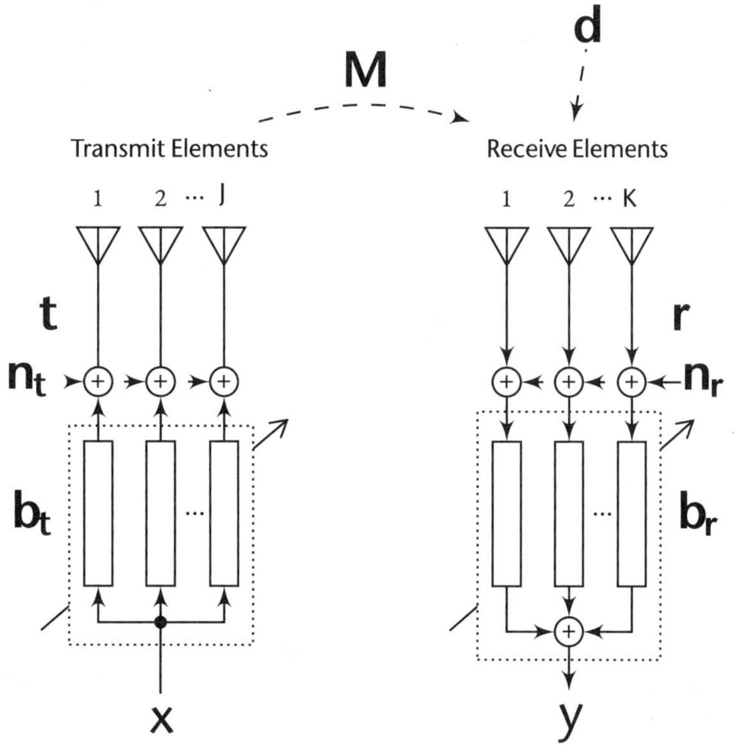

Figure 13.2 Basic STAR phased array signal model.

where $\boldsymbol{x}_t$ and $\boldsymbol{y}_t$ are the x- and y-positions of each transmitting antenna element in the array plane.

On receive, the signal $\boldsymbol{r}(n) \in \mathbb{C}^{K\times 1}$ incident at each receiving element is

$$\boldsymbol{r}(n) = \boldsymbol{d}(n) + \boldsymbol{i}(n) + \boldsymbol{M}\boldsymbol{t}(n), \tag{13.8}$$

where $\boldsymbol{d}(n) \in \mathbb{C}^{K\times 1}$ represents an external desired signal, $\boldsymbol{i}(n) \in \mathbb{C}^{K\times 1}$ represents any external undesired signal(s), and $\boldsymbol{M} \in \mathbb{C}^{K\times J}$ describes the mutual coupling from transmitting elements to receiving elements. If transmit and receive channels share common radiating elements via a three-port network such as a circulator (as discussed in Chapter 4), the mutual coupling $\boldsymbol{M}$ and antenna gain $g(\theta, \phi)$ are defined to include the response of the three-port network.

Individual received signals are combined to form a beam $y(n)$

$$y(n) = \boldsymbol{b}_r^H \left[\boldsymbol{d}(n) + \boldsymbol{i}(n) + \boldsymbol{M}\boldsymbol{t}(n) + \boldsymbol{n}_r(n)\right], \tag{13.9}$$

where $\boldsymbol{b}_r \in \mathbb{C}^{K\times 1}$ is the vector of beamforming coefficients with $\boldsymbol{b}_r^H \boldsymbol{b}_r = 1$, and $\boldsymbol{n}_r(n)$ is the receiver noise with covariance $N_r \triangleq E\left[\mathbf{n}_r \mathbf{n}_r^H\right]$. The far-field receive array gain for a signal incident from angle (θ_r, ϕ_r) is given by

$$G_R(\theta_r, \phi_r) = g(\theta_r, \phi_r)\boldsymbol{b}_r^H \boldsymbol{q}_r(\theta_r, \phi_r)\boldsymbol{q}_r^H(\theta_r, \phi_r)\boldsymbol{b}_r, \tag{13.10}$$

where $\boldsymbol{q}_r(\theta_r, \phi_r)$ is the receive steering vector given by

$$\boldsymbol{q}_r(\theta_r, \phi_r) = \exp\left[j\frac{2\pi}{\lambda}\left(\boldsymbol{x}_r cos(\phi_r) sin(\theta_r) + \boldsymbol{y}_r \sin(\phi_r)\sin(\theta_r)\right)\right]. \tag{13.11}$$

The received power of the desired signal arriving from angle (θ_d, ϕ_d) is then

$$P_D = P_d^{iso} G_R(\theta_d, \phi_d) \tag{13.12}$$

Provided $x(n)$, $\boldsymbol{d}(n)$, $\boldsymbol{i}(n)$, $\boldsymbol{n}_t(n)$, and $\boldsymbol{n}_r(n)$ are all mutually uncorrelated, the output SINR of the received beam (without additional SI mitigation) is

$$\text{SINR}_o = \frac{P_D}{P_I + P_X + P_{NT} + P_{NR}}, \tag{13.13}$$

where P_I is the external interference power, $P_X = \boldsymbol{b}_r^H \boldsymbol{M} \boldsymbol{b}_t \boldsymbol{b}_t^H \mathbf{M}^H \boldsymbol{b}_r$ is the SI power due to the transmitted signal x, $P_{NT} = \boldsymbol{b}_r^H \boldsymbol{M} \boldsymbol{N}_t \boldsymbol{M}^H \boldsymbol{b}_r$ is the received SI power due to transmitter noise, and $P_{NR} = \boldsymbol{b}_r^H \mathbf{N}_r \boldsymbol{b}_r$ is the total receiver noise power.

The transmit and receive noise covariance matrices may be expressed as

$$\boldsymbol{N}_t = \eta_t^{-1} Diag\left(\boldsymbol{b}_t \boldsymbol{b}_t^H\right) \tag{13.14}$$

$$N_r = \eta_r^{-1} Diag\left(E\left[\boldsymbol{r}\boldsymbol{r}^H\right]\right) + \sigma_{nr}^2 \boldsymbol{I}, \tag{13.15}$$

where η_t and η_r are, respectively, the maximum SNR of the transmit and receive channels under high-power excitation, and σ_{nr}^2 is the receiver thermal noise power under low-power excitation. Here, we are assuming that each receive channel has sufficient gain control so as to avoid saturation and can maintain maximum SNR in the presence of high-power signals.

To characterize EII, we consider a scenario without external interference and where the desired signal power is small relative to SI. Under these conditions, the receiver noise covariance becomes

$$N_r = \eta_r^{-1} Diag\left(\boldsymbol{M}\boldsymbol{b}_t\boldsymbol{b}_t^H\boldsymbol{M}^H\right) + \eta_r^{-1}\eta_t^{-1} Diag\left(\boldsymbol{M} Diag\left(\boldsymbol{b}_t\boldsymbol{b}_t^H\right)\boldsymbol{M}^H\right) + \sigma_{nr}^2 \boldsymbol{I}. \tag{13.16}$$

The total receiver noise power is then

$$P_{NR} = P_{NR}^{x} + P_{NR}^{nt} + \sigma_{nr}^2, \tag{13.17}$$

where $P_{NR}^{x} = \eta_r^{-1}\boldsymbol{b}_r^H Diag\left(\boldsymbol{M}\boldsymbol{b}_t\boldsymbol{b}_t^H\boldsymbol{M}^H\right)\boldsymbol{b}_r$ represents the elevated receiver noise power caused by the high-power transmit signal and $P_{NR}^{nt} = \eta_r^{-1}\eta_t^{-1}\boldsymbol{b}_r^H Diag\left(\boldsymbol{M} Diag\left(\boldsymbol{b}_t\boldsymbol{b}_t^H\right)\boldsymbol{M}^H\right)\boldsymbol{b}_r$ represents the elevated receiver noise caused by the high-power transmitter noise.

From (13.3) and (13.13), we can now assemble an overall expression for the EII of a STAR phased array

$$EII = \frac{EIRP(\theta_t,\phi_t)G_R(\theta_d,\phi_d)}{P_X + P_{NT} + P_{NR}^{x} + P_{NR}^{nt} + \sigma_{nr}^2}. \tag{13.18}$$

The residual SI and noise power terms in the denominator of (13.18) are summarized in Table 13.1. Up to this point, we have not discussed any specific design features to mitigate SI. In the following section, we will introduce several such techniques.

Table 13.1 SI and Noise Components in a STAR Phased Array

Term	*Description*	*SI and Noise Power*
P_X	Residual transmit signal in receive beam	$b_r^H M b_t b_t^H M^H b_r$
P_{NT}	Residual transmit noise in receive beam	$\eta_t^{-1} b_r^H M Diag\left(b_t b_t^H\right) M^H b_r$
P_{NR}^{x}	Elevated receive noise due to high-power transmit signal	$\eta_r^{-1} b_r^H Diag\left(M b_t b_t^H M^H\right) b_r$
P_{NR}^{nt}	Elevated receive noise due to high-power transmit noise	$\eta_r^{-1}\eta_t^{-1} b_r^H Diag\left(M Diag\left(b_t b_t^H\right) M^H\right) b_r$
σ_{nr}^2	Baseline receiver thermal noise floor	σ_{nr}^2

13.4 Self-Interference Mitigation in Phased Arrays

In multichannel arrays, correlated signals and uncorrelated noise experience somewhat different net coupling mechanisms and an effective design must take both into account. Although the simple AWGN model developed in the previous section does not account for more complex effects such as nonlinear distortion and phase noise, such RF impairments may often be mitigated in the same manner as uncorrelated AWGN. In this section, we discuss application of SIC and spatial cancellation via beamforming for STAR phased arrays.

13.4.1 Self-Interference Cancellation

SIC is perhaps the primary technique used by most STAR/IBFD systems to mitigate SI and is discussed extensively elsewhere in this book. This technique combines a filtered copy of the interfering signal with the received signal in order to achieve cancellation via destructive interference. The accuracy of the SIC system determines how well the cancellation channel can mimic the SI coupling channel, and thus establishes the maximum level of realizable cancellation. The dynamic range of the SIC system determines the SI power handling and amount of added noise or distortion generated by the system itself. Finally, the efficiency of the SIC system characterizes additional losses introduced in the transmit and/or receive signal paths due to SIC implementation. Whereas residual SI resulting from limited accuracy may in principle be removed by subsequent cancellation stages, impairments due to reduced efficiency and finite dynamic range are generally difficult or impossible to repair. The development of effective SIC for phased arrays requires consideration of all of these design factors, compounded by the multiplicity of transmit and receive channels interacting within a complex coupling environment. Next, we discuss how analog, digital, and hybrid SIC architectures may be used for STAR phased arrays.

13.4.1.1 Analog SIC

A multichannel analog cancellation network $\boldsymbol{H}_a \in \mathbb{C}^{K \times J}$ may by integrated within a STAR phased array, as shown in Figure 13.3, and is similar to those discussed in Chapters 8 and 9. By coupling a portion of each transmitted signal, applying appropriate filtering and recombining with the received signals, SI can, in principle, be cancelled. The received signal model from (13.9) is modified to be

$$y(n) = \boldsymbol{b}_r^H \left[\boldsymbol{d}(n) + \boldsymbol{i}(n) + \boldsymbol{Mt}(n) + \boldsymbol{H}_a \boldsymbol{t}(n) + \boldsymbol{H}_a \boldsymbol{n}_a(n) + \boldsymbol{n}_r(n) \right], \tag{13.19}$$

where $\boldsymbol{n}_a(n)$ represents added noise due to the finite dynamic range η_a of the cancellation electronics, with covariance $\mathbf{N}_a \triangleq E\left[\boldsymbol{n}_a \boldsymbol{n}_a^H\right] = \eta_a^{-1} Diag\left(\boldsymbol{b}_t \boldsymbol{b}_t^H\right)$. With analog SIC present, the net coupling becomes $\boldsymbol{M} + \boldsymbol{H}_a$ and all of the SI terms in Table 13.1 then depend on this modified net coupling. An additional SI term $P_{NA} = \boldsymbol{b}_r^H \boldsymbol{H}_a \mathbf{N}_a \boldsymbol{H}_a{}^H \boldsymbol{b}_r$ is also introduced to represent canceller noise. This model does not explicitly account for reduced efficiency due to SIC coupling, and so any loss or phase shift introduced into the transmit and receive paths should be incorporated into the mutual coupling $\boldsymbol{M}$ and element gain $g(\theta, \phi)$.

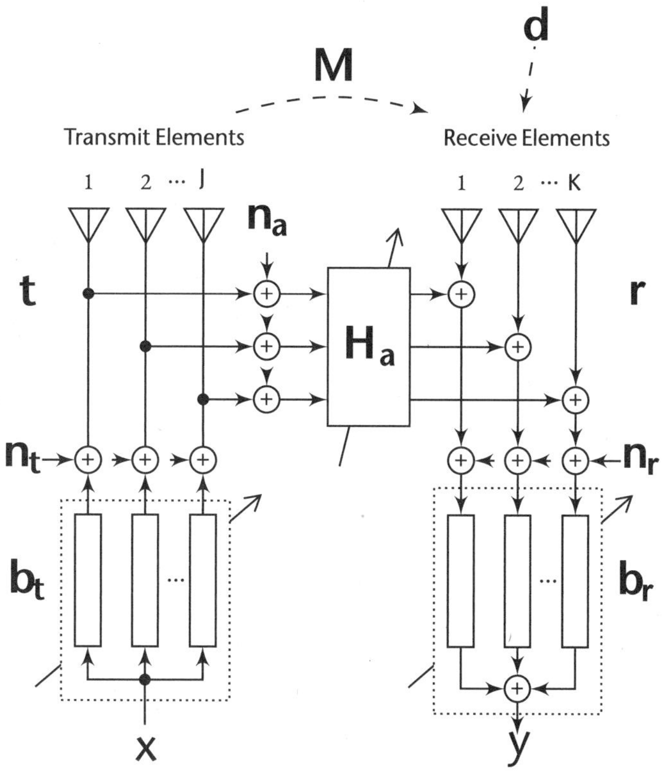

Figure 13.3 STAR phased array signal model with analog SIC network Ha.

If the canceller network could be tuned to perfectly match the full mutual coupling matrix, then all SI terms would be cancelled, leaving only the canceller noise P_{NA} and thermal noise σ_{nr}^2. We refer to this matching objective as SIC Type A, which represents a theoretically ideal analog SIC strategy.

SIC Type A:

$$\boldsymbol{H}_a \rightarrow -\boldsymbol{M} \tag{13.20}$$

Various attempts have been made to approximate such networks directly within the array feed itself by inserting additional coupling paths between neighboring elements [2–4]. Mutual coupling compensation networks (MCCNs) are generally passive and therefore do not generate additional noise (i.e., $P_{NA} = 0$). However, these simple static networks generally have poor accuracy and narrow bandwidth, resulting in limited isolation improvement. SIC Type A cancellation networks have also been proposed for small STAR arrays using multichannel active tunable analog filters [5]. This approach of directly matching the pairwise coupling between all elements in the array quickly becomes infeasible, as the array grows beyond a handful of elements.

Rather than attempting to perfectly reconstruct the complete coupling matrix $\boldsymbol{M}$, lower-rank approximations of (13.20) may be implemented with simpler SIC networks. By restricting $\boldsymbol{H}_a$ to a diagonal form, each receive channel requires only a single coupling path from a corresponding transmit channel. For ELSTAR arrays, such a network may be implemented using local canceller circuits that are self-contained within each element. Two possible methods for optimizing such designs are discussed below.

A reasonable goal for optimizing H_a would be to minimize the transmitted signal x at the input to each receive channel. This goal is satisfied by the following matching objective, which we denote SIC Type B:

SIC Type B:

$$\boldsymbol{H}_a \boldsymbol{b}_t \rightarrow -\boldsymbol{M}\boldsymbol{b}_t \tag{13.21}$$

SIC Type B is a rank-1 constraint, which for $J = K$ admits a diagonal solution,

SIC Type B, diagonal form:

$$\boldsymbol{H}_a \rightarrow -Diag(\boldsymbol{b}_t)^{-1} Diag(\boldsymbol{M}\boldsymbol{b}_t) \tag{13.22}$$

As $\boldsymbol{H}_a$ is tuned to satisfy (13.22), the transmitted signal x coupling into each receive channel will be cancelled. This is equivalent to matching the active reflection coefficient at each element [6]. Unfortunately, the transmitted noise may not be similarly cancelled and terms P_{NT} and P_{NR}^{nt} may remain high.

Rather than attempting to cancel the incident transmit signal at the individual receiver inputs, the canceller can instead be tuned to reduce the overall SI at the output of the receive beamformer. This objective will be satisfied under a third matching objective, referred to as SIC Type C.

SIC Type C:

$$\boldsymbol{b}_r^H \boldsymbol{H}_a \rightarrow -\boldsymbol{b}_r^H \boldsymbol{M} \tag{13.23}$$

Under this condition, the net SIC coupling will match the net SI at the output of the receive beamformer, resulting in reduction of both the transmitted signal and the uncorrelated transmitter noise. As with SIC Type B, (13.23) admits a diagonal solution, which requires only a local cancellation circuit within each element of an ELSTAR array.

SIC Type C, diagonal form:

$$\boldsymbol{H}_a \rightarrow -Diag(\boldsymbol{b}_r^H)^{-1} Diag(\boldsymbol{b}_r^H \boldsymbol{M}) \tag{13.24}$$

Although the signal model assumes linear AWGN channels, SIC Type C will also cancel transmitter nonlinearities and distortion products, which in practice may significantly exceed random noise. The drawback is that, unlike SIC Type B, the incident power is not minimized into each receive channel. Thus, although the interference power terms P_X and P_{NT} may theoretically be cancelled, the receiver noise terms P_{NR}^{x} and P_{NR}^{nt} may remain elevated.

Finally, we note that SIC Types B and C are equivalent for the symmetrical case where $\boldsymbol{M} = \boldsymbol{M}^T$ and $\boldsymbol{b}_r^H = \boldsymbol{b}_t^T$. This condition is met, for example, by an ELSTAR array with transmit and receive beams pointing in the same direction. In this case, a simple diagonal canceller can simultaneously cancel the signal at each receiver as well as the total interference and noise in the receive beam. The only residual SI terms will then be P_{NR}^{x}, the elevated receiver noise due to high-power transmit noise, and P_{NA}, the noise generated by the cancellation network itself. The three analog SIC types discussed above are summarized in Table 13.2.

Simulated results for a 50-element ELSTAR array described in Figure 13.4 are provided in Figure 13.5. For each SIC type, we assume the analog SIC system has perfect accuracy and efficiency but finite dynamic range η_a of 80 dB. These simulations therefore represent a theoretical upper bound on the possible performance of each of the analog SIC types for this array configuration and dynamic range. We observe that, even with perfect cancellation and very high dynamic range, the receiver is still significantly desensitized by SI. In practice, implementation of analog SIC networks with high accuracy, efficiency, dynamic range, and sufficiently small SWaP for integration within phased array electronics remains a substantial engineering challenge, and practical analog SIC for arrays with more than a handful of elements has yet to be demonstrated. An alternate approach of performing SIC within the digital domain is discussed in the following section.

13.4.1.2 Digital SI Cancellation

Whereas analog SIC is implemented using physical couplers and filtering circuits, digital SIC uses digital filtering to combine the transmitted waveform $x(n)$ with the digital received signal,

$$y(n) = \boldsymbol{b}_r^H \left[\boldsymbol{d}(n) + \boldsymbol{i}(n) + \boldsymbol{M}\boldsymbol{b}_t x(n) + \boldsymbol{M}\boldsymbol{n}_t(n) + \boldsymbol{n}_r(n) \right] + h_d x(n), \tag{13.25}$$

If the filter coefficient h_d is tuned such that $h_d \rightarrow -\boldsymbol{b}_r^H \boldsymbol{M}\boldsymbol{b}_t$, then the transmitted signal $x(n)$ will be cancelled[2]. This may be implemented using any number of adaptive digital filtering algorithms such as least mean squares (LMS) or recursive least squares (RLS) [7]. Digital filtering enables precise, fast, repeatable, and adaptive cancellation far beyond what is generally possible with analog circuits. Adaptive filtering algorithms can compensate in real time to dynamic changes to the coupling and scattering environment. Digital distribution and combining of signals do not

Table 13.2 Residual SI and Noise Power Terms for STAR Phased Array Using Analog SIC Network H_a with Dynamic Range η_a

		SIC Type A	*SIC Type B*	*SIC Type C*
Term	*Residual SI Power under Analog SIC*	$\boldsymbol{H}_a \rightarrow -\boldsymbol{M}$	$\boldsymbol{H}_a\boldsymbol{b}_t \rightarrow -\boldsymbol{M}\boldsymbol{b}_t$	$\boldsymbol{b}_r^H\boldsymbol{H}_a \rightarrow -\boldsymbol{b}_r^H\boldsymbol{M}$
P_X	$\boldsymbol{b}_r^H(\boldsymbol{M}+\boldsymbol{H}_a)\boldsymbol{b}_t\boldsymbol{b}_t^H(\boldsymbol{M}+\boldsymbol{H}_a)^H\boldsymbol{b}_r$	Cancelled	Cancelled	Cancelled
P_{NT}	$\eta_t^{-1}\boldsymbol{b}_r^H(\boldsymbol{M}+\boldsymbol{H}_a)Diag(\boldsymbol{b}_t\boldsymbol{b}_t^H)(\boldsymbol{M}+\boldsymbol{H}_a)^H\boldsymbol{b}_r$	Cancelled	Cancelled if symmetrical	Cancelled
P_{NR}^x	$\eta_r^{-1}\boldsymbol{b}_r^H Diag\left((\boldsymbol{M}+\boldsymbol{H}_a)\boldsymbol{b}_t\boldsymbol{b}_t^H(\boldsymbol{M}+\boldsymbol{H}_a)^H\right)\boldsymbol{b}_r$	Cancelled	Cancelled	Cancelled if symmetrical
P_{NR}^{nt}	$\eta_r^{-1}\eta_t^{-1}\boldsymbol{b}_r^H Diag\left((\boldsymbol{M}+\boldsymbol{H}_a)Diag(\boldsymbol{b}_t\boldsymbol{b}_t^H)(\boldsymbol{M}+\boldsymbol{H}_a)^H\right)\boldsymbol{b}_r$	Cancelled	Not cancelled	Not cancelled
P_{NA}	$\eta_a^{-1}\boldsymbol{b}_r^H\boldsymbol{H}_a Diag(\boldsymbol{b}_t\boldsymbol{b}_t^H)\boldsymbol{H}_a{}^H\boldsymbol{b}_r$	Introduced	Introduced	Introduced

2. If analog and digital SIC are applied simultaneously, then $\boldsymbol{M}$ should be replaced with $\boldsymbol{M} + \boldsymbol{H}_a$, for digital SIC.

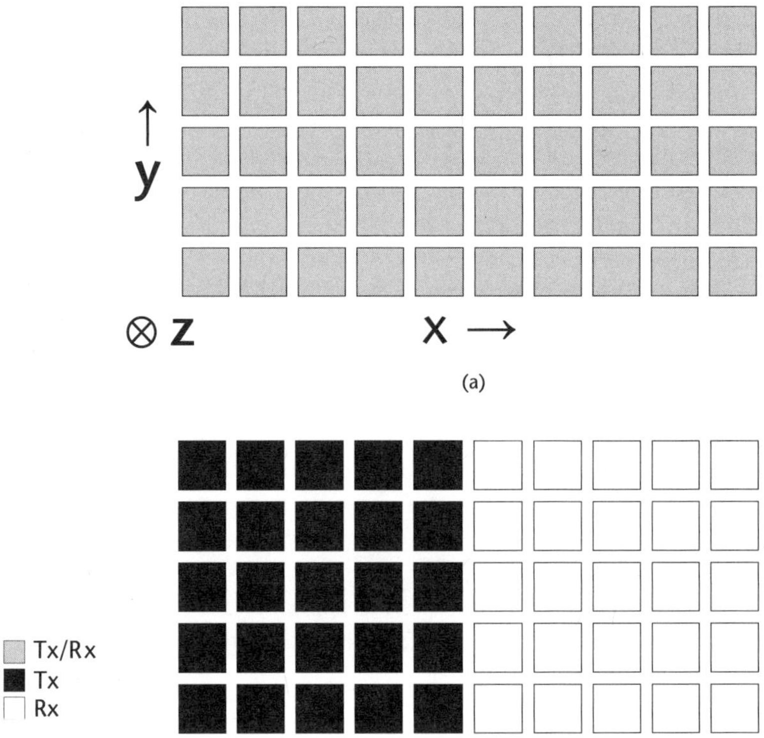

(a)

(b)

Figure 13.4 (a) ELSTAR configuration and (b) ALSTAR. Two example configurations for 50-element STAR phased array of half-wavelength spaced patches. The array model assumes 30-dBm transmit power per element with $\eta_t = 30$ dB, $\eta_r = 60$ dB, and $\sigma_r^2 = -90$ dBm. Array mutual coupling was simulated in Ansys HFSS.

incur losses or reduction in SNR as in analog implementations, and thus neither efficiency nor dynamic range is a limitation. For wideband signals, the scalar cancellation coefficient h_d may be replaced with a multitap digital filter to enable digital cancellation of wideband signals, and to compensate for distant scatterers with long propagation delays. Digital SIC implementations may also leverage nonlinear channel modeling to provide some mitigation of distortion products, as discussed in Chapter 11. Overall, digital cancellation provides significantly superior channel modeling and effective cancellation of linear, coherent SI components.

A purely digital SIC implementation cannot remove random transmitter noise (P_{NT}), nor can it mitigate effects of finite receiver dynamic range under high power SI (P_{NR}^{x} and P_{NR}^{nt}). The resulting residual SI terms under digital SIC are summarized in Table 13.3.

13.4.1.3 Hybrid SIC with Observation Receivers

A hybrid of analog and digital techniques offers some additional flexibility that may overcome some of the limitations of purely analog and digital SIC implementations. Analog SIC can mitigate both correlated signal and uncorrelated noise but generally suffers from limited matching accuracy, low efficiency and bandwidth, and high SWaP. Conversely, digital SIC provides high accuracy, efficiency, and bandwidth,

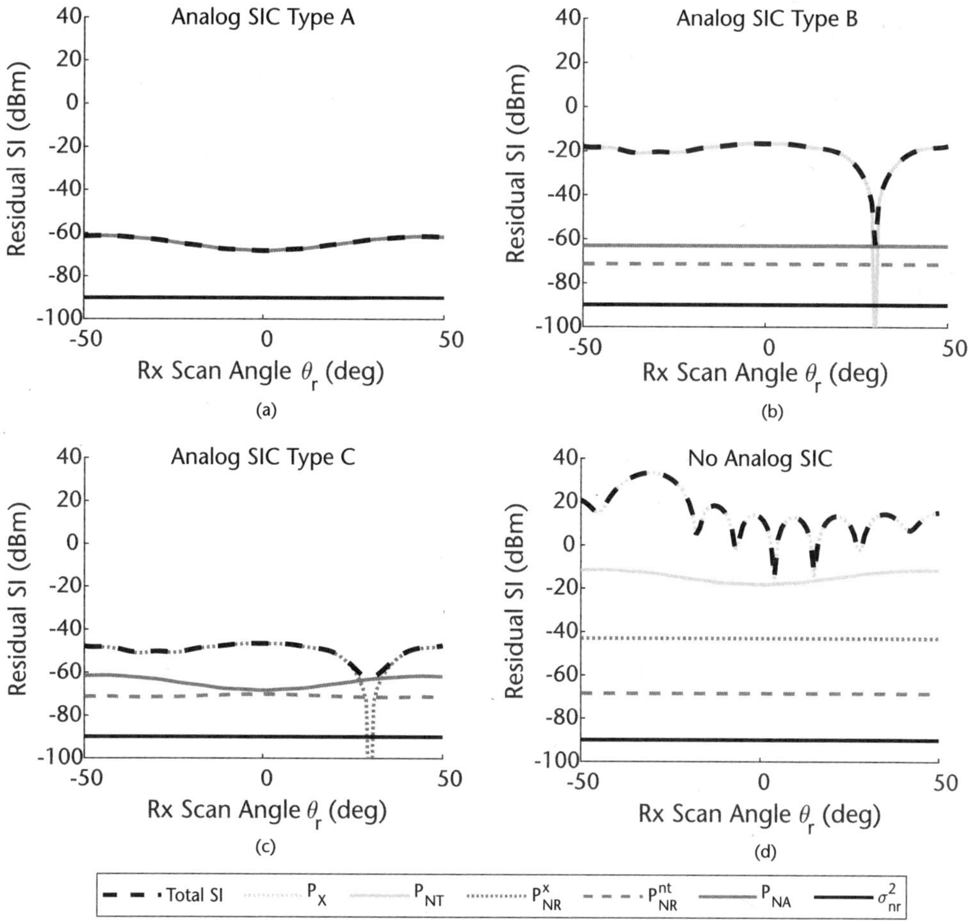

Figure 13.5 Simulated residual SI for the ELSTAR array depicted in Figure 13.4(a), under ideal analog SIC with 80-dB dynamic range. The beams are scanned in the x-z plane, with the receive beam swept from –50° to +50° while the transmit beam is fixed at +30°. Note the isolation improvement when $\theta_r = \theta_t$ for SIC Types B and C.

Table 13.3 Residual SI Under Ideal Digital SIC with Cancellation Coefficient h_d

Term	*Residual SI Power under Digital SIC*	$h_d \to -b_r^H M b_t$
P_X	$\left(b_r^H M b_t + h_d\right)^2$	Cancelled
P_{NT}	$\eta_t^{-1} b_r^H M Diag\left(b_t b_t^H\right) M^H b_r$	Unaffected
P_{NR}^x	$\eta_r^{-1} b_r^H Diag\left(M b_t b_t^H M^H\right) b_r$	Unaffected
P_{NR}^{nt}	$\eta_r^{-1} \eta_t^{-1} b_r^H Diag\left(M Diag\left(b_t b_t^H\right) M^H\right) b_r$	Unaffected

but cannot mitigate SI due to transmitter noise. By introducing additional observation channels that measure and digitize the physical transmitted signals as discussed in Chapter 12 and shown in Figure 13.6, hybrid SIC can realize the advantages of both approaches.

For ELSTAR arrays that must transmit and receive on the same elements, the addition of such observation channels requires a secondary digital receiver at each element, which may be impractical. For ALSTAR arrays, existing (otherwise unused) receive channels may be repurposed as observation channels, provided that the transmitted signal at each active element can be coupled back into the corresponding receiver at an appropriate level [8].

The observation signals $\boldsymbol{o}(t) \in \mathbb{C}^{J\times 1}$ are given by

$$\boldsymbol{o}(t) = \boldsymbol{H}_o \boldsymbol{t}(n) + \boldsymbol{H}_o \boldsymbol{n}_o(n), \tag{13.26}$$

where $\boldsymbol{H}_o \in \mathbb{C}^{J\times J}$ is the diagonal matrix representing the coupling of each transmit channel to the corresponding observation channel and $n_o(n)$ is AWGN due to limited observation channel dynamic range η_o, with noise covariance $\mathbf{N}_o \triangleq E\left[\boldsymbol{n}_o \boldsymbol{n}_o^H\right] = \eta_o^{-1} diag\left(\boldsymbol{b}_t \boldsymbol{b}_t^H\right)$. Passing $\boldsymbol{o}(t)$ through a multichannel digital filter $\boldsymbol{h}_d \in \mathbb{C}^{J\times 1}$ and combining with the received signal, we have

$$y(n) = \boldsymbol{b}_r^H \left[\boldsymbol{d}(n) + \boldsymbol{i}(n) + \boldsymbol{n}_r(n) + \boldsymbol{M}\boldsymbol{t}(n)\right] + \boldsymbol{h}_d^H \boldsymbol{H}_o \boldsymbol{t}(n) + \boldsymbol{h}_d^H \boldsymbol{H}_o \boldsymbol{n}_o(n). \tag{13.27}$$

The SI $\boldsymbol{t}(n)$ will then be cancelled if

$$\boldsymbol{h}_d^H \to -\boldsymbol{b}_r^H \boldsymbol{M} \boldsymbol{H}_o^{-1}. \tag{13.28}$$

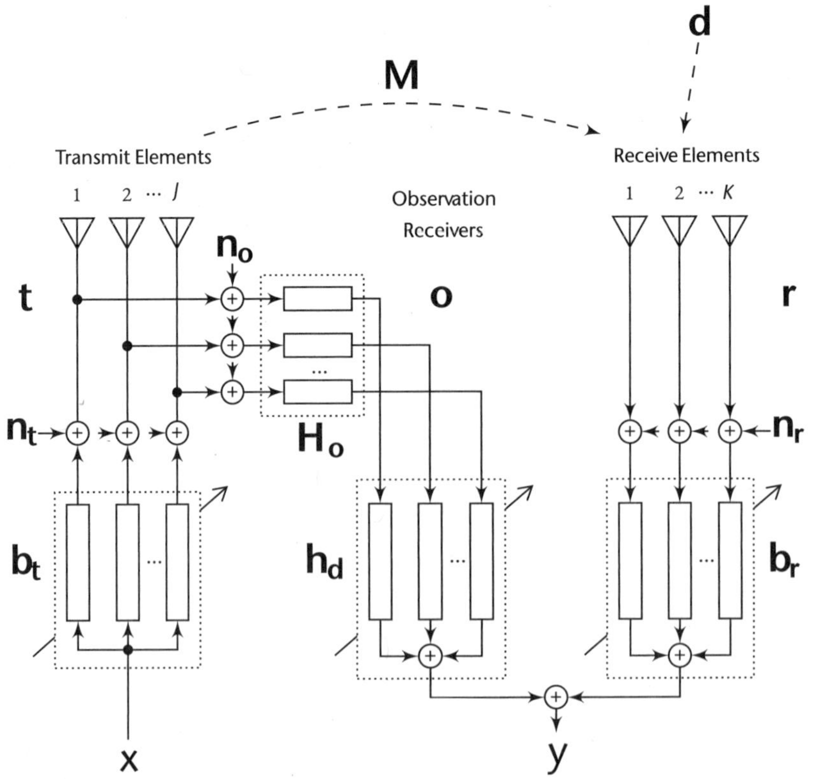

Figure 13.6 STAR phased array model with hybrid SIC using observation receivers.

Table 13.4 summarizes the residual interference terms using this technique. Comparing to simple digital cancellation, we see that both the transmit signal P_X and transmit noise P_{NT} may be cancelled at the cost of adding an additional observation noise term P_{NO}. The use of observation receivers for hybrid SIC is therefore useful in cases where the observation channel dynamic range η_o is significantly greater than the transmitter dynamic range η_t. Neither digital nor hybrid SIC can protect the receiver front end from high-power SI, and thus P_{NR}^{x} and P_{NR}^{nt} are unaffected. For this reason, digital and/or hybrid SIC may be best used as a complement to other cancellation techniques, for example, to clean up residual SI remaining after an analog SIC stage with limited accuracy, such as those covered in Chapters 8 and 9. Hybrid SIC is also especially effective in digital ALSTAR arrays when combined with adaptive beamforming, which is discussed in the following section.

13.4.2 Spatial Cancellation with Beamforming

In addition to SIC, phased arrays may also mitigate SI by controlling their radiation patterns to further improve isolation via spatial cancellation [8–11]. We may write the total effective isolation as

$$\text{EII} = \text{EIRP}(\theta_t, \phi_t) \cdot g(\theta_r, \phi_r) \cdot \frac{\boldsymbol{b}_r^H \boldsymbol{q}_r(\theta_r, \phi_r) \boldsymbol{q}_r^H(\theta_r, \phi_r) \boldsymbol{b}_r}{\boldsymbol{b}_r^H \mathbb{M}_r \boldsymbol{b}_r}, \tag{13.29}$$

where all noise and interference terms have been collected into a single term $\boldsymbol{b}_r^H \mathbb{M}_r \boldsymbol{b}_r = P_X + P_{NT} + P_{NR}^{x} + P_{NR}^{nt} + P_{NR}^{nr} + P_{NA} + P_{NO}$. The matrix $\mathbb{M}_r$ represents the effective covariance of the overall received interference and noise. The expansion of $\mathbb{M}_r$ depends on the particular SIC cancellation architecture (refer to Tables 13.1 to 13.4). For hybrid SIC tuning (per (13.28)), we find

$$\begin{aligned} \mathbb{M}_r = {} & \eta_r^{-1} Diag\left(\boldsymbol{M}\boldsymbol{b}_t\boldsymbol{b}_t^H\boldsymbol{M}^H\right) + \eta_r^{-1}\eta_t^{-1} Diag\left(\boldsymbol{M} Diag\left(\boldsymbol{b}_t\boldsymbol{b}_t^H\right)\boldsymbol{M}^H\right) + \\ & \eta_o^{-1}\boldsymbol{M} Diag\left(\boldsymbol{b}_t\boldsymbol{b}_t^H\right)\boldsymbol{M}^H + \sigma_{nr}^2 \boldsymbol{I} \end{aligned}. \tag{13.30}$$

Table 13.4 Residual SI Under Ideal Hybrid SIC with Observation Network H_o and Cancellation Filter h_d with Dynamic Range η_o

Term	*Residual SI Power under Hybrid SIC*	$b_d^H \rightarrow -b_r^H M H_o^{-1}$
Px	$\left(b_r^H M + b_d^H H_o\right) b_t b_t^H \left(M^H b_r + H_o^H b_d\right)$	Cancelled
P_{NT}	$\eta_t^{-1}\left(b_r^H M + b_d^H H_o\right) Diag\left(b_t b_t^H\right)\left(M^H b_r + H_o^H b_d\right)$	Cancelled
P_{NR}^{x}	$\eta_r^{-1} b_r^H Diag\left(M b_t b_t^H M^H\right) b_r$	Unaffected
P_{NR}^{nt}	$\eta_r^{-1}\eta_t^{-1} b_r^H Diag\left(M diag\left(b_t b_t^H\right) M^H\right) b_r$	Unaffected
P_{NO}	$\eta_o^{-1} b_d^H H_o Diag\left(b_t b_t^H\right) H_o^H b_d$	Introduced

The effective isolation as expressed in (13.29) contains a generalized Rayleigh quotient in terms of the receive beamformer vector $\boldsymbol{b}_r$. EII can therefore be maximized by choosing

$$\boldsymbol{b}_r \rightarrow \alpha\left(\mathbb{M}_r + \gamma_r \boldsymbol{I}\right)^{-1} \boldsymbol{q}_r(\theta_r, \phi_r), \tag{13.31}$$

where α is a scalar constant chosen to maintain $\boldsymbol{b}_r^H \boldsymbol{b}_r = 1$. A regularization parameter γ_r enables the designer to balance improvement in EII against main beam distortion. Choosing $\gamma_r > 0$ corresponds to maximizing EII while also constraining the loss of receive gain $G_R(\theta_r, \phi_r)$ [12]. Equation (13.31) represents an adaptive beamforming approach that is similar to that used for external interference nulling and may be generalized to provide reduction of both SI and external interferences.

The matrix $\mathbb{M}_r$ depends on the transmit beamformer, and (13.29) can also be refactored as a generalized Rayleigh quotient in terms of $\boldsymbol{b}_t$. Applying the identity $\boldsymbol{a}^H Diag\left(\boldsymbol{b}\boldsymbol{b}^H\right)\boldsymbol{a} = \boldsymbol{b}^H Diag\left(\boldsymbol{a}\boldsymbol{a}^H\right)\boldsymbol{b}$ to the individual noise and interference terms in Tables 13.1 through 13.4, we may effectively turn $\boldsymbol{b}_r^H \mathbb{M}_r \boldsymbol{b}_r$ inside out into $\boldsymbol{b}_t^H \mathbb{M}_t \boldsymbol{b}_t$ and rewrite (13.29) as

$$\text{EII} = G_R(\theta_r, \phi_r) \cdot g(\theta_t, \phi_t) \cdot \frac{\boldsymbol{b}_t^H \boldsymbol{q}_t(\theta_t, \phi_t) \boldsymbol{q}_t^H(\theta_t, \phi_t) \boldsymbol{b}_t}{\boldsymbol{b}_t^H \mathbb{M}_t \boldsymbol{b}_t}, \tag{13.32}$$

where $\mathbb{M}_t$ behaves like a virtual covariance matrix of the interference and noise from the perspective of the transmit beamformer. For ideal hybrid SIC,

$$\begin{aligned} \mathbb{M}_t = {} & \eta_r^{-1} \boldsymbol{M}^H Diag\left(\boldsymbol{b}_r \boldsymbol{b}_r^H\right)\boldsymbol{M} + \eta_r^{-1}\eta_t^{-1} Diag\left(\boldsymbol{M}^H Diag\left(\boldsymbol{b}_r \boldsymbol{b}_r^H\right)\boldsymbol{M}\right) + \\ & \eta_o^{-1} Diag\left(\boldsymbol{M}^H \boldsymbol{b}_r \boldsymbol{b}_r^H \boldsymbol{M}\right) + \frac{\sigma_{nr}^2}{P_t}\boldsymbol{I} \end{aligned} \tag{13.33}$$

As above with the receive beamformer, the transmit beamformer that maximizes EII is given by

$$\boldsymbol{b}_t \rightarrow \beta\left(\mathbb{M}_t + \frac{\gamma_t}{P_t}\boldsymbol{I}\right)^{-1} \boldsymbol{q}_t(\theta_t, \phi_t), \tag{13.34}$$

where β is a scalar constant chosen to maintain $\boldsymbol{b}_t^H \boldsymbol{b}_t = P_t$, and γ_t is a regularization parameter that constrains the loss of main-beam EIRP.

Because $\mathbb{M}_r$ is a function of $\boldsymbol{b}_t$ and $\mathbb{M}_t$ is a function of $\boldsymbol{b}_r$, (13.31) and (13.34) are optimal only relative to a fixed choice for the other beamformer. In practice, these equations may be solved iteratively, alternately recalculating the residual covariance matrices with each iteration. Although this iterative approach is not strictly guaranteed to converge to the global maximum, it has been shown to provide near-optimal solutions for ALSTAR arrays [9].

The residual SI for the simulated digital ALSTAR array of Figure 13.4(b) using digital and hybrid SIC, with uniform and adaptive beamforming, is shown in

Figure 13.7, and the corresponding array gain and isolation are shown in Figure 13.8. Regularization was used to constrain the reduction in transmit and receive gain to 1 dB at all scan angles. For this array, the combination of hybrid SIC and adaptive transmit and receive beamforming provides 166-dB EII at broadside and >160 dB when scanning within ±35° of broadside.

Optimizing EII via adaptive beamforming involves a trade-off between the desired far-field response and the undesired near-field transmit-receive coupling. Each beamformer has a finite number of spatial degrees of freedom (DOF) that may be consumed to satisfy near-field or far-field design constraints. If the array coupling matrix were random and full rank, maintaining a deep null to isolate K receiving elements would consume K DOF from the transmit beamformer. Likewise, maintaining a deep null on each of the J transmitting elements would consume J DOF from the receive beamformer. In this case, no remaining DOF would be available to form the desired far-field beams. Fortunately, the coupling matrix is highly structured in most arrays due to the physical array geometry, and significant isolation may often

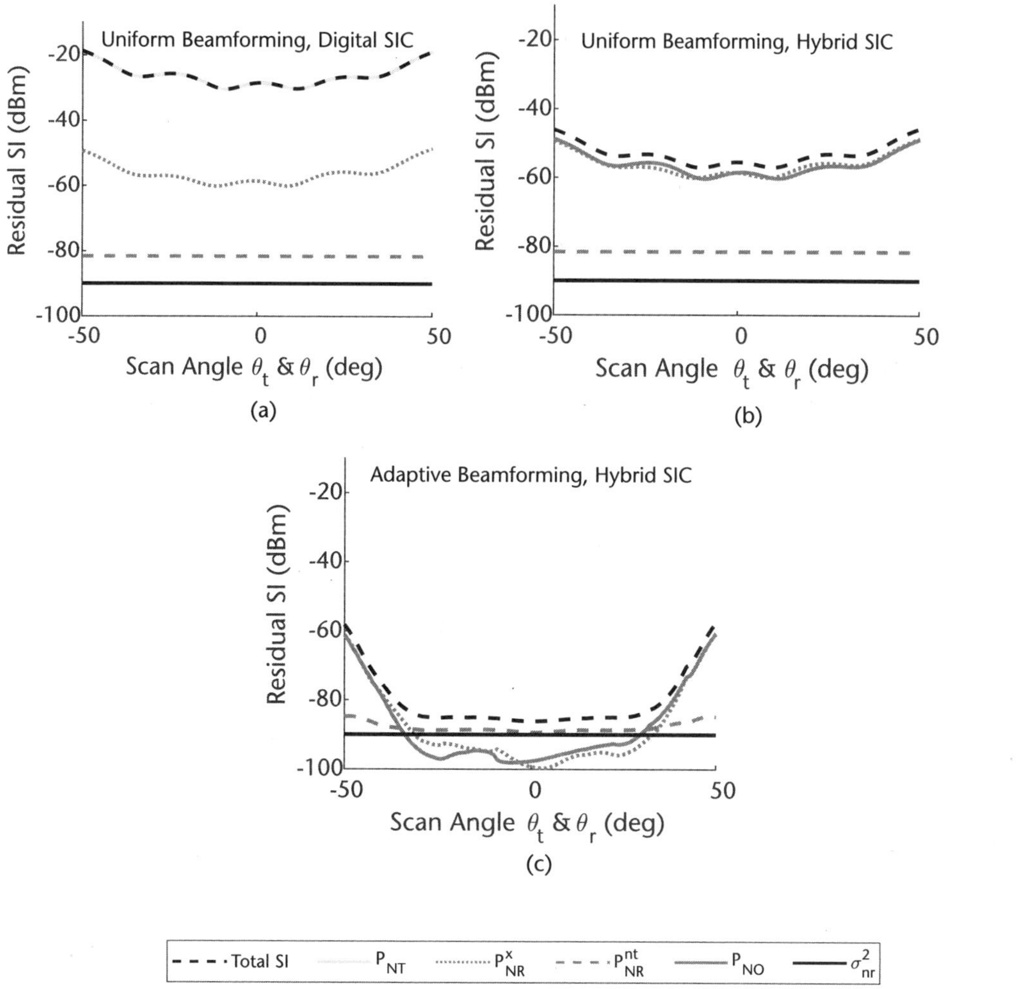

Figure 13.7 Simulated residual SI and noise of the ALSTAR array of Figure 13.4(b) using digital and hybrid SIC with uniform and adaptive beamforming. Both transmit and receive beams are scanned together from –50° to +50° in the x-z plane.

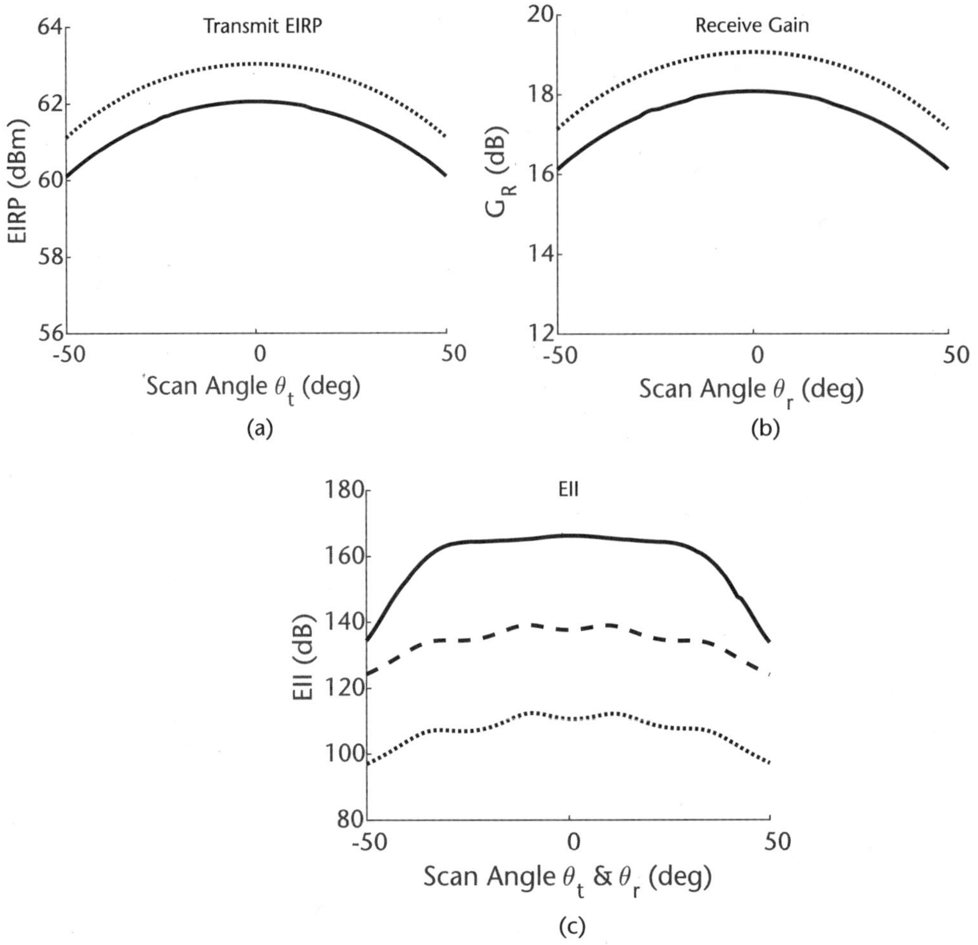

Figure 13.8 Main-beam response and EII for the simulated ALSTAR results of Figure 13.7.

be achieved with relatively few DOFs. This is especially true for separate apertures or side-by-side ALSTAR arrays, as shown below.

To illustrate the relationship between beamforming, mutual coupling, and array geometry, the covariance matrices $\mathbb{M}_r$ and $\mathbb{M}_t$ for the ELSTAR and ALSTAR arrays from Figure 13.4 were computed and their eigenvalue spectra are plotted in Figure 13.9. Large eigenvalues correspond to propagation modes that produce high SI, while small eigenvalues correspond to low interference modes. The ALSTAR configuration is dominated by several strong coupling modes, suggesting that isolation improvement requires consumption of a correspondingly few DOF, and the remaining DOF may be allocated to the desired far-field pattern. In contrast, the ELSTAR configuration requires many more DOF for isolation improvement, significantly reducing the DOF available for the main beams. Allocation of DOF for isolation versus far-field beams can be managed in practice via the regularization parameters γ_r and γ_t.

This difference between the relative effectiveness of beamforming for ALSTAR and ELSTAR configurations is intuitive. For an ALSTAR array with separate transmit and receive regions, the elements to be isolated are located off to one side, away

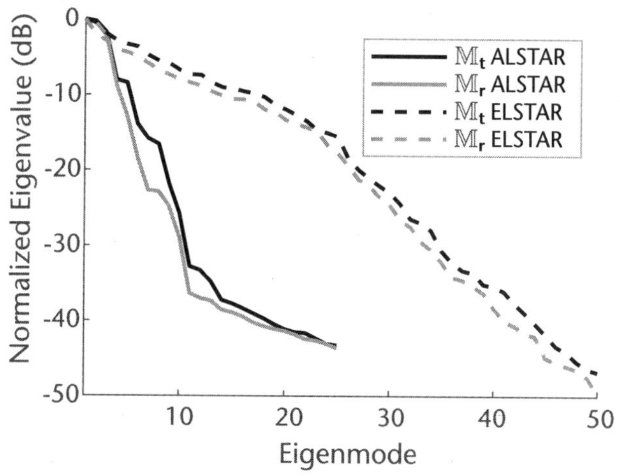

Figure 13.9 Eigenvalue spectra of $\mathbb{M}_r$ and $\mathbb{M}_t$ for the ELSTAR and ALSTAR configurations of the array described in Figure 13.4.

from the main beam direction. Thus, the desired main beam can be formed without exciting the undesired coupling modes. Moreover, because the elements to be isolated are all coplanar and generally located in the same direction, a single spatial null may effectively maintain isolation to multiple elements. Conversely, for an ELSTAR array, there is no safe direction to steer the main beam, as the active elements and elements to be isolated are colocated. The result is that although beamforming is quite effective for improving isolation in side-by-side ALSTAR arrays, it may be less so for ELSTAR arrays.

13.5 Practical Considerations

As has been discussed, the maturity of analog SIC solutions for STAR phased arrays is still quite low due to performance limitations and integration challenges. At the same time, element-level digitization of arrays is an emerging capability that has received significant recent investment from government and commercial sources [13]. There is presently a significant opportunity to leverage new fully digital array architectures for STAR. The complementary techniques of adaptive digital beamforming and hybrid SIC using digital observation channels may enable practical ALSTAR digital arrays without requiring specialized RF cancellation hardware. By implementing SI mitigation entirely in the digital domain, the physical design and RF performance of a STAR array need not be compromised by the limitations of analog SIC circuits. Moreover, fully digital ALSTAR arrays can easily switch between STAR and non-STAR operational modes as required by the application without performance penalty.

A small proof-of-concept fully digital ALSTAR array was constructed and demonstrated by MIT Lincoln Laboratory and maintained 140-dB EII over 100-MHz instantaneous bandwidth, as seen in Figure 13.10 [14]. Several practical challenges encountered in the development and testing of this array are highlighted. The performance of both the beamforming and digital cancellation was primarily limited

(a)

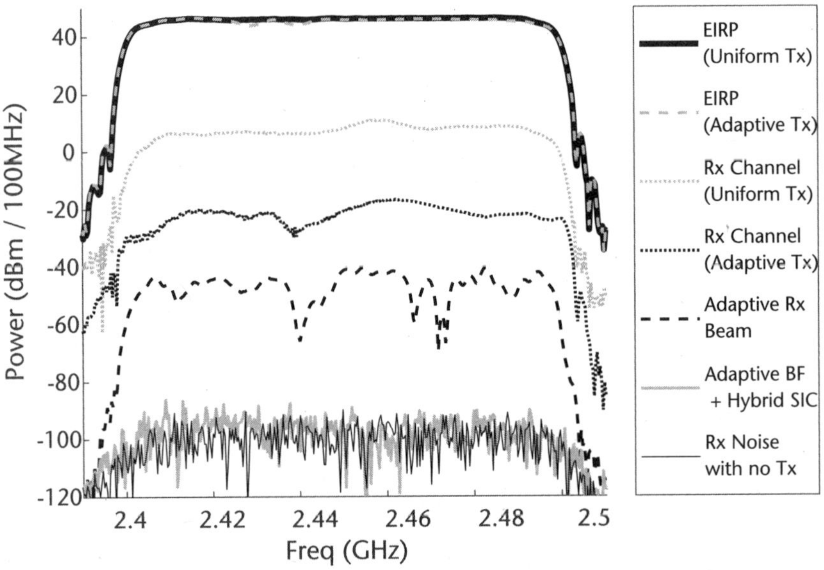

(b)

Figure 13.10 (a) MIT Lincoln Laboratory 8-element ALSTAR array prototype (4 transmitting, 4 receiving, 1 unused element), and (b) measured results demonstrating 140-dB EII over 100 MHz. (Adapted from [14].)

by the respective transmit and receive channel linearity, and careful attention had to be given to the cascaded system design. Moreover, effective transmit beamforming requires the PAs to operate in their linear region, which may reduce overall system efficiency. Also of practical concern are multipath effects in the testing environment. Even a good anechoic chamber may introduce reflections that become significant when attempting to measure >120-dB isolation. For very high-isolation systems, outdoor testing and/or larger anechoic chambers may be required to fully characterize performance.

13.6 Conclusion

13.6.1 Summary

Although STAR phased arrays have much in common with single-channel STAR/IBFD architectures, there are many specific challenges and opportunities that arise in developing high-isolation arrays. In this chapter, we have introduced EII as a useful metric for characterizing systems with nontrivial antennas. A simple signal model was provided to quantify SI and EII in STAR phased arrays and formed the basis for a discussion of analog, digital, and hybrid SIC and spatial cancellation via adaptive beamforming. Simulated results were provided for a 50-element array under both ELSTAR and ALSTAR configurations, and measured results were provided for a small, 8-element ALSTAR prototype. These preliminary results suggest that while practical ELSTAR arrays may continue to represent a significant engineering challenge for the foreseeable future, recent advances in fully digital array architectures may enable practical ALSTAR arrays capable of extremely high isolation without requiring complex and lossy cancellation circuits.

13.6.2 Future Directions

STAR phased arrays represent an exciting technological frontier that promises significant opportunities for fundamentally new phased array capabilities and applications. There are many important areas for further investigation and development that will be essential to advancing the practicality and performance of STAR architectures for phased arrays. First, nonlinear channel modeling and equalization should be incorporated to improve mitigation of distortion products introduced by the receiver or analog SIC electronics. The use of digital predistortion (DPD) techniques and/or phase-only beamforming may allow the transmitters to operate closer to saturation while still benefiting from adaptive transmit beamforming. Another important challenge relates to the digital processing architecture. As the array size grows, signal distribution and adaptive processing becomes impractical with a centralized processor. Alternate processing architectures should be explored to take advantage of distributed networking and real-time computation across multiple processing nodes. Finally, full ELSTAR arrays will almost certainly require development of novel highly integrated analog SIC solutions that can support high bandwidth and efficiency while maintaining very high dynamic range at the high power levels required by many phased array applications, such as MIMO communications, discussed in Chapter 14.

References

[1] Doane, J. P., "Isolation Metrics for Single-Channel and Multi-Channel Simultaneous Transmit and Receive Systems," *2020 IEEE AP-S Symposium on Antennas and Propagation*, July 2020.

[2] Hannan, P., D. Lerner, and G. Knittel, "Impedance Matching a Phased-Array Antenna over Wide Scan Angles by Connecting Circuits," *IEEE Transactions on Antennas and Propagation*, Vol. 13, No. 1, January 1965, pp. 28–34.

[3] Amitay, N., "Improvement of Planar Array Match by Compensation Through Contiguous Element Coupling," *IEEE Transactions on Antennas and Propagation*, Vol. 14, No. 5, September 1966, pp. 580–586.

[4] Duffy, S. M., G. A. Brigham, and J. S. Herd, "Integrated Compensation Network for Low Mutual Coupling of Planar Microstrip Antenna Arrays," *2007 IEEE Antennas and Propagation Society International Symposium*, Honolulu, HI, 2007, pp. 1389–1392.

[5] Bharadia, D., and S. Katti, "Full Duplex MIMO Radios," *Proc. 11th USENIX Symp. NSDI*, 2014, pp. 359–372.

[6] Hannan, P., "The Element-Gain Paradox for a Phased-Array Antenna," *IEEE Transactions on Antennas and Propagation*, Vol. 12, No. 4, July 1964, pp. 423–433.

[7] Haykin, S., *Adaptive Filter Theory*, 5th ed., Boston, MA: Pearson, 2013.

[8] Doane, J. P., K. E. Kolodziej, and B. T. Perry, "Simultaneous Transmit and Receive with Digital Phased Arrays," *Proc. 2016 IEEE International Symposium on Phased Array Systems and Technology*, Waltham, MA, October 2016.

[9] Cummings, I. T., et al., "Aperture-Level Simultaneous Transmit and Receive with Digital Phased Arrays," *IEEE Transactions on Signal Processing*, Vol. 68, 2020, pp. 1243–1258.

[10] Everett, E., et al., "SoftNull: Many-Antenna Full-Duplex Wireless via Digital Beamforming," *IEEE Transactions on Wireless Communications*, Vol. 15, No. 12, December 2016, pp. 8077–8092.

[11] Dastjerdi, M. B., et al., "Full Duplex Circulator-Receiver Phased Array Employing Self-Interference Cancellation via Beamforming," *2018 IEEE Radio Frequency Integrated Circuits Symposium (RFIC)*, Philadelphia, PA, 2018, pp. 108–111.

[12] Boyd, S., and L. Vandenberghe, *Convex Optimization*, New York: Cambridge University Press, 2004.

[13] Fulton, C., et al., "Digital Phased Arrays: Challenges and Opportunities," *Proceedings of the IEEE*, Vol. 104, No. 3, March 2016, pp. 487–503.

[14] Doane, J. P., K. E. Kolodziej, and B. T. Perry, "Simultaneous Transmit and Receive Performance of an 8-Channel Digital Phased Array," *2017 IEEE International Symposium on Antennas and Propagation & USNC/URSI National Radio Science Meeting*, San Diego, CA, 2017, pp. 1043–1044.

CHAPTER 14

MIMO Array Systems

Mahmood Baraani Dastjerdi, Tingjun Chen, Gil Zussman, and Harish Krishnaswamy

14.1 Introduction

Multi-antenna systems, including phased arrays and MIMO radios, are expected to play an essential role in next-generation wireless networks. As discussed in Chapter 13, phased arrays can provide coherent beamforming gain and provide spatial interference rejection, as well as increase the communication range. However, MIMO technology promises to significantly enhance system performance in coverage, capacity and data rates through beamforming or diversity/capacity gains that are challenging from both the Tx power handling and Rx noise perspectives in a multipath environment. MIMO systems can also simultaneously serve multiple users (i.e., multi-user MIMO (MU-MIMO)), which is vital for femtocell base station and access point (AP) deployments [1–4]. This chapter discusses MIMO arrays developed for the IBFD operation, as highlighted within the IBFD transceiver diagram of Figure 14.1(a) and the SIC techniques tree of Figure 14.1(b).

Because IBFD wireless can provide double data rates at the physical layer and reduced communication latency, integrating IBFD capability in multi-antenna systems has the potential to further boost the achievable data rates and spectrum efficiency, as well as to enhance the communication distance (see Figure 14.2). However, enabling IBFD operation in multi-antenna systems is challenging because there are two sources of SI: the SI from each Tx to its own Rx, and the cross-talk interference (XI) that exists between every pair of Tx and Rx elements.

14.2 Challenges with Multi-Antenna Technologies

14.2.1 Power Handling

Tx power handling remains an open problem in IBFD radios, particularly in ones with integrated shared antenna interfaces, such as those presented in Chapter 4. IBFD multi-antenna radio power handling further suffers from XI in addition to SI (Figure 14.3). For each Rx channel in an N-element IBFD phased array transceiver,

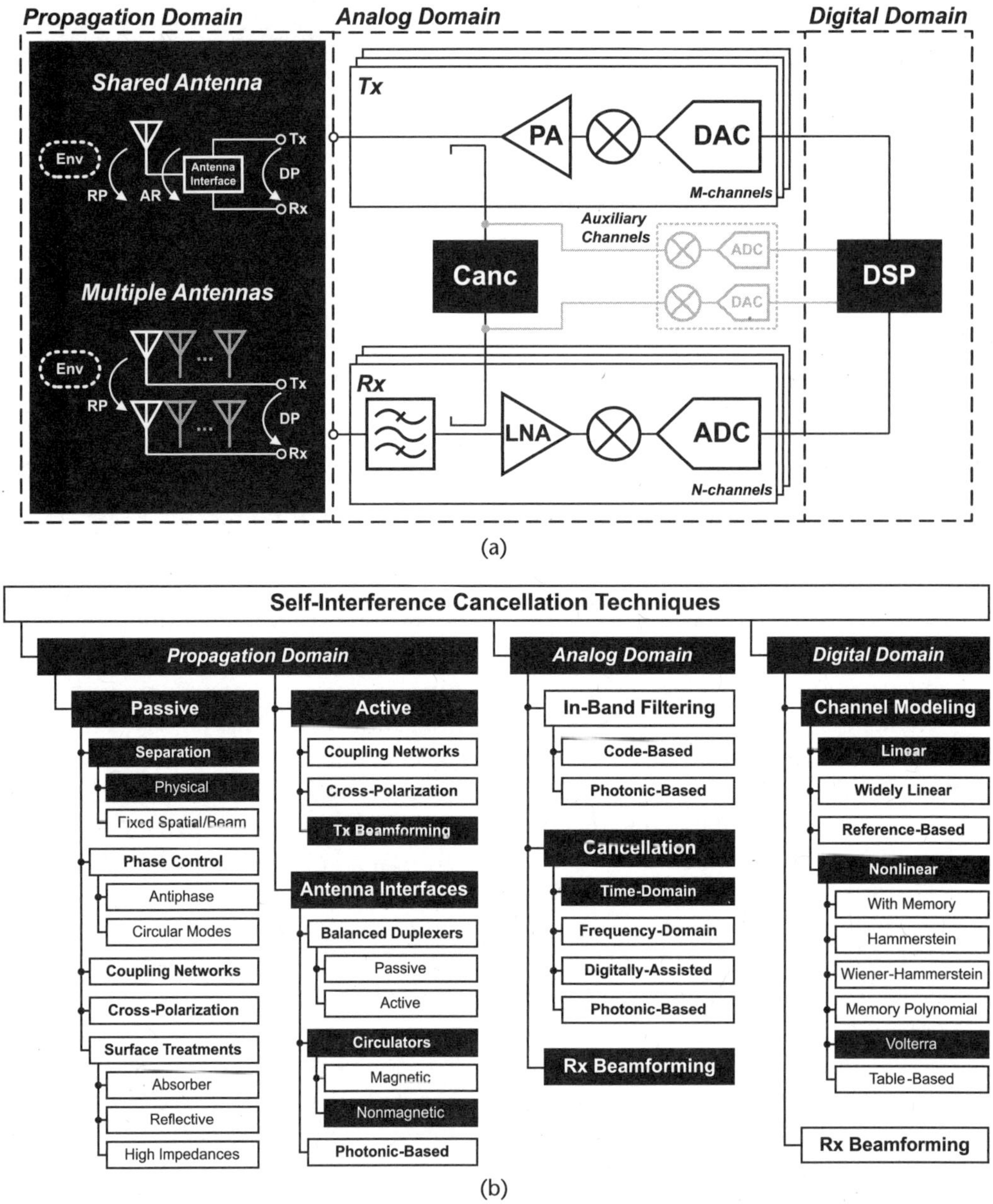

Figure 14.1 MIMO array systems and specific chapter focus highlighted within (a) a generic IBFD transceiver and (b) a SIC techniques tree diagram.

the SI and XI can possibly add up constructively to increase by an amount equal to the Tx array gain (N^2) and then add up constructively after Rx beamforming to increase by an amount equal to the Rx array gain (N^2), resulting in a total increase of N^4 relative to a single-element transceiver. Similarly, in an N-element IBFD MIMO transceiver, assuming orthogonal coding, the total SI and XI power can be N times higher than the single-element case (assuming similar power levels for the SI and XI across all elements). This substantially limits the IBFD phased array/MIMO transceiver power handling compared to the single-element case and, therefore, alleviates benefits gained from deploying such multi-antenna systems.

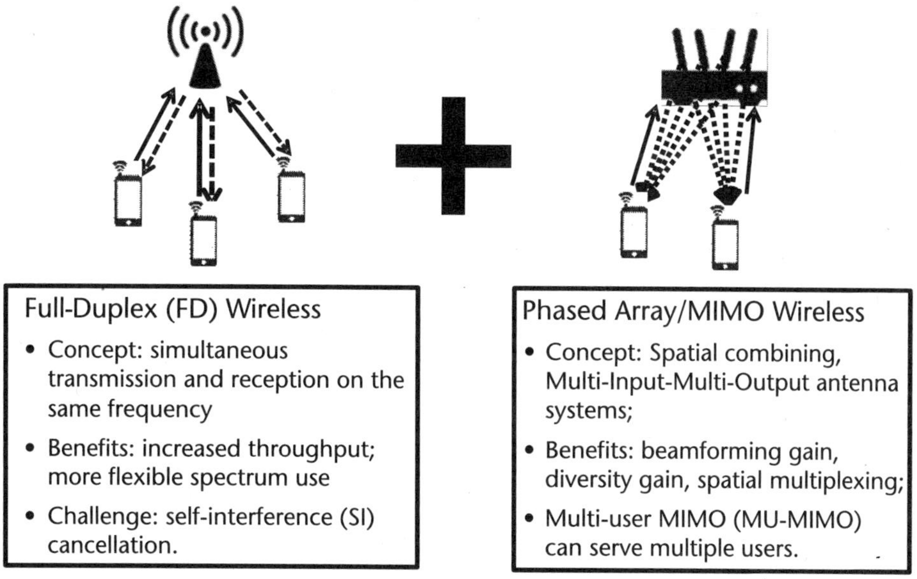

Figure 14.2 Integrating full-duplex (FD) operation with MIMO technology can significantly improve the data rates and spectrum efficiency for the next generation of wireless communication.

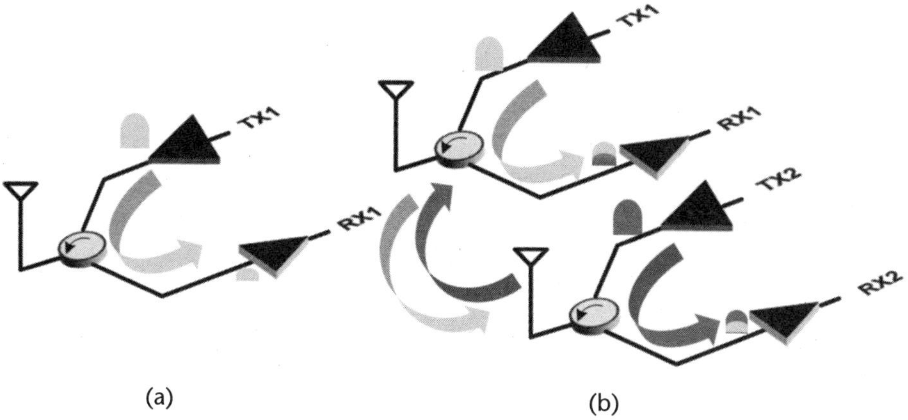

Figure 14.3 SI and XI in full-duplex (FD) wireless: (a) in a shared-antenna transceiver; and (b) in a multi-antenna transceiver.

14.2.2 Canceller Complexity

In an IBFD MIMO transceiver, XI between each Tx-Rx pair needs to be suppressed. As depicted in Figure 14.4, a cancellation path, H_{ij}, is required from each Tx to all other Rx's. An N-element IBFD MIMO transceiver, therefore, requires N^2 cancellation paths, which largely increases the complexity of the canceller in terms of associated area, power dissipation and noise penalties. This is both area-hungry and power-hungry and is not feasible for radios with a large number of elements. Furthermore, the noise from SI and XI cancellation paths can accumulate and further degrades the radio sensitivity level.

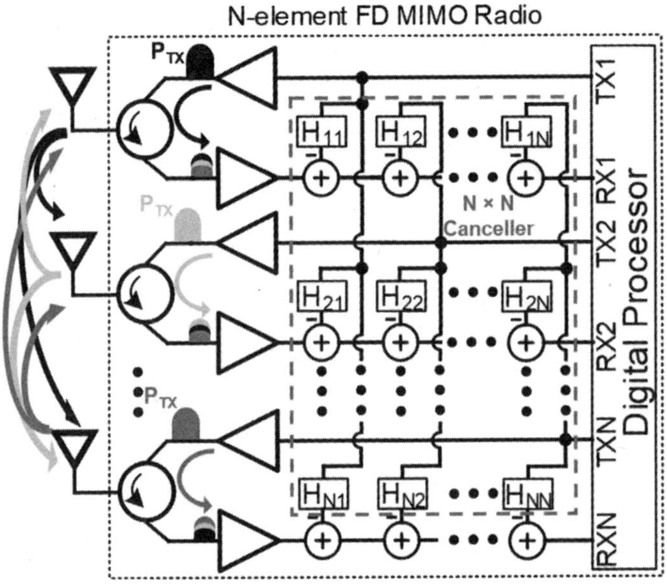

Figure 14.4 In an N-element full-duplex (FD) MIMO radio, a cancellation path for each Tx-Rx pair is required, resulting in a total canceller complexity of O(N^2). (© 2019 IEEE. Reprinted, with permission, from: [5].)

14.3 Prior Art

There are a few state-of-the-art IBFD multi-antenna radios on the system-level using off-the-shelf discrete components. In [4], the correlation between SI and XI is exploited to share cancellation delay taps between SI and XI cancellation paths. This is based on the idea that XI signals experience slightly higher delay compared to the SI. This enables the IBFD MIMO canceller complexity to increase linearly with the number of elements (i.e., O(N)). This radio uses bulky delay lines, and there is no phase control on the delay taps. SoftNull [6] is another technique that divides the number of available antennas between Tx and Rx and sacrifices MIMO degrees of freedom (DOF) on digital Tx beamforming to achieve SIC. This work does not employ a shared antenna interface and does not feature analog cancellers (as discussed in Chapters 8 through 10), with SI and XI cancellation only achieved through digital Tx beamforming. In [7], an IBFD 2-element MIMO radio is presented for LTE applications using four multitap cancellers to suppress the SI and XI in the RF domain. This radio does not address the canceller complexity challenge, requires bulky delay cells, and does not use a shared antenna interface.

14.4 Phased Arrays Based on SIC via Beamforming

14.4.1 System Considerations

14.4.1.1 Array SIC

An IBFD link consisting of two N-element IBFD phased arrays is depicted in Figure 14.5. Although the SIC after beamforming depends heavily on the antenna response, the SI and XI, in the worst case, can add up constructively in each Rx

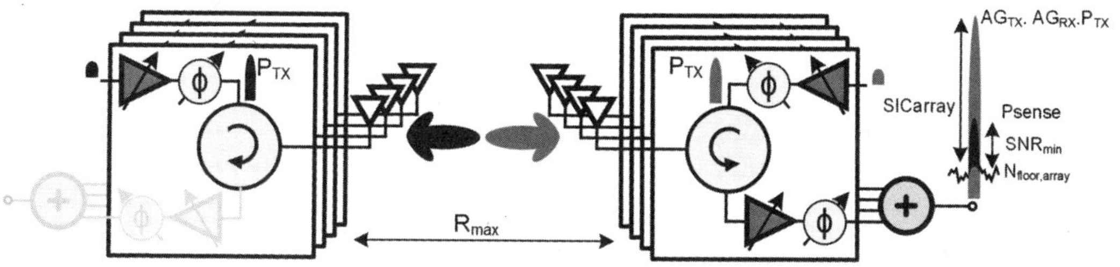

Figure 14.5 IBFD phased array link budget.

channel to increase by an amount equal to the Tx array gain (N^2) and then add up constructively after Rx beamforming to increase by an amount equal to the Rx array gain (N^2), resulting in a total increase of N^4 relative to a single-element transceiver. Hence, while phased array beamforming provides N^2 increase in the array gain on both Tx/Rx sides, it can also potentially increase the total SI at the Rx output after beamforming by the same amount. Therefore, we define the array SIC similar to the single-element counterpart with the addition of Tx/Rx array gain

$$SIC_{array} = P_{TX} \bullet AG_{TX} \bullet AG_{RX} / SI, \qquad (14.1)$$

where P_{TX} is the per-element Tx power level, SI is the residual SI power level, and $AG_{Tx/Rx}$ is the Tx/Rx array gain.

In a single-input single-output (SISO) radio, the required SIC to suppress the SI level down to the noise level is given by $SIC_{SISO} = P_{TX}/N_{floor,SISO}$, where $N_{floor,SISO} = kT \cdot BW \cdot NF$. In an N-element IBFD phased array, the required SIC is given by $SIC_{array} = P_{TX} \cdot AG_{TX} \cdot AG_{RX}/N_{floor,array}$, where $N_{floor,array} = kT \cdot BW \cdot NF \cdot N$. It can be seen that the required SIC in an N-element IBFD phased array is $AG_{TX} \cdot AG_{TX}/N$ times (which could be as high as N^3) higher than the SISO case.

14.4.1.2 Link Budget Calculations

The desired received signal power in an N-element IBFD phased array link after Tx and Rx beamforming is given by

$$P_{RX,array} = \frac{P_{TX} \bullet AG_{TX} \bullet G_{TX} \bullet AG_{RX} \bullet G_{RX}}{FSPL \bullet IL}, \qquad (14.2)$$

where $G_{TX/RX}$ is the Tx/Rx antenna gains, $FSPL$ is the free-space propagation loss equal to $(4\pi \cdot R \cdot f/c)^2$, and IL is a margin considered for practical implementation losses (R, f, and c are the range, the carrier frequency, and the speed of light, respectively). The minimum required SNR determines the radio sensitivity level ($P_{sense} = 2 \cdot SNR \cdot N_{floor,array}$), where the factor of 2 covers the SNR degradation due to the residual SI. The maximum communication range that the link can support (R_{max}) can thus be calculated as

$$R_{max} = \frac{\lambda}{4\pi}\sqrt{P_{TX} \bullet AG_{TX} \bullet G_{TX} \bullet AG_{RX} \bullet G_{RX} \bullet IL / P_{sense}}. \quad (14.3)$$

This clearly shows the benefit of array gain at large link distances. Table 14.1 summarizes the link budget calculation of our proposed system. In particular, for an 8-element, 730-MHz array with +1-dBm Tx power per element (P_{TX}), 6-dBi antenna gain, 15-dB Tx and Rx array gains (AG_{TX} and AG_{RX}, 3 dB degraded from the ideal 18-dB array gain due to the need to achieve sufficient amount of SIC), 16.25-MHz bandwidth (BW), 5-dB Rx noise figure (NF), 20-dB required SNR, and 10-dB implementation losses, one can establish an IBFD link over a distance of 2.6 km. This shows that phased arrays can substantially enhance range in silicon-based IBFD transceivers that are limited in power handling, making them suitable for both space-constrained Wi-Fi APs and small-cell base stations.

14.4.2 RF Self-Interference Cancellation via Joint Tx and Rx Beamforming

An N-element phased array with a shared antenna interface and amplitude and phase controls on each Tx and Rx element features overall $2(N - 1)$ complex-valued DOF on Tx and Rx sides ($(N - 1)$ DOF on each side). These DOF represent the complex-valued weights (amplitudes and phases) of each element as it compares to that of the first element. Typically, these DOF are used to form the beams towards desired spatial directions and also minimize interference to/from nearby radios by aiming nulls towards them or suppressing the radiation pattern's side-lobes. Alternatively, a few beamforming DOF at the Tx and Rx can be repurposed so the total SI is suppressed after Rx beamforming at the expense of some Tx and Rx beam characteristics, such as a few nulls and/or some gain loss in the beam-pointing direction(s).

Table 14.1 IBFD Phased Array Link Budget

Metric	*Calculation*	*Value*
Frequency (f)	—	730 MHz
No. of ANT elements (N)	—	8
Tx power per elements (P_{TX})	—	1 dBm
Tx/Rx array gain ($AG_{TX/RX}$)	$20 \cdot \log_{10}(N) - 3$	15 dB
Tx/Rx ANT gain ($G_{TX/RX}$)	—	6 dBi
Bandwidth (BW)	—	16.26 MHz
Rx noise figure (NF)	—	5 dB
Rx array noise floor referred to ANT input ($N_{floor,array}$)	$kT \cdot BW \cdot NF \cdot N$	–88 dBm
Required SNR	—	20 dB
Rx array sensitivity referred to ANT input (P_{sense})	$2 \cdot SNR \cdot N_{floor,array}$	–65 dBm
Implementation losses (IL)	—	10 dB
Supported range (R_{max})	$\frac{\lambda}{4\pi}\sqrt{P_{TX} \cdot AG_{TX} \cdot G_{TX} \cdot AG_{RX} \cdot G_{RX} \cdot IL / P_{sense}}$	2.6 km
Required array SIC (SIC_{array})	$P_{TX} \cdot AG_{TX} \cdot AG_{RX}/N_{floor,array}$	119 dB

Figure 14.6(a) depicts our implementation of a 2 × 4 rectangular array of slot loop antennas at 730 MHz with $\lambda/2$ spacing, whose SI channel matrix in the frequency domain is denoted by $\mathbf{H}^{SI}(f) = \left[H^{SI}_{(m',n'),(m,n)}(f)\right] \in \mathbb{C}^{8\times 8}$. Let $x(f)$ be the transmit signal in the frequency domain, and $\mathbf{w}^{TX}$ and $\mathbf{w}^{RX}$ be the complex-valued Tx

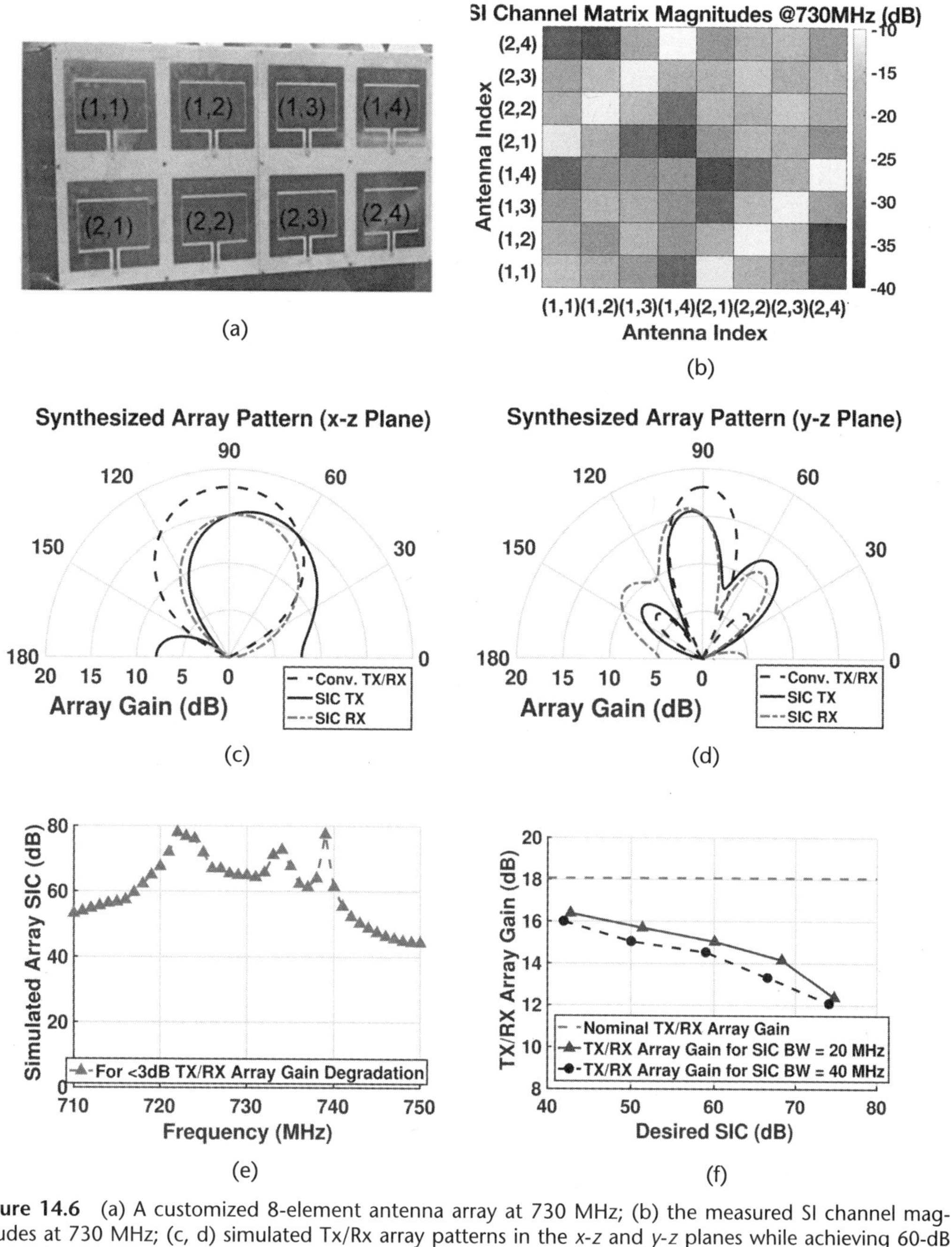

Figure 14.6 (a) A customized 8-element antenna array at 730 MHz; (b) the measured SI channel magnitudes at 730 MHz; (c, d) simulated Tx/Rx array patterns in the *x-z* and *y-z* planes while achieving 60-dB array SIC across 20 MHz with 3-dB array gain degradation in the Tx/Rx broadside beam-pointing directions; (e) simulated array SIC where at least 60-dB SIC is guaranteed to be achieved across 720 to 740 MHz; and (f) simulated Tx/Rx array gain for a requested array SIC based on the optimization problem. (© 2018 IEEE. Reprinted, with permission, from [8].)

and Rx beamforming weight vectors (i.e., beamformers), respectively. Then the phased array SI after Tx and Rx beamforming is given by $x^{SI}(f) = (\mathbf{w}^{RX})^T \cdot \mathbf{H}^{SI}(f) \cdot \mathbf{w}^{TX} \cdot x(f)$.

Consider a 3-D coordinate system where the 2 × 4 rectangular array is located on the x-y plane. We denote the Tx and Rx beamforming direction by the azimuth and elevation angles (θ, ϕ) in a horizontal coordinate system. We use $\mathbf{s}^{TX/RX}(\theta,\phi) = \left[s_{(m,n)}^{TX/RX}(\theta,\phi)\right] = \left[\exp\left(j\pi\left[(m-1)\cos\phi\cos\theta + (n-1)\sin\phi\cos\theta\right]\right)\right]$ to denote the Tx/Rx steering vector in the spatial direction (θ, ϕ). The goal is ultimately to achieve wideband SIC in the near field with minimal penalty in the Tx/Rx beamforming gains in the far field.

We formulate an optimization challenge with the goal of maximizing the Tx and Rx array gains, subject to the constraint under which a desired array SIC is achieved after Tx and Rx beamforming, that is,

$$\begin{aligned}
&\text{max: } AG_{TX/RX} \\
&\text{subject to: } \left|\left(\mathbf{s}^{TX}\left(\theta^{TX},\phi^{TX}\right)\right)^T \bullet \mathbf{w}^{TX}\right|^2 \geq AG_{TX} \\
&\left|\left(\mathbf{s}^{RX}\left(\theta^{RX},\phi^{RX}\right)\right)^T \bullet \mathbf{w}^{RX}\right|^2 \geq AG_{RX} \\
&\left|\left(\mathbf{w}^{RX}\right)^T \bullet \mathbf{H}^{SI}(f) \bullet \mathbf{w}^{TX}\right|^2 \leq \chi, \forall f \\
&\left|w_{(m,n)}^{TX}\right|^2 \leq 1, \left|w_{(m,n)}^{RX}\right|^2 \leq 1, \forall m,n .
\end{aligned}$$

In particular, the first two constraints ensure minimal degradation in the Tx and Rx array gain in the desired Tx and Rx beam-pointing directions, (θ^{TX}, ϕ^{TX}) and (θ^{RX}, ϕ^{RX}) (compared to the maximal Tx and Rx array gains without sacrificing any DOFs). The third constraint guarantees at least $\chi_{dB} = 10 \cdot \log_{10}(\chi)$ dB SIC across the desired bandwidth so that the total RF SIC is ($\chi_{dB} + AG_{TX} + AG_{RX}$). The last constraint sets the normalization on each Tx/Rx beamforming weight.

We evaluate this idea through simulations using the measured array SI channel Figure 14.6(b)), where the optimization problem is solved using thc MATLAB nonlinear optimization solver. We set $\chi = 10^{-3}$ so that the total array SIC is given by ($30 + AG_{TX} + AG_{RX}$) dB. The results are summarized in Figure 14.6(c, d). In our simulation, the Tx/Rx array gains are maximized for broadside beamforming (i.e., $\theta = 90°$) subject to the requirement that at least 60-dB array SIC is achieved between 720 and 740 MHz. Figure 14.6(c, d) show that an array gain degradation of only 3 dB compared with the maximal array gain of $N^2 = 18$ dB can be maintained while achieving 60-dB array SIC across 20 MHz. Moreover, Figure 14.6(f) shows the trade-off between the maximum achievable Tx/Rx array gain and different array SIC performance for 20-MHz and 40-MHz bandwidth.

In general, we would like to highlight four important features: (1) SI suppression is essentially achieved in the spatial domain through a trade-off between near-field SI nulling and far-field beamforming without any explicit cancellers, similar to Chapter 13; (2) the SI suppression is wideband as different antenna coupling paths are cancelling each other, rather than having an integrated circuit (IC) canceller duplicate the antenna coupling path's frequency characteristics; (3) the beamforming-IBFD trade-off can be dynamically adapted in the field, with the number of DOF sacrificed being dependent on the required SIC, bandwidth, external interferers that must be nulled; and (4) the trade-off between IBFD and beamforming will become more favorable for larger arrays. In [8], we further investigate this problem and develop efficient algorithms for cases with large-scale arrays (e.g., 36 or 72 elements).

14.4.3 Integrated Circuit Implementation

In [8], we designed and implemented a scalable 4-element integrated IBFD circulator receiver (C-Rx) phased array at 730 MHz using 65-nm CMOS technology, whose circuit diagram and chip photo are shown in Figure 14.7. It leverages the N-path-filter-based combined-circulator-Rx concept described in [9], which provides an integrated shared antenna interface, as discussed in Chapter 4. The availability

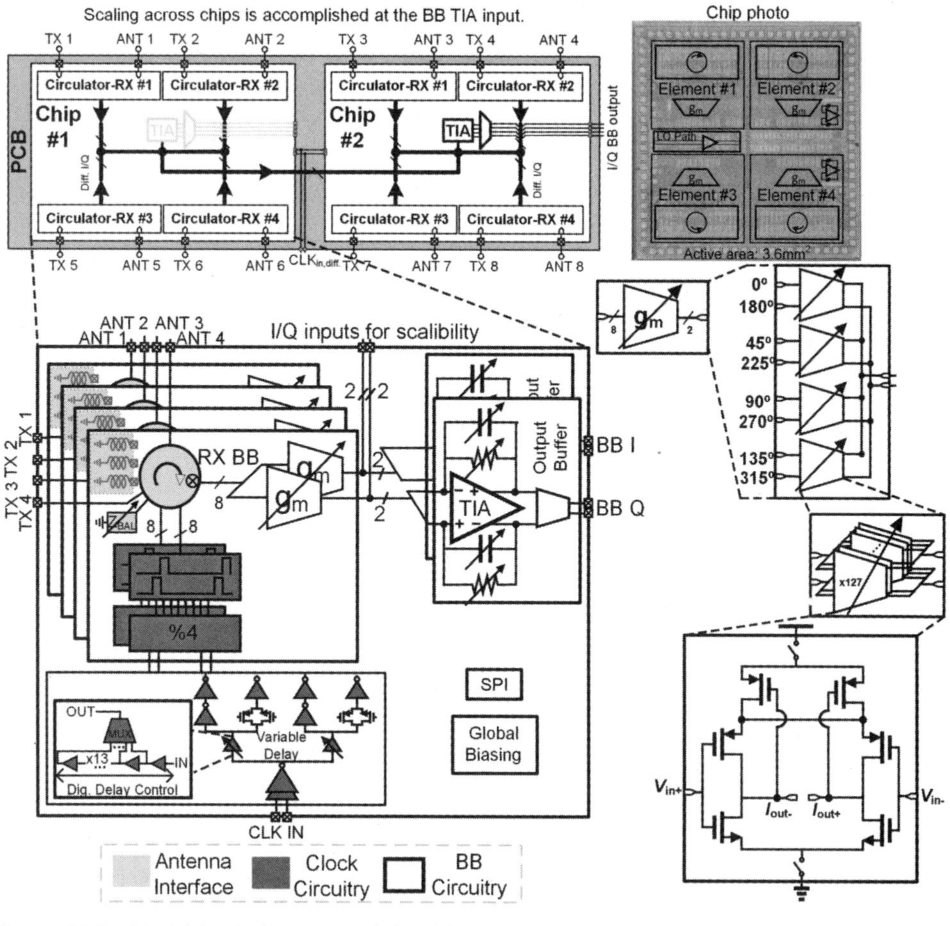

Figure 14.7 Block/circuit diagram and chip photo of the 65-nm CMOS 730-MHz, 8-element IBFD circulator receiver (C-Rx) phased-array system employing scalable 4-element ICs. (© 2018 IEEE. Reprinted, with permission, from [8].)

of 8-phase baseband nodes within each C-Rx further simplifies the Rx beamforming, as 7-bit programmable g_m-cells are implemented for each baseband node in each C-Rx and later combined in the current domain across all 4 elements into low-input impedance IQ transimpedance amplifiers (TIAs) (implemented using two-stage op-amps) to allow for Cartesian beamforming. The low input impedance combining point is also brought over to pads to facilitate scaling across chips. To tile two chips, the TIAs in the second chip would first be turned off, and the current from its g_m-cells would be combined into the TIAs of the first chip, with the low impedance making the board trace capacitance more resilient.

Each C-Rx exhibits 1.7-dB Tx-ANT loss, +28-dBm Tx-ANT IIP3, 41-dB nominal single-element ANT-baseband conversion gain, –31 dBm/+22.5 dBm in-band/out-of-band ANT-baseband IIP3, and 5-dB single-element noise figure. IBFD phased array measurements are performed with the 2 × 4 rectangular array of slot loop antennas described earlier (Figure 14.6(a)). Two ICs are tiled on a customized printed circuit board (PCB) to realize an 8-element IBFD C-Rx phased array. We also implemented a custom 8-element phased-array Tx using discrete components on a PCB. The isolation of each C-Rx is around 15 dB. When the Tx and Rx arrays are configured to perform nominal beamforming in the broadside, the average array SIC over 16.25 MHz is only 23 dB, thus preventing the array from operating in the IBFD mode. When the Tx and Rx arrays are configured to repurpose beamforming DOFs for SIC based on the optimization code described earlier while allowing 3-dB Tx and Rx array gain loss, 40.7 dB array SIC is achieved over 16.25-MHz bandwidth. The SIC profile is very wideband and is very similar to the simulated profile shown in Figure 14.6(e), albeit at somewhat lower SIC levels, due to practical factors that are not considered in the simulations (e.g., internal isolation of the circulator, quantization of beamforming weights). Lastly, custom-designed tuners are integrated with the antennas, and cooptimizing the tuners (which are configured identically across all elements) to yield 50-dB array SIC over 16.25 MHz. The Rx and Tx array gains were measured across frequency for broadside excitation with these beamforming weights optimized for SIC, and the 3-dB array gain loss was verified. The synthesized Tx and Rx array patterns for these weights are illustrated in Figure 14.6(c, d).

With the beamforming and SIC configured and the Tx array power ($P_{TX} \cdot AG_{TX}$) swept, the next step is to monitor the gain imparted to a weak in-band signal radiated towards the array. The output 1-dB compression point of the weak in-band signal occurs at +16.5 dBm, which can be interpreted as the Tx array power handling. Lastly, two-tone Tx tests are performed with beamforming and SIC configured, and the Rx baseband outputs are also monitored. The effective IIP3 referred to the Tx array power is +17.5 dBm, and at +16.7-dBm average Tx array power, nonlinear Volterra-series-based digital SIC is able to suppress the residual total SI and its associated third-order intermodulation distortion (IM3) to below –84 dBm, indicating 100-dB total array SIC. Another 19 dB of SIC is needed to suppress the SI to the array noise floor (on the effective IIP3 graph, the noise floor would be at N_{floor}/AG_{RX} = –103 dBm) and can potentially be achieved with additional analog SIC.

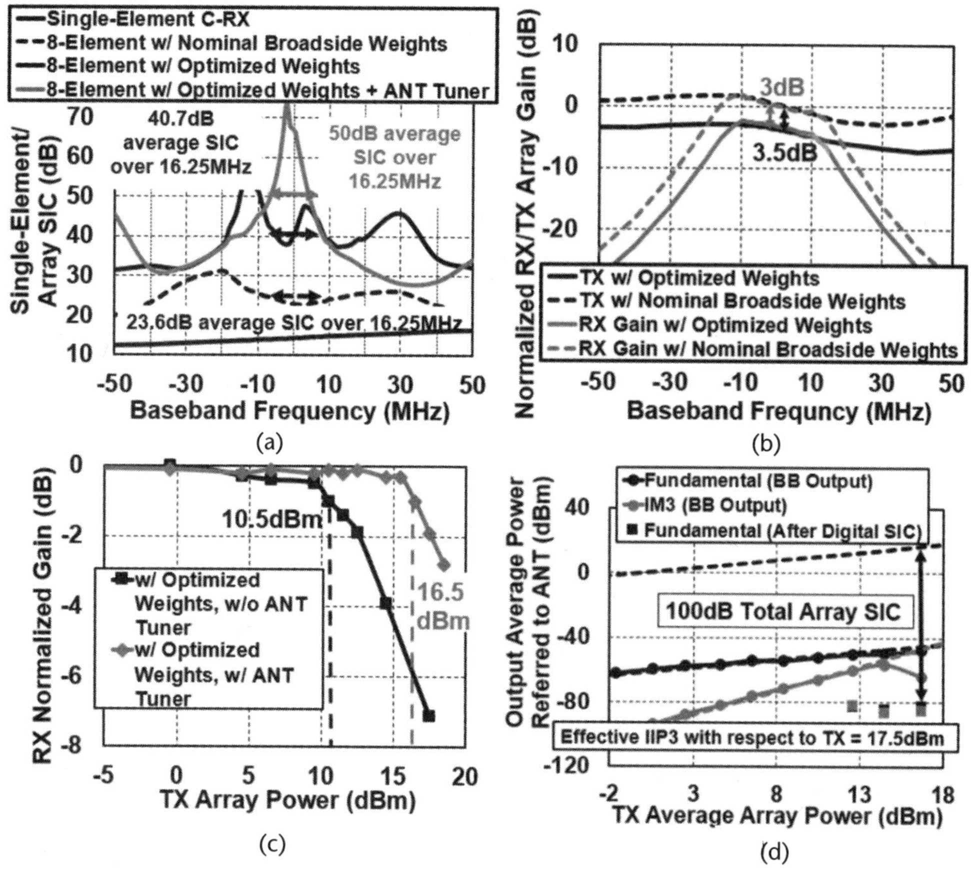

Figure 14.8 Measured IBFD phased-array performance across 8 elements (tiling of 2 ICs): (a) array SIC; (b) impact of optimized weights to achieve SIC on the Tx/Rx array gain; (c) gain compression of a small received signal under the influence of Tx power with optimized weights both with and without the antenna tuner; and (d) two-tone Tx test that tracks the Tx total SI and its third-order intermodulation distortion (IM3) products at the receiver output with digital SIC. (© 2018 IEEE. Reprinted, with permission, from [8].)

14.5 MIMO Arrays Based on Shared-Delay Cancellation

14.5.1 System Considerations

14.5.1.1 Dual-Injection MIMO Canceller

Ideally, cancellers should be low-noise, highly linear and with minimum area and power overhead. The cancellation signal can be injected at different stages of the Rx chain. Injecting before the LNA in the Rx chain using passive noiseless components can substantially enhance the power handling with minimum noise figure penalty (as discussed in Chapters 8 and 9). Achieving wideband cancellation through the implementation of true-time delay at RF is very challenging and is typically achieved at baseband through the use of noisy active delay cells [9]. In this work, we pursue a dual-injection MIMO cancellation technique, depicted in Figure 14.9, where passive noiseless cancellers provide initial SI and XI cancellation at RF before amplification and relax Rx dynamic range requirements. Then FIR-based active baseband cancellers further knock down SI after the baseband amplifier with minimum noise figure degradation.

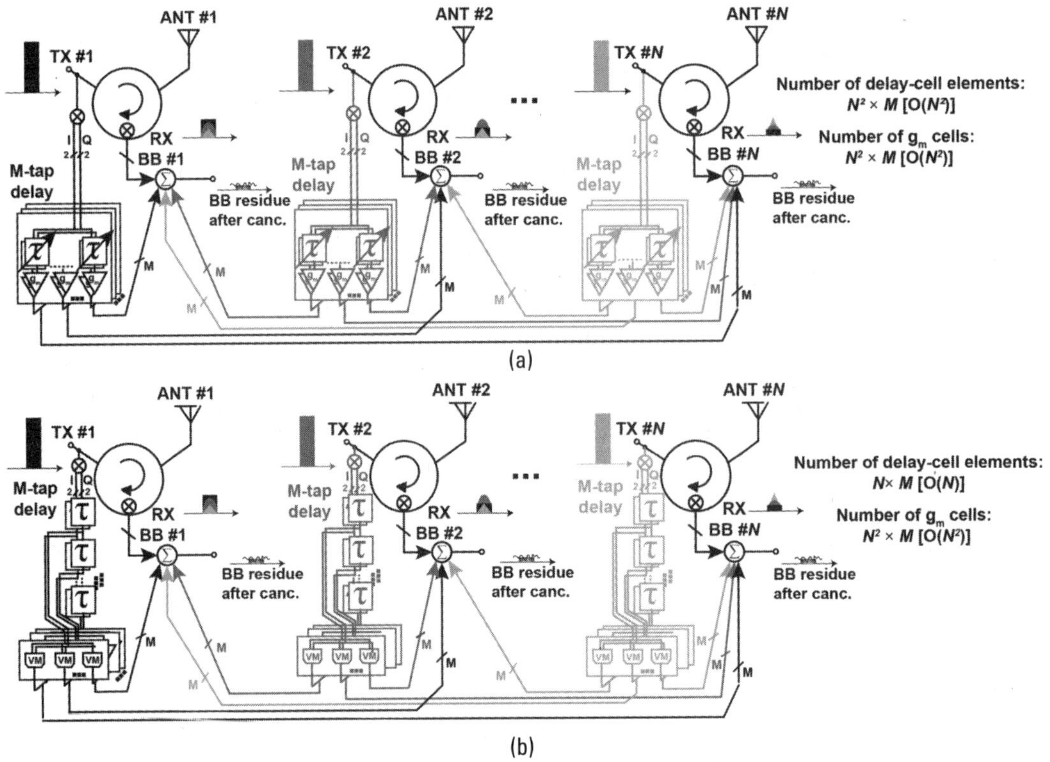

Figure 14.9 MIMO baseband (BB) canceller structure and power consumption distribution: (a) typical MIMO canceller; and (b) proposed shared-delay BB canceller architecture. (© 2019 IEEE. Reprinted, with permission, from [5].)

As discussed in Chapter 1, several leakage paths contribute to the SI and XI, such as direct coupling from the integrated circulator; on-chip routing as well as PCB traces; reflection at the circulator antenna port due to mismatch caused by the wirebond, package parasitics, and off-chip balun; reflection due to antenna impedance variation; direct coupling between adjacent antennas; and environmental reflections from nearby objects [9]. The overall channel response is the combination of all these leakages and can consist of a large group delay on the order of tens of nanoseconds. Delay cells with large group delays are therefore necessary to achieve wideband cancellation in IBFD radios, as mentioned in Chapter 8.

14.5.1.2 Shared-Delay Baseband Canceller Architecture

A traditional baseband canceller architecture employs parallel programmable delay cells with gain control to emulate the SI channel through the implementation of FIR filters [10]. Figure 14.10 presents an extension of such an approach to MIMO. Parallel delay cells are necessitated in [10] by the use of transmission lines on board to realize delay, making it challenging to tap the signal from intermediary points. Active delay elements in an integrated setting enable cascaded delay lines with intermediary taps [9]. In [9, 10], the ability to control phase shift independent of delay is absent. In our implementation, we employ FIR filters operating on both I and Q components of the Tx baseband signal (obtained through I/Q downconversion of

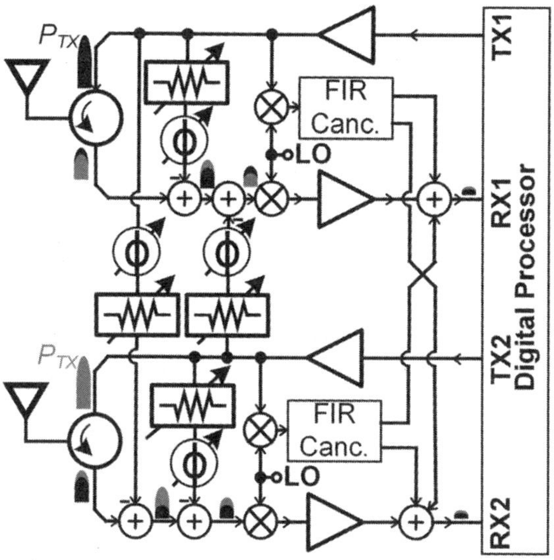

Figure 14.10 Proposed dual-injection MIMO (RF and baseband (BB)) canceller to address power handling and complexity challenges of IBFD MIMO implementation. (© 2019 IEEE. Reprinted, with permission, from [5].)

the Tx output, which has the advantage of capturing PA nonlinearities as well as enabling mitigation of LO phase-noise-induced residual SI assuming a common LO [11]) and injecting into both I and Q paths of the Rx, thus enabling independent control of phase and delay through IQ vector modulation, as discussed in Chapter 8. Furthermore, MIMO operation additionally complicates matters as the number of delay cells and gain controls scale quadratically with the number of elements and are area-hungry and power-hungry (Figure 14.10). In this work, a shared-delay-cell architecture, shown in Figure 14.10, is proposed where g_m-cell based vector modulators (VMs) tap from the discrete delay steps of each Tx element and inject into all Rx baseband paths to approximate the SI channel. Because each Tx path has a single delay line whose taps are used in common for all Rx paths, the number of required delay lines increases only linearly with the number of MIMO elements. Although the number of required g_m-cell based VMs still scales quadratically with the number of elements, the canceller power and area consumption are dominated by the delay elements. Therefore, shared-delay baseband cancellation in conjunction with passive MIMO RF cancellation directly addresses the $O(N^2)$ power consumption challenge.

14.5.1.3 Baseband Canceller Noise Penalty

The noise penalty associated with the proposed baseband cancellation can be analyzed by considering the noise of active delay cells.

$$\Delta F_{i,SIC} = \frac{v_{n,out,d}^2 ISO_{ii}^2}{A_{delay,i}^2 k_B T \times R_S \times CG_i^2} + \sum_{j \neq i}^{N} \frac{v_{n,out,d}^2 ISO_{ij}^2}{A_{delay,j}^2 k_B T \times R_S \times CG_j^2}, \quad (14.4)$$

where ISO_{ij} represents the ANT-to-ANT isolation between elements i and j (including the RF cross-talk interference cancellation (XIC) and circulator conversion gain) and $A_{delay,j}$ represents the lumped voltage gain of the jth delay element.

Figure 14.11 shows the circuit model used for the analysis of noise degradation, where ISO_{ii} represents intrinsic Tx-Rx isolation in the ith C-Rx in addition to RF SIC provided by feed capacitors. Note that the C-Rx downconverts the Tx signal to baseband, and thus ISO_{ii} also includes the intrinsic downconversion gain that the circulator has due to its architecture. The signal path transconductance following the circulator in the ith-element is represented by $g_{m,RX,i}$. For cancellation of the residual SI in the Rx baseband, (delayed) copies of the downconverted Tx signal must be injected into the Rx following $g_{m,RX,i}$. Assuming a single dominant delay in the SI channel, a voltage gain of $A_{delay,i}$ from the PA output to the corresponding tap in the delay line, and a weight of $g_{m,SIC}$ in the subsequent g_m-cell, baseband SIC requires

$$g_{m,SIC} = \frac{ISO_{ii} \times g_{m,RX}}{A_{delay,i}}. \tag{14.5}$$

The noise factor degradation in the system due to active baseband SIC is then given by

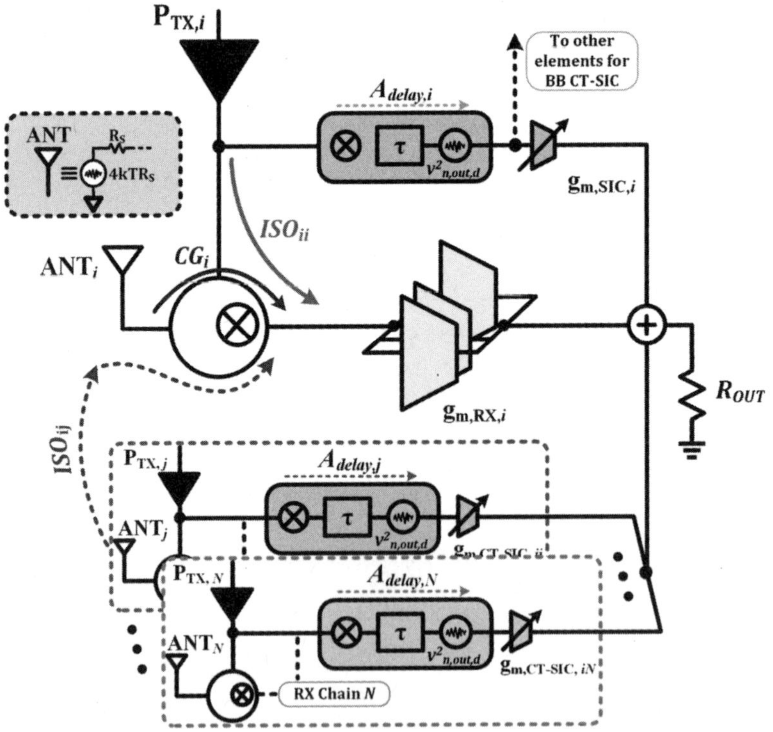

Figure 14.11 Model for the analysis of noise degradation due to active delay-based baseband (BB) cancellation. (© 2019 IEEE. Reprinted, with permission, from [5].)

$$\Delta F_{SIC} = \frac{v^2_{n,out,d} ISO^2_{ii}}{A^2_{delay,i} k_B T \times R_S \times CG^2_i}, \tag{14.6}$$

where $V^2_{n,out,d}$ represents the total output noise of the delay element. Note that the noise associated with the delay element is considered but not the SIC g_m-cell, as the delay element's noise contribution dominates in our implementation (the g_m-cell input-referred noise power is around 6 aV^2/Hz while delay-cell output-referred noise power is 52 aV^2/Hz). In the MIMO case, the baseband cancellers must cancel both the residual and XI. Using similar constraints for XIC as used for SIC in (14.6), the overall degradation in the noise figure in the MIMO case is given by

$$\Delta F_{i,SIC} = \frac{v^2_{n,out,d} ISO^2_{ii}}{A^2_{delay,i} k_B T \times R_S \times CG^2_i} + \sum_{j \neq i}^{N} \frac{v^2_{n,out,d} ISO^2_{ij}}{A^2_{delay,j} k_B T \times R_S \times CG^2_j}, \tag{14.7}$$

where ISO_{ij} represents the ANT-to-ANT isolation between elements i and j (including the RF XIC and circulator conversion gain) and $A_{delay,j}$ represents the lumped voltage gain of jth delay element.

In order to illustrate the impact of delay cell noise, we consider (14.7) assuming per differential delay tap (based on our implementation). Figure 14.12 plots the degradation in the noise figure (assuming a nominal noise figure of 6.5 dB) as a function of ISO_{ii} for a different number of MIMO elements. Note that MIMO cross-talk isolation levels (ISO_{ij}) are assumed to be identical to ISO_{ii}. Also, note that a higher RF SIC in decibels on the x-axis is equivalent to a smaller absolute value for ISO_{ii}. Several trade-offs become apparent. As expected, higher RF SIC/XIC reduces the noise penalty of active delay-based baseband cancellation, and at lower RF SIC/XIC levels, a higher MIMO order results in more severe penalties. A higher-voltage gain to the output of the delay cells achieved, for instance, by

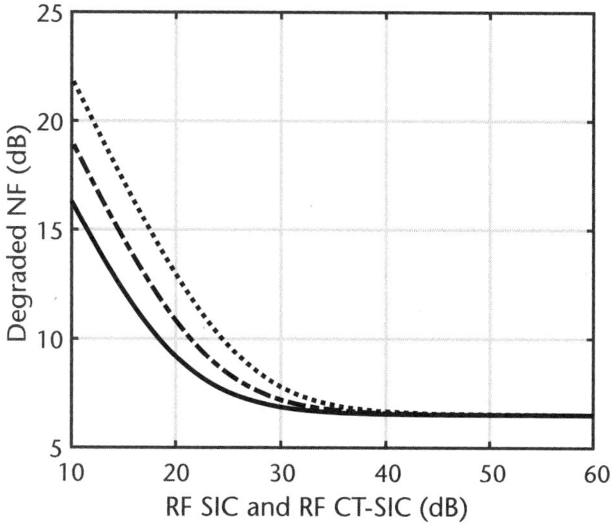

Figure 14.12 Impact of the baseband (BB) active delay cells on noise figure (NF) as a function of RF SIC and XI. (© 2019 IEEE. Reprinted, with permission, from [5].)

coupling a stronger portion of the Tx signal, also alleviates the noise penalty, at the expense of the need for higher linearity and power handling in the cancellation path.

14.5.2 Integrated Circuit Implementation

Figure 14.13 presents block and circuit diagrams of the 2.2-GHz IBFD 2-element MIMO C-Rx prototype using 65-nm CMOS technology [12]. The N-path-filter-based C-Rx described in [9] is deployed in a differential configuration as the shared antenna interface, which merges a nonmagnetic circulator with a downconverting mixer, directly providing baseband signals at its output. Programmable feed capacitors are connected to both sides of the *N*-path filter inside the integrated circulator to provide initial passive RF SI and XI cancellation [5]. Each of the feed capacitors for the VM RF SI cancellers are implemented using 6 bits of control. Two differential capacitor banks (6 bits) are added to the Tx and ANT ports to compensate for

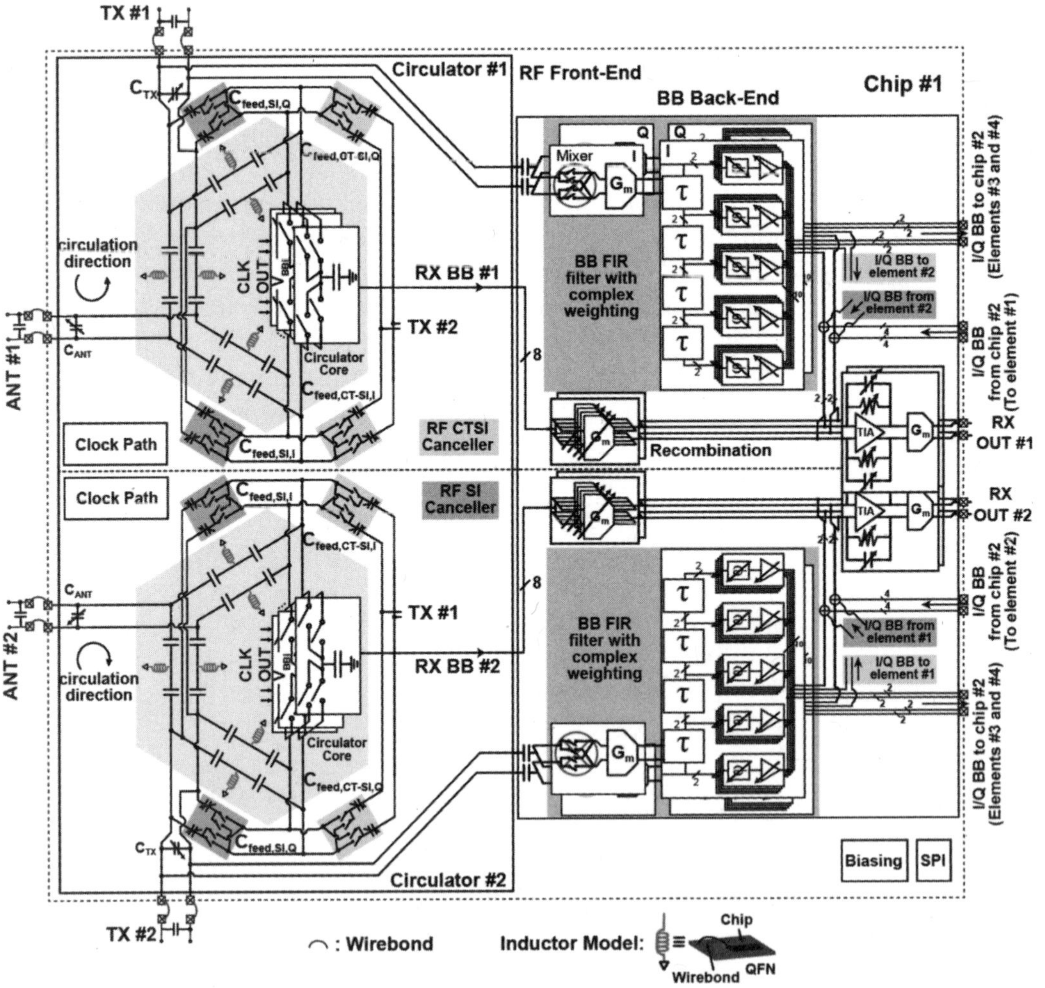

Figure 14.13 Block and circuit diagram of the 65-nm CMOS 2.2-GHz IBFD 2-element MIMO circulator-Rx (C-Rx) with high Tx power handling exploiting MIMO RF and shared-delay baseband (BB) SIC. (© 2019 IEEE. Reprinted, with permission, from [5].)

package parasitics and to further improve the antenna VSWR coverage. Furthermore, MIMO shared-delay baseband cancellation using 4-stage shared delay lines is implemented to achieve further SI and XI cancellation. Each Tx signal is coupled through a small capacitor to a 4-path passive mixer to downconvert the Tx signal for baseband cancellation. Delay cells are implemented with 4 taps using a topology similar to [9] but with a complementary implementation to enhance the maximum voltage swing at the output resulting in less noise contribution from the baseband canceller and less power consumption in the g_m-cells. A single IQ delay line is used for each Tx, with g_m-cells injecting into each Rx IQ path. Therefore, the delay line complexity only grows as $O(N)$ while the g_m-cell complexity grows as $O(N^2)$. A two-stage TIA is used for each element to provide a low-input-impedance node into which the output of all g_m-cell based VMs of the baseband cancellers can be combined with the Rx side g_m cells while preserving high linearity. This low-impedance node can also be used to tile ICs to increase the MIMO order and achieve XIC across ICs.

The chip microphotograph of the 65-nm CMOS IBFD 2-element MIMO C-Rx array is shown in Figure 14.14(a). It has a total active area of 5.6 mm^2, is packaged in an 80-pin, 12 mm × 12 mm quad-flat no-leads (QFN) package, and is mounted on an FR-4 PCB for all measurements. The performance of the feed-capacitor-based RF SI/XI cancellers are shown in Figure 14.14(c, d). RF SI capacitors, in addition to the capacitor banks at Tx and ANT ports, can be used to cancel SI due to antenna reflection (AR). Measurements are done on 1,024 sample points showing full coverage of antenna impedance inside the VSWR = 2 circle, and partial coverage of VSWR < 3. Similarly, cross-element RF passive XI canceller measurements show cancellation of up to –18.5 dB of cross-antenna coupling with arbitrary phase. Note that RF passive cancellation consumes no power and can both enhance IBFD power handling substantially and reduce the noise figure penalty of subsequent baseband cancellers.

Figure 14.14(b) shows the setup for IBFD measurements to characterize both SIC and XIC performance. Figure 14.14(e) shows measurement results corresponding to SIC provided by: (1) circulator internal isolation (average ~15 dB), (2) RF SIC using the feed-forward capacitors, and (3) baseband SI cancellation. Over a bandwidth of 40/20 MHz around 2.2 GHz, the average total SIC with RF SIC turned on was ~28/30 dB. With both RF SIC and baseband SIC, the average total was ~35/45 dB. Similarly, Figure 14.14(f) shows the measurement results corresponding to the XIC provided by: (1) ~30-dB average ANT-to-ANT isolation, (2) RF XIC using the feed-forward capacitors, and (3) baseband SIC. The peak XIC provided with RF XIC (along with inherent ANT isolation) was up to ~60 dB. In addition, the total SIC (ANT-ANT isolation, RF XIC, and baseband XIC combined) was on an average ~42/53 dB over 40/20-MHz bandwidth around 2.2-GHz carrier frequency. The SIC and XIC performance are achieved with 1.5-dB and <2.1-dB penalty in terms of noise figure. It is also critical to measure the noise figure degradation in the presence of the Tx signal. Measuring the Tx-induced ANT-baseband noise figure using a regular Y-factor method is not feasible as the noise source is sensitive to the Tx power, and thus the gain method is used. For Tx operation power of +8 dBm (6 dB backed off from the +14-dBm Tx-induced Rx 1-dB compression point to account for the peak-to-average power ratio (PAPR) of a modulated signal), the increase in

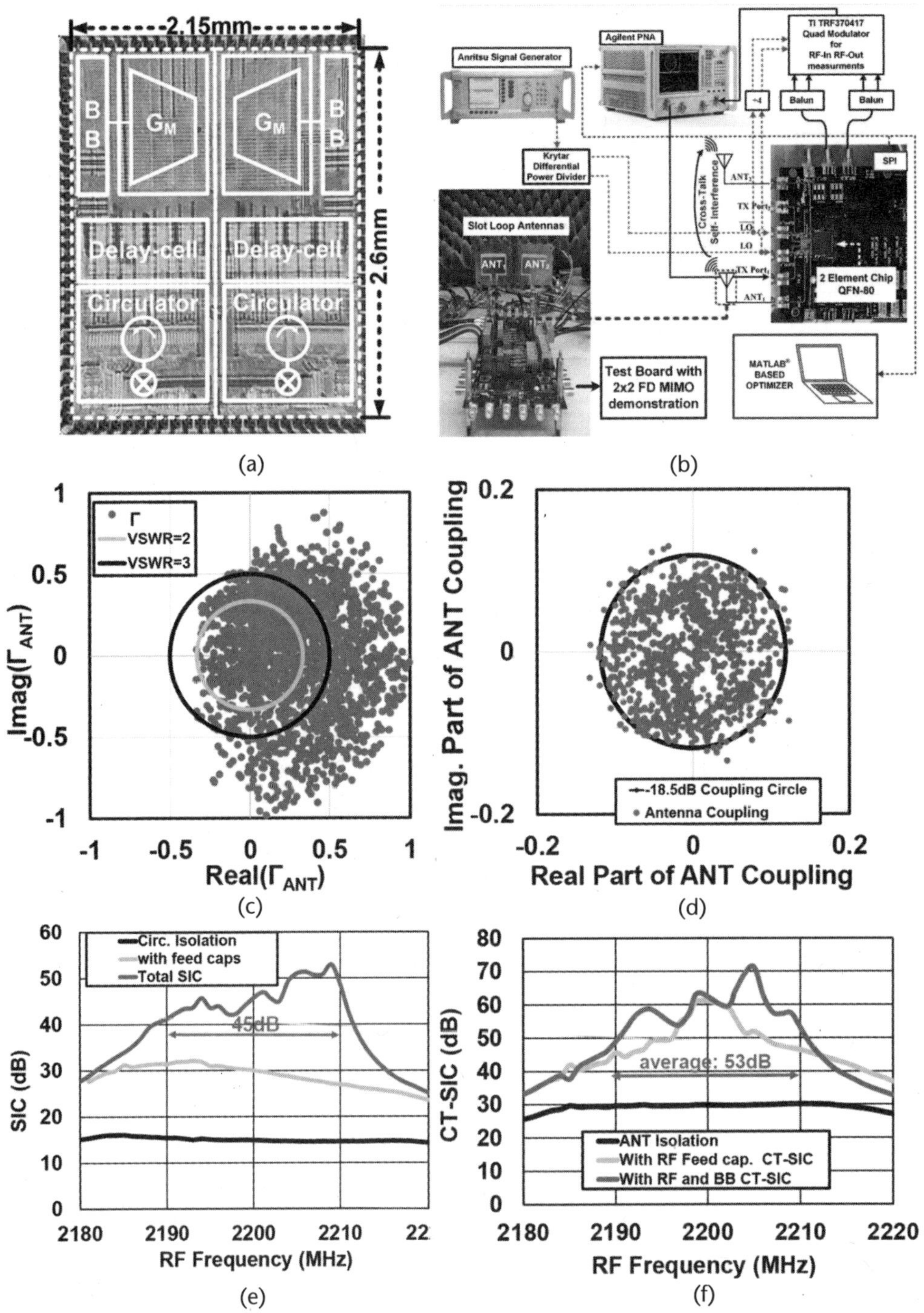

Figure 14.14 (a) Chip microphotograph of the 65-nm CMOS IBFD 2-element MIMO circulator-Rx (C-Rx); (b) IBFD MIMO wireless SIC and XIC measurement setup. Measured RF SIC/XIC performance; (c) SIC antenna VSWR coverage; and (d) XIC antenna coupling coverage. Wireless IBFD measurements: (e) single-element IBFD SIC measurement results; and (f) two-element IBFD XIC measurement results. (© 2019 IEEE. Reprinted, with permission, from [5].)

noise figure over the noise figure in the absence of a Tx signal is ~3 dB, while the increase at that same power level is ~14 dB without RF and baseband cancellation.

14.6 Practical Considerations

Design of IBFD multi-antenna radios is challenging due to the XI in addition to SI, which requires highly linear IC implementations. Our approach combines leakage signals and canceller signals in the current domain and deploys inverter-based g_m-cells to achieve maximum power handling. Noise contribution of delay elements in the IBFD MIMO cancellation paths is a strong function of $g_{m,sic}$ and has to handle a larger Tx signal to reduce these delay elements, causing us to use delay elements in complementary structures to minimize their noise contribution. The maximum delay required in a baseband canceller is a function of SI and XI. Thus, an evaluation of ANT-ANT leakage in the antenna array is critical to finding the range that is needed to be covered by delay cells to achieve an appropriate SIC and XIC profile.

14.7 Conclusion

14.7.1 Summary

In this chapter, we demonstrated how IBFD capability can be integrated with phased array and MIMO systems, which can substantially improve wireless spectral efficiency [13]. Although the combination of these two technologies can result in higher data rate and spectrum efficiency, it is challenging since a large number of SI and XI paths need to suppressed/cancelled.

We first presented a scalable 8-element IBFD phased array C-Rx that combines an IBFD operation with phased arrays. A joint optimization of Tx and Rx beamforming is employed for simultaneous beamforming and RF SIC with a minimum penalty on array gains. The beamforming is implemented with a minimum overhead by exploiting the available baseband outputs of N-path-filter-based integrated C-Rx [14, 15]. A 730-MHz scalable 8-element IBFD phased array C-Rx prototype was presented in 65-nm CMOS, which achieved up to 50 dB of array SIC over +16.25 MHz with only less than 3.5/3-dB degradation in Tx and Rx array gain while handling +16.5-dBm Tx array power. In conjunction with digital SIC, our IBFD phased array achieved an overall 100 dB of array SIC.

We then discussed how IBFD capability can be integrated with MIMO radios using RF and baseband SI/XI cancellers. MIMO operation increases the SIC requirements by a factor of N^2, resulting in severe area, power consumption, and noise penalty. We presented the detailed analysis, design, and implementation of RF/baseband cancellers that address this challenge, as well as a technique to enhance the power handling of the integrated C-Rx. A 65-nm CMOS scalable 2-element MIMO IBFD C-Rx array with integrated N-path-filter-based nonmagnetic circulators was described that exploits techniques in the circulator N-path-filters to enhance Tx power handling by 8 dB, and features area-efficient and power-efficient passive RF and active baseband wideband MIMO cancellation with shared-delay elements to address the $O(N^2)$ cancellation challenge. The prototype exhibits up to 35/45-dB average SIC across 40/20-MHz bandwidth, more than 42/53-dB average XIC across 40/20-MHz bandwidth with <2.1-dB degradation in Rx noise figure, and an overall Tx power handling of +14 dBm.

14.7.2 Future Directions

One future research direction is to develop an efficient scheme for obtaining the optimized RF and baseband canceller weights based on SI and XI channels, which is important for practical, real-time IBFD MIMO systems. Furthermore, a digital canceller is required in addition to an analog one to establish an IBFD link. As mentioned in Chapter 11, the digital cancellers' performance depends heavily on how accurate the SI channel estimation is, and new algorithms are required for the IBFD MIMO radio in which both SI and XI should be suppressed down to the noise floor.

References

[1] Naqvi, A. H., and S. Lim, "Review of Recent Phased Arrays for Millimeter-Wave Wireless Communication," *Sensors*, Vol. 18, No. 10, 2018.

[2] Hong, W., et al., "Study and Prototyping of Practically Large-Scale mmWave Antenna Systems for 5G Cellular Devices," *IEEE Communications Magazine*, Vol. 52, No. 9, September 2014, pp. 63–69.

[3] Hassan, N., and X. Fernando, "Massive MIMO Wireless Networks: An Overview," *Electronics*, Vol. 6, No. 3, 2017.

[4] Zhou, J., et al., "Integrated Full-Duplex Radios," *IEEE Communications Magazine*, Vol. 55, No. 4, April 2017, pp. 142–151.

[5] Baraani Dastjerdi, M., et al., "Analysis and Design of a Full-Duplex Two-Element MIMO Circulator-Receiver with High Tx Power Handling Exploiting MIMO RF and Shared-Delay Baseband Self-Interference Cancellation," *The IEEE Journal of Solid-State Circuits (JSSC)*, Vol. 54, No. 12, December 2019, pp. 3525–3540.

[6] Everett, E., et al., "SoftNull: Many-Antenna Full Duplex Wireless via Digital Beamforming," *IEEE Transactions on Wireless Communications*, Vol. 15, 2016, pp. 8077–8092.

[7] Zhang, Z., et al., "Full Duplex 2x2 MIMO Radios," *Proc. WCSP'14*, 2014.

[8] Baraani Dastjerdi, M., et al., "Full Duplex Circulator-Receiver Phased Array Employing Self-Interference Cancellation via Beamforming," *2018 IEEE Radio Frequency Integrated Circuits Symposium (RFIC)*, Philadelphia, PA, 2018, pp. 108–111.

[9] Reiskarimian, N., et al., "Analysis and Design of Commutation-Based Circulator-Receivers for Integrated Full-Duplex Wireless," *IEEE Journal of Solid-State Circuits (JSSC)*, Vol. 53, No. 8, August 2018, pp. 2190–2201.

[10] Bharadia, D., et al., "Full Duplex Radios," *Proc. of the ACM SIGCOMM'13*, 2013.

[11] Zhou, J., and H. Krishnaswamy, "A System-Level Analysis of Phase Noise in Full-Duplex Wireless Transceivers," *IEEE Transactions on Circuits and Systems II: Express Briefs*, Vol. 65, No. 9, September 2018, pp. 1189–1193.

[12] Baraani Dastjerdi, M., et al., "28.6 Full-Duplex 2×2 MIMO Circulator-Receiver with High Tx Power Handling Exploiting MIMO RF and Shared-Delay Baseband Self-Interference Cancellation," *IEEE International Solid-State Circuits Conference (ISSCC)*, San Francisco, CA, 2019, pp. 448–450.

[13] Bharadia, D., and S. Katti, "Full Duplex MIMO Radios," *Proc. USENIX, NSDI'14*, 2014.

[14] Chen, T., et al., "Wideband Full-Duplex Phased Array with Joint Transmit and Receive Beamforming: Optimization and Rate Gains," *Proc. ACM MobiHoc'19*, 2019.

[15] Zhang, T., et al., "Wideband Dual-Injection Path Self-Interference Cancellation Architecture for Full-Duplex Transceivers," *IEEE Journal of Solid-State Circuits (JSSC)*, Vol. 53, No. 6, 2018, pp. 1563–1576.

CHAPTER 15

Photonic-Based Cancellation Systems

Eric C. Blow and Paul R. Prucnal

15.1 Introduction

The increasing severity of the wireless spectral deficit has accelerated the transition from LTE to 5G standards [1, 2]. 5G requires an order of magnitude increase in capacity per link and expansion of operating frequencies [3]. This necessitates new communication systems to be capable of flexible spectral resource allocation and wideband tunable signal processing to maintain relevance. Traditional RF electronic technology struggles to meet this demand [4]. Dynamically processing high-bandwidth signals over all existing communication channels is a task perfect for microwave photonics [5], as mentioned in Chapters 5 and 7.

Significant alleviation of the spectral deficit would result from using current RF spectrum more efficiently. In this chapter, we discuss the significance of using microwave photonic signal processing techniques to enable IBFD communications, as highlighted within the IBFD transceiver diagram of Figure 15.1(a) and the SIC techniques tree of Figure 15.1(b).

As depicted in Figure 15.2, a microwave photonic processor requires four primary components: electro-optic (E-O) conversion (modulation), optical signal processing, transmission medium, and optic-electro (O-E) conversion (detection). The optical processing is computed by tunable wideband optical elements that can simultaneously process tens of gigahertz [5]. With recent advances in silicon photonics, photonic integrated circuits (PICs) have demonstrated commercially available modulators [6] and photodetectors [7] operating from 0 to 40 GHz as well as research-grade modulators operating up to 100 GHz [8]. In a microwave photonic system, the entirety of the RF spectrum becomes upconverted to a 193-THz optical carrier and therefore becomes narrowband relative to the fractional bandwidth of the carrier and the flat frequency response of the tunable optical elements. For example, the LTE band number 5 is a 25-MHz signal, which has a dramatically higher fractional bandwidth when modulated onto an 850-MHz carrier in comparison to a 193-THz optical carrier. Photonic systems can further increase processing capacity by leveraging wavelength division multiplexing (WDM), where RF signals

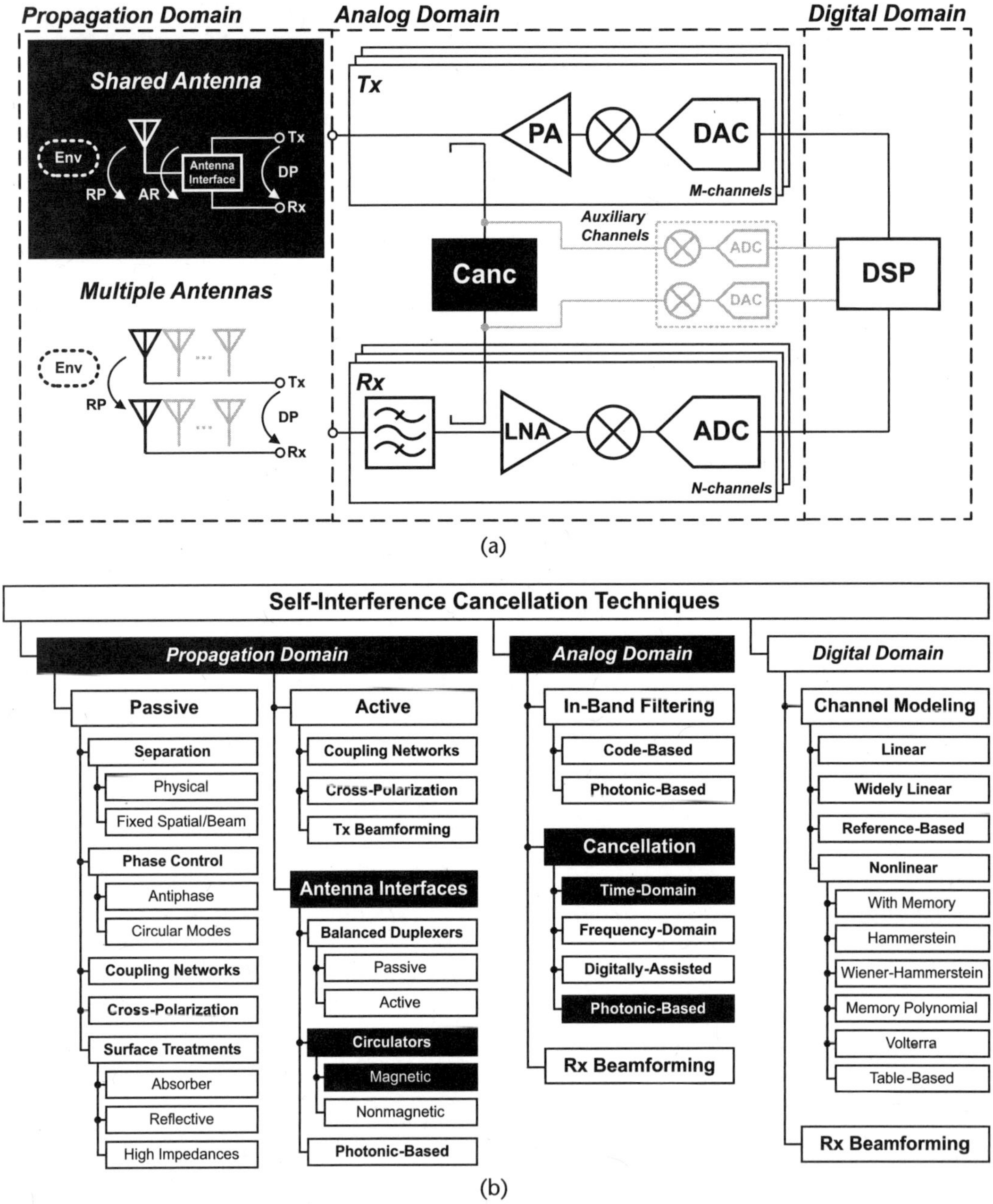

Figure 15.1 Photonic-based cancellation systems and specific chapter focus highlighted within (a) a generic IBFD transceiver and (b) a SIC techniques tree diagram.

are modulated on unique wavelengths of light that can be simultaneously processed on a single waveguide.

While RF electronic processing is narrowband and often fixed to single operating points, microwave photonic links allow for wideband analog signal processing with extremely high fidelity. These systems leverage the physics of optics to overcome fundamental limitations of RF analog processing and achieve novel functionality. Microwave photonic systems offer wide-bandwidth operation, high

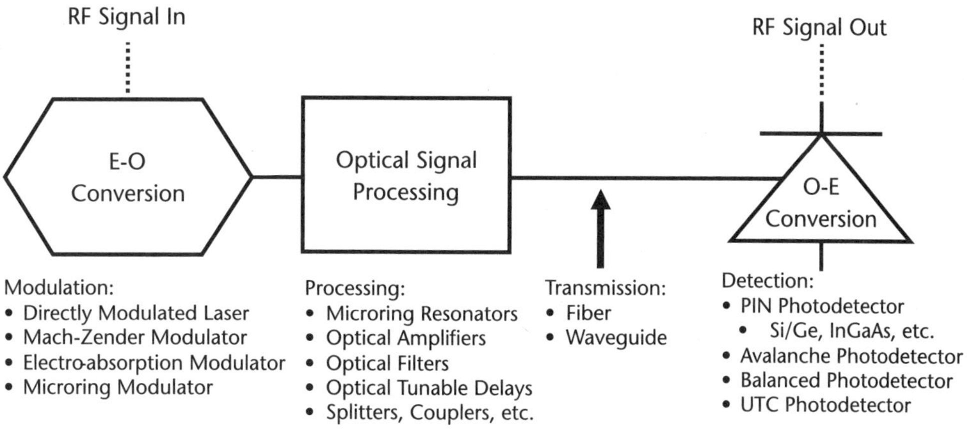

Figure 15.2 Generalized microwave photonic link, including modulations, processing, transmission, and detections. Each required element includes corresponding optical devices for electro-optic (E-O) and optic-electro (O-E) conversion.

dynamic range, low susceptibility to electromagnetic interference, reconfigurability, and low latency tunability [5, 8–10]. For accurate comparison across technological platforms, photonic systems should be analyzed with the same RF metrics: link loss, noise figure, and dynamic range. While dynamic range is intrinsically high in microwave photonic, link loss, and noise figure are limited by the electro-optic-electro (E-O-E) conversion and are typically poor in comparison to RF electronics [11]. Research focused on improving RF metrics in microwave photonic systems has been the focus of photonic engineers in recent years and has proven successful [11–14].

As briefly discussed in Chapter 8, these microwave photonic links can be constructed to create microwave photonic cancellers (MPCs) for IBFD. These designs typically require three fundamental elements: tunable optical delays for phase matching, tunable optical weights for amplitude matching, and a balanced photodetector (BPD) for subtraction. MPCs have been developed significantly over the past decade, including new architectures, system integration, and RF performance optimization [15–31]. The continued development of MPCs offers a pathway forward to photonic-enabled full-duplex communication in next-generation wireless networks.

15.2 Discrete Microwave Photonic Cancellers

RF electronic cancellation systems are generally a favorable method of interference isolation when system requirements are narrowband and fixed operation, as discussed in Chapters 8, 9, and 10. RF cancellers have demonstrated active analog cancellation of 47 dB over 80 MHz operating at 2.4 GHz [32] and multitap cancellation of 30 dB over 30 MHz operating at 2.4 GHz [33]. With the implementation of the 5G standard, network requirements are increased to wider bandwidth, higher-frequency operating points, and dynamic spectral allocation [34]. In these future environments, microwave photonic-enabled technology thrives.

A generalized schematic of a one-tap, single-ended canceller, shown in Figure 15.3, consists of two optical paths. The top optical path is modulated by the received signal, which is an unknown combination of the signal-of-interest (SOI) and the interference signal (IS). This optical path is terminated in the positive port of a BPD. The bottom optical path is modulated by the interference reference signal, which in an IBFD system is the transmitted signal. The reference modulated carrier is then processed by an optical linear filter composed of tunable optical weights and delays that matches the amplitude and phase of the reference signal to the received IS. Following the matched filter, the optical signal is terminated in the negative port of a BPD. The two optical signals are detected, and the modulated signals are subtracted by the BPD. When matched properly, the interference is cancelled and the uncorrupted SOI is detected.

The cancellation is equivalent to the interference power ratio over a defined bandwidth, Δf, with (P_c) and without (P_{nc}) cancellation enabled, according to

$$Cancellation = \frac{\int_{\Delta f} P_c(f)}{\int_{\Delta f} P_{nc}(f)}. \tag{15.1}$$

The amount of cancellation achievable by a canceller determines the maximum difference in power between the transmitted and received signal in an IBFD communication link. The amount of cancellation achievable by a given canceller is dependent on how accurately and precisely the matching mimics the transfer function of the wireless channel, as discussed in Chapter 8. This relationship can be shown by modeling the canceller without the SOI ((15.2) through (15.4)). When modeling, removing the presence of the SOI simplifies the objective of the cancellers to minimizing the output. Therefore, the inputs of the system are two equal interferers: $X(\omega)$, and then the output of the canceller $Y(\omega)$ will be equal to the difference in the transfer functions between the two optical paths, H_{Rx} and H_{Tx}, which represent both the difference in the optical paths and the difference between the matched filter and the wireless channel. The transfer functions of the Rx and Tx branches are represented by attenuation and time delay, α and $e^{-j\omega t}$, respectively. Therefore, the transfer function of the canceller can be represented by (15.3) and highlights the performance limitations of the cancellation achievable to two metrics: amplitude and phase mismatch, as shown in (15.4).

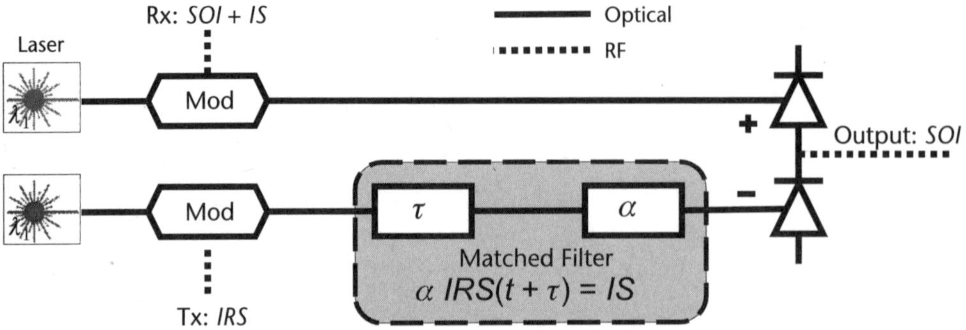

Figure 15.3 Generalized schematic of one-tap single-ended MPC, indicating a SOI, IS, interference reference signal (IRS), weight α, and delay τ.

$$Y(\omega) = X(\omega)H_{Rx} - X(\omega)H_{Tx} \tag{15.2}$$

$$\left|\frac{Y(\omega)}{X(\omega)}\right|^2 = \alpha_{Rx}^2 * \left[1 + \left(\frac{\alpha_{Tx}}{\alpha_{Rx}}\right) - 2\frac{\alpha_{Tx}}{\alpha_{Rx}}\cos\left(i\omega\left(\tau_{Tx} - \tau_{Rx}\right)\right)^2\right] \tag{15.3}$$

$$\begin{aligned} &\text{Amplitude Mismatch: } \Delta\alpha = \frac{\alpha_{Tx}}{\alpha_{Rx}}; \\ &\text{Phase Mismatch: } \Delta\phi = \left(\tau_{Rx} - \tau_{Tx}\right) \end{aligned} \tag{15.4}$$

An MPC developed in 2013 demonstrated narrowband 10-kHz cancellation of over 65 dB and broadband 40-MHz cancellation over 30 dB at an operating frequency of both 900 MHz and 2.4 GHz [17]. This system was composed of electro-absorption modulators (EAMs) for modulation, microelectromechanical systems (MEMS) optical attenuators for amplitude matching and a mirror-based optical delay line for phase matching. The experimental results shown in Figure 15.4 illustrate the cancellation of both narrowband and wideband ISs. The received signal without cancellation is plotted with a dashed line. This curve shows that the SOI is buried by the stronger IS. The solid line shows that when cancellation is enabled and optimized, the IS sufficiently cancelled and the SOI is recovered. By simply tuning the weights and delays of the system, the matched filter can be controlled to operate at various frequency ranges. When tuning the MPC, the system

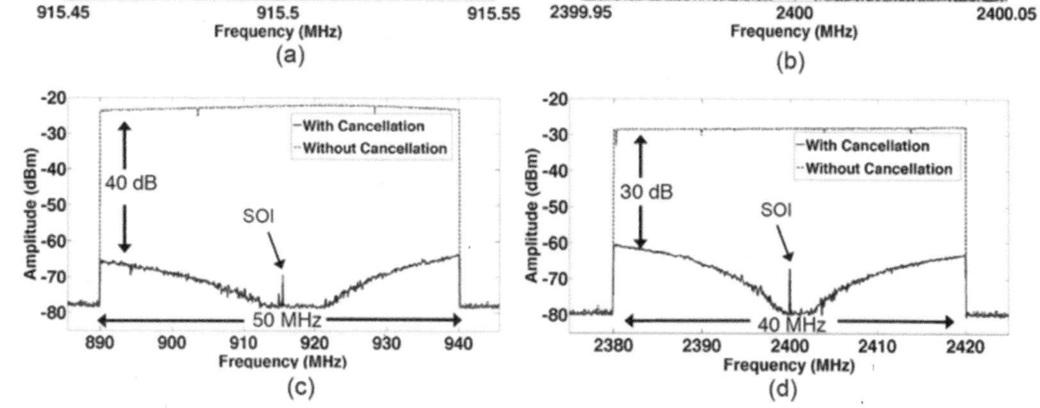

Figure 15.4 Experimental cancellation results of EAM-based MPC. Cancellation of narrowband interferers (top) and broadband interferers (bottom) at 900 MHz (left) and 2.4 GHz (right). (© 2013 IEEE. Reprinted, with permission, from [17].)

is accounting for frequency-dependent differences in the transfer functions of the two optical paths. When tuning the system to increase the instantaneous cancellation bandwidth, optimizing for minimization of average mismatch is prioritized over maximum cancellation at a single operating point. In the narrowband case, the transfer functions are well-matched, and the system can achieve higher cancellation depths.

The architecture of an MPC can be easily scaled to include additional interference reference taps. The addition of multiple taps enables cancellers to address the multipath wireless environments discussed in Chapter 1. In these conditions, SI signals can reflect off surfaces and take several paths when coupling from the transmitter to the receiver. These multipath effects result in a combination of amplitude and phase-shifted copies of the original interferer. In a single-tap cancellation system, the interference power of the second-strongest interferer would limit the cancellation depth achievable. MPCs can leverage the advantages of WDM to scale the number of taps of the filter to simultaneously cancel multipath interferers. This concept is demonstrated in the dashed box of Figure 15.5, where two optical sources with unique wavelengths are used to cancel two interferers.

For a system with n optical taps, the modeling equations can be expanded to include this increase in complexity ((15.5) and (15.6)). The trade-off for this increased functionality is increased complexity of control. The summation of amplitude and phase mismatch from all taps now limit the possible cancellation depth.

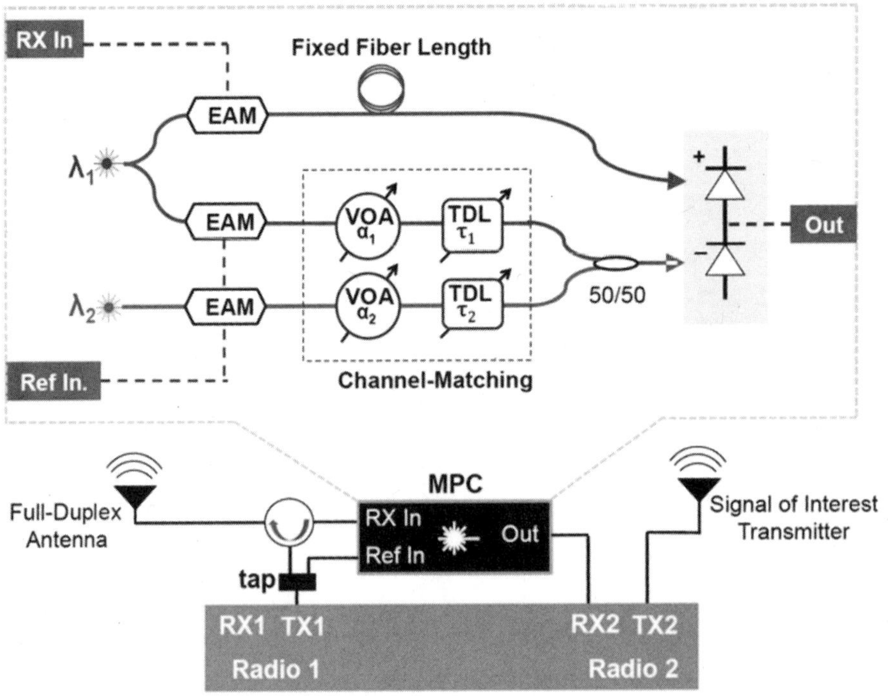

Figure 15.5 Experimental optical and RF schematic of MPC in full-duplex communication link with EAM, variable optical attenuators (VOAs) and tunable delay lines (TDLs). (© 2016 IEEE. Reprinted, with permission, from [20].)

$$\left|\frac{Y(\omega)}{X(\omega)}\right|^2 = \sum_{n=1}^{N} \alpha_{Rx,n}^2 * \left[1+\left(\frac{\alpha_{Tx,n}}{\alpha_{Rx,n}}\right) - 2\frac{\alpha_{Tx,n}}{\alpha_{Rx,n}}\cos\left(i\omega\left(\tau_{Tx,n} - \tau_{Rx,n}\right)\right)^2\right] \quad (15.5)$$

$$\text{Amplitude Mismatch: } \Delta\alpha_n = \frac{\alpha_{Tx,n}}{\alpha_{Rx,n}};$$
$$\text{Phase Mismatch: } \Delta\phi_n = \left(\tau_{Rx,n} - \tau_{Tx,n}\right) \quad (15.6)$$

In 2016, a multitap MPC in an IBFD wireless communication link was demonstrated and validated [20]. The system, shown in Figure 15.5, utilized two software-defined radios to generate pseudorandom binary sequences that were transmitted using orthogonal frequency-division multiplexing (OFDM) with a QPSK modulation type. The radio used a single antenna, which was duplexed via an RF magnetic circulator while the SOI was transmitted through an additional antenna. To measure the impact of the canceller on the full-duplex link, the bit error rate (BER) of the SOI and SI were measured from the reference of the receiver as a function of SI transmitter RF power. The resultant measurements in Figure 15.6 show the MPC tuned to two operating frequencies. With optical cancellation enabled, the radio showed an improvement of 10-dB interference rejection at 2.2 GHz and over 15 dB at 2.46 GHz.

15.3 RF Performance Metrics

While the operational functionality of MPCs with respect to bandwidth and cancellation is tremendously promising, these cancellers must be evaluated in terms of RF performance metrics to ensure these microwave photonic systems are compatible and viable with IBFD radio communication networks. The three primary RF metrics to be considered are link loss, noise figure, and dynamic range. In general, simplified microwave photonic links (Figure 15.2), and as implemented into a canceller

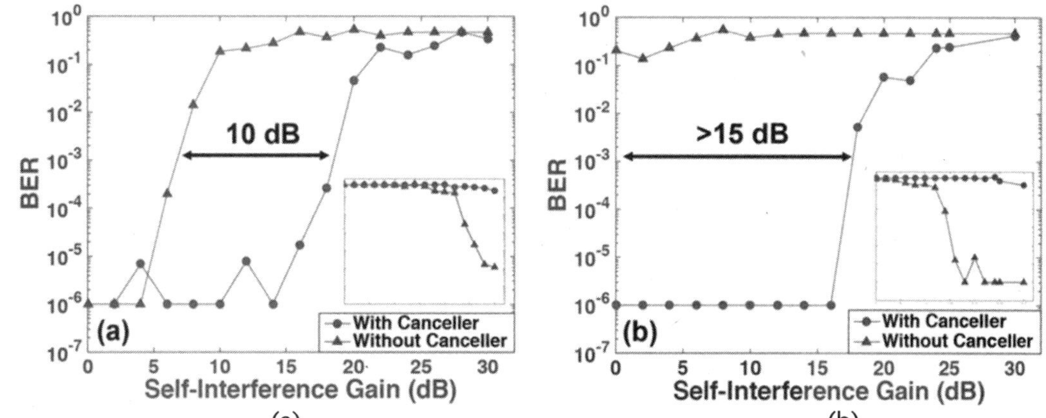

(a) (b)

Figure 15.6 SOI BER with and without MPC with signals centered at (a) 2.2 GHz and (b) 2.46 GHz. Inserts show the corresponding SI BER under same conditions. (© 2016 IEEE. Reprinted, with permission, from [20].)

(Figure 15.3), often demonstrate poor RF performance upwards of a 40-dB link loss and a 50-dB noise figure [11, 15]. The minimization of these two metrics is a consistent goal in microwave photonics [14].

15.3.1 Link Loss

The link loss of an MPC is defined as the RF degradation in power and can be attributed to four primary contributions: impedance matching, E-O conversion, optical gain, and O-E conversion. First, for this analysis, passive impedance matching between the load impedance and photodetector is assumed, which results in a one-quarter reduction of gain. The E-O conversion is further separated into two architecture types: direct modulation (15.7) and external modulation (15.8). For a directly modulated MPC, the gain of the E-O conversion is dependent only on the square of the slope efficiency of the laser, s_{LD}. For externally modulated MPCs, the E-O conversion is proportional to the square of the optical power in, P_{in}, as well as the optical transmission and source resistance of the modulator, T_{mod} and R_S, while inversely proportional to the square of the modulator sensitivity, V_π. The RF gain of an MPC is proportional to the square of the optical gain, G_{opt}, and therefore minimizing optical loss is of critical importance when designing microwave photonic systems. Lastly, the gain of O-E conversion is proportional to the square of the responsivity of the photodiode, r_{PD}. Combined, these four sources of loss result in (15.7) and (15.8) for directly and externally modulated MPCs, respectively.

$$G_{DM} = \frac{1}{4}\left(s_{LD}\right)^2 \left(G_{opt}\right)^2 \left(r_{PD}\right)^2 \tag{15.7}$$

$$G_{EM} = \frac{1}{4}\left(\frac{\pi P_{in} T_{mod} R_S}{2V_\pi}\right)^2 \left(G_{opt}\right)^2 \left(r_{PD}\right)^2 \tag{15.8}$$

These parameters that determine the loss of the photonic link can be characterized into two categories: device and architecture. Device parameters such as the sensitivity of the modulator and responsivity of the photodiode have been developed for decades and the optimal components should be selected. While this is important, there is a great deal of performance to be gained from exploring and developing novel architectures for these systems. Developing higher-optical power-handling circuits, reduction in optical loss, and impedance matching are all architectural design choices that can be furthered developed to significantly improve MPCs. The decision of modulation type is critical when optimizing an MPC; the most important distinction between the two equations is the dependency on optical input power.

15.3.2 Noise Figure

The noise figure is a powerful performance metric that encapsulates the effect of a system on RF signal fidelity. The noise factor, the linear equivalent of noise figure, is defined in (15.9) and is equal to the SNR before and after the device under test. This expression can be simplified to three terms: N_{in} is the input noise, N_{out} is the output noise, and G is the RF link loss, as described in the previous section. The input noise of an MPC is simply the thermal noise. The output noise is a combination of the three noise sources of an optical link: thermal noise, shot noise, and relative intensity noise (RIN). The inclusion and expansion of these terms result in the final expression of (15.9). Additional terms include: R_L for load resistance, k for Boltzmann constant, T for temperature, and electron charge q.

$$F = \frac{SNR_{in}}{SNR_{out}} = \frac{N_{out}}{GN_{in}} = 1 + \frac{1}{G} + \frac{qP_{in}r_{PD}R_L}{2GkT} + \frac{P_{in}^2 r_{PD}^2 R_L RIN}{4GkT} \tag{15.9}$$

Equation (15.9) is expanded to (15.10) and (15.11) by substituting the link loss equation for both modulation architectures for the RF link loss term. The resultant equations shed light on the MPC noise factor's dependency on optical power with respect to the three noise sources. For directly modulated MPCs, the noise factor is only increased with optical power because the gain of the system is independent, while shot noise is proportional to optical power, and RIN is proportional to the square of optical power. Therefore, operating at relatively low optical powers is beneficial for directly modulated links.

In contrast, for external modulated MPCs, the gain and total noise both have unique dependencies on optical power. The link loss improves proportional to the square of optical power, as shown in (15.8); therefore, the noise factor due to the three-noise source will behave as such when increasing optical power: thermal noise contribution will decease inversely proportional to the square of optical power, shot noise contribution will decrease inversely proportional to optical power, and the RIN contribution will remain constant. This results in three operating regions for externally modulated MPC: low optical power thermal noise dominates, medium optical power shot noise dominates, and high optical power RIN noise dominates.

$$F_{DM} = 1 + \frac{4}{s_{LD}^2 G_{opt}^2 r_{PD}^2} + \frac{2qP_{in}R_L}{s_{LD}^2 G_{opt} r_{PD} kT} + \frac{P_{in}^2 R_L RIN}{s_{LD}^2 kT} \tag{15.10}$$

$$F_{EM} = 1 + \frac{16V_\pi^2}{\pi^2 P_{in}^2 G_{opt}^2 r_{PD}^2 R_L^2} + \frac{16qV_\pi^2}{\pi^2 P_{in} G_{opt} r_{PD} R_L kT} + \frac{4V_\pi^2 RIN}{\pi^2 R_L kT} \tag{15.11}$$

Further mathematical analysis of microwave photonics links and improving RF performance can be found in [11–14, 36], and MPC-specific RF analysis can be found in [21, 23, 35].

15.4 Balanced Microwave Photonic Cancellers

Traditional MPCs, as shown in Figure 15.5, demonstrate poor RF performance when compared to alternative technology, specifically a link loss of –44.9 dB and 49.6 dB, respectively [17]. With reducing noise figure as a primary goal, novel balanced architectures were developed to suppress noise within the MPC [21, 25, 30]. This architecture, inspired by differential detection techniques of analog photonic links [37–39], leverages the same optical processing responsible for interference cancellations to simultaneously cancel the common-mode of the link, RIN.

The optical schematic shown in Figure 15.7 highlights the operation of this system. The critical element in achieving simultaneous cancellation is the use of a dual-output Mach-Zehnder modulator (MZM) to generate two complementary modulated signals on the optical carriers. As an example, focus only on the two outside branches colored in black, which are the Rx branches. When these two phase-matched and amplitude-matched optical paths are subtracted at the BPD, two operations will occur. First, the two complementary modulated signals will sum when subtracted. Second, the common noise of both optical paths will subtract. The same process will hold true with respect to the Tx branches (the four lighter gray branches), although the matched filter from the traditional MPC architecture is required on both complementary Tx branches. Similar to the Rx branches, the modulated signals will sum, while the RIN of the laser sources will subtract. The key difference between the Rx and Tx branches is that the BPD terminations are inverted with respect to one another. This results in the Rx and Tx modulated signals subtracting, such as in the single-end MPC.

As seen in Figure 15.8, the experimental cancellation results for the system are truly wideband, including multipath cancellation of 25 dB over 1-GHz instantaneous bandwidth with tunability between 2 and 9 GHz [30]. The primary trade-off in this architecture is that the MPC's matching control complexity doubles, resulting in difficulty achieving narrowband high-cancellation depths.

The RF simulation and experimental results of the balanced MPC demonstrates the improved RF performance achievable via implementation of this architecture. The noise figure as a function of input optical power plot, shown in

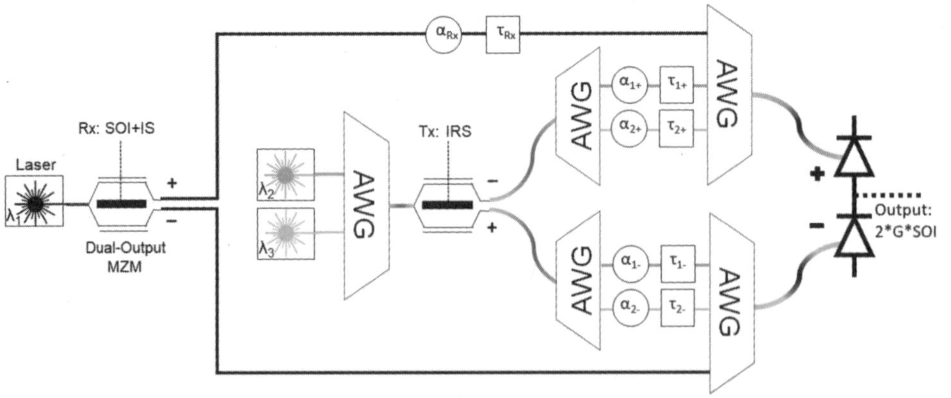

Figure 15.7 Schematic for MPC implemented with a balanced architecture for noise figure (NF) reduction, showing a SOI, IS, interference reference signal (IRS), arrayed waveguide grating (AWG), weights α, and delays τ. (© 2020 IEEE. Reprinted, with permission, from [30].)

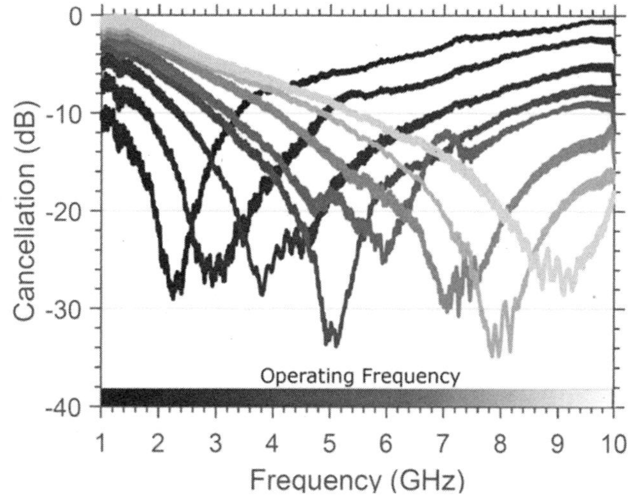

Figure 15.8 Experimental cancellation results for two-tap balanced MPC tuned from 2 to 9 GHz, shown by the 8 cancellation curves (simultaneous cancellation of two interferers with 25 dB over 1-GHz instantaneous bandwidth). (© 2020 IEEE. Reprinted, with permission, from [30].)

Figure 15.9 (left), highlights the mathematical expression of (15.11). While operating at +16-dBm optical power, the link loss and noise figure of the balanced MPC was measured to be –23.2 dB and 32.2 dB, respectively [31]. This was a significant improvement over the single-ended architecture and was limited to maximum laser output power. The three lighter-shaded solid curves in Figure 15.9 (left) highlight the effects of nonidealities when matching for RIN suppression. Mismatch during this process results in a higher RIN noise power, which saturates the achievable noise figure reduction at lower optical. This is best demonstrated by plotting the slope of the noise figure curve as a function of optical power (Figure 15.9, right). As mathematically described in (15.11), the MPC has three optical power regions

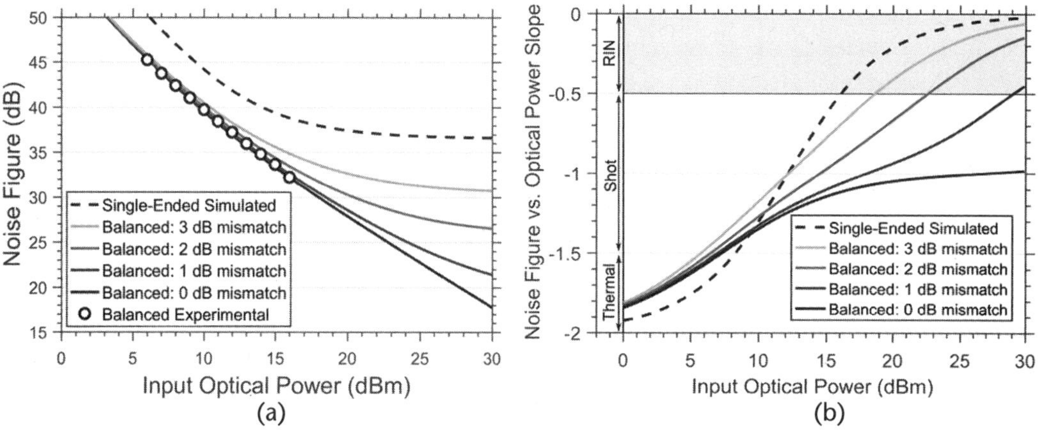

Figure 15.9 Simulation and experimental results for noise figure (NF) analysis of a balanced MPC, with NF as a function of input optical power (left) and a derivative of NF versus optical power as a function of input optical power, highlighting three noise-limiting regions (right). Dashed curves indicate single-ended architectures. Solid lines indicate balanced architecture with varying amounts of the amplitude/phase mismatch. Circles are experimental results. (© 2020 IEEE. Reprinted, with permission, from [30].)

in which a different noise source is dominant. In the thermal and shot noise region, increasing optical power reduces the noise figure, but once a system enters the RIN region (highlighted in gray), both the gain and noise begin to increase equivalently, limiting the noise figure reduction. This analysis provides further practical insights on the state of the MPC giving an operating optical power in addition to the limits of RF performance providing a maximum optical power-handling capability.

15.5 Integrated Microwave Photonic Cancellers

In parallel to the development of novel MPC architectures, recent advancements in PICs have enabled enhanced microwave photonic systems with respect to performance and complexity, which was previously intangible using discrete optics. Integrated optics offers significant improvements in performance, cost, size, weight, and power (SWaP) [40–46]. The performance improves as shrinking devices reduce energy requirements, latencies, and component-to-component coupling loss. The cost of optical processing has continually decreased with economies of scale, as large industrial demands have led to the establishment of more photonic foundries. The convergence of two simultaneously maturing fields generates a promising future for the continued development of MPCs and the potential ubiquitous implementation of this technology in 5G IBFD communication networks.

There are three primary technology platforms for integrated photonics: III-V semiconductors, silicon nitride, and silicon photonics. The benefit of using a III-V semiconductor platform such as indium phosphide is the availability of active optical devices-on-chip, including lasers and amplifiers. Although this increase in functionality is advantageous, these platforms suffer from high propagation loss, high fabrication cost, and low yield [41, 42].

Silicon nitride platforms offer the lowest loss properties of any PICs and therefore can yield microwave photonic circuits with optimal RF performance metrics [45]. The trade-off for this technology is the reduction of functionality and maturity; silicon nitride integration only supports passive optical devices.

Lastly, over recent years, silicon photonics has dominated commercial research and development. This technology leverages fabrication compatibility with traditional digital electronic circuit, CMOS fabrication. This compatibility enables acceleration in fabrication maturity as CMOS has been optimized for decades as well as enables the revolutionary prospect of state-of-the-art microelectronics and photonics cointegrated on the same integrated circuit [46]. There are no active devices on silicon photonics, but doping allows for the creation of PN junctions, which enables modulators and photodetectors [46].

While silicon integration approaches offer a promising future for this technology, previous research into integrated MPCs were focused on indium phosphide approaches [22, 23, 25]. This technology was first developed largely due to the benefit of fully monolithic integration, where all light generation, processing, and detection was contained on-chip, only requiring electrical inputs and output. This greatly reduced the complexity with regards to packaging of this technology.

The first integrated MPC is shown on the top of Figure 15.10. This system is a directly modulated, single-ended, single-tap architecture [28]. The system used directly modulated distributed feedback lasers for the modulator, semiconductor

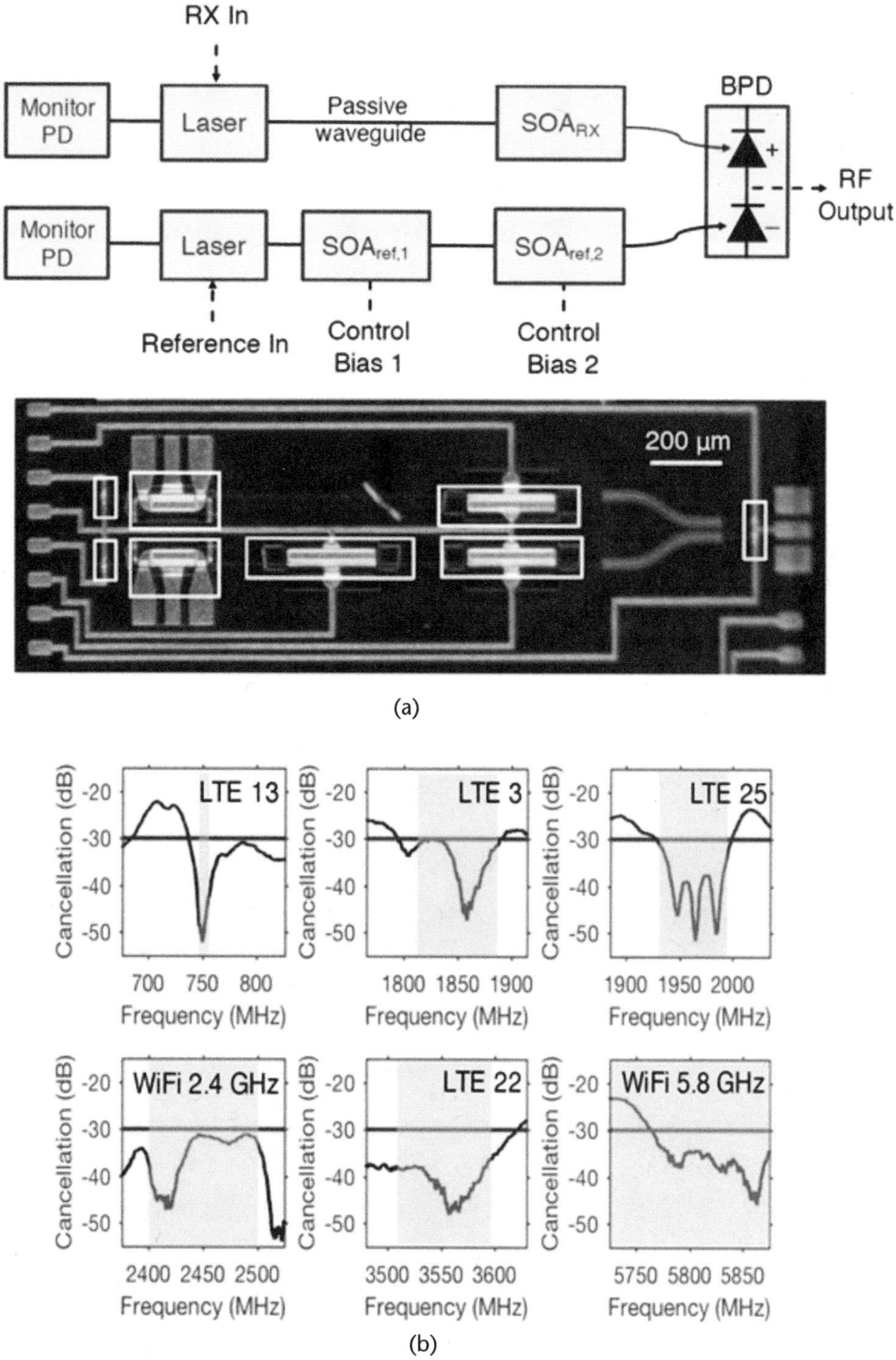

Figure 15.10 Indium phosphide, single-ended, single-tap integrated MPC with SOAs and a balanced photodiode (BPD) (a). Experimental cancellation results for various LTE and Wi-Fi bands (b). Horizontal line indicates 30 dB of cancellation. Gray zone indicates the operational bandwidth of the wireless channel. (© 2017 IEEE. Reprinted, with permission, from [22].)

optical amplifiers (SOAs) for amplitude matching, and phase matching via slow and fast light tunable delays. The experimental cancellation results, shown on the bottom of Figure 15.10, demonstrated acceptable cancellation of greater than 30 dB over various Wi-Fi and LTE bands, with tunable instantaneous bandwidth dependent on wireless channel requirements. The verification of this simple MPC architecture led to the development of more advanced integrated MPC architectures, such as balanced external modulated MPC [25, 30] (Figure 15.11 (top)), and three-tap externally modulated MPC (Figure 15.11 (bottom)). While these systems achieved acceptable cancellation results, they suffered from poor RF performance due to a maximum output laser power of 6 dBm [25, 30]. At this constrained optical power, directly modulated systems perform better.

15.6 Practical Considerations

When designing an MPC for an IBFD link, there are three primary practical considerations to consider: optical power of source, additional sources of cancellation, and inclusion of LNAs.

The maximum optical power handling of the MPC has a significant impact that determines the capabilities of the system. For thermal noise-dominated,

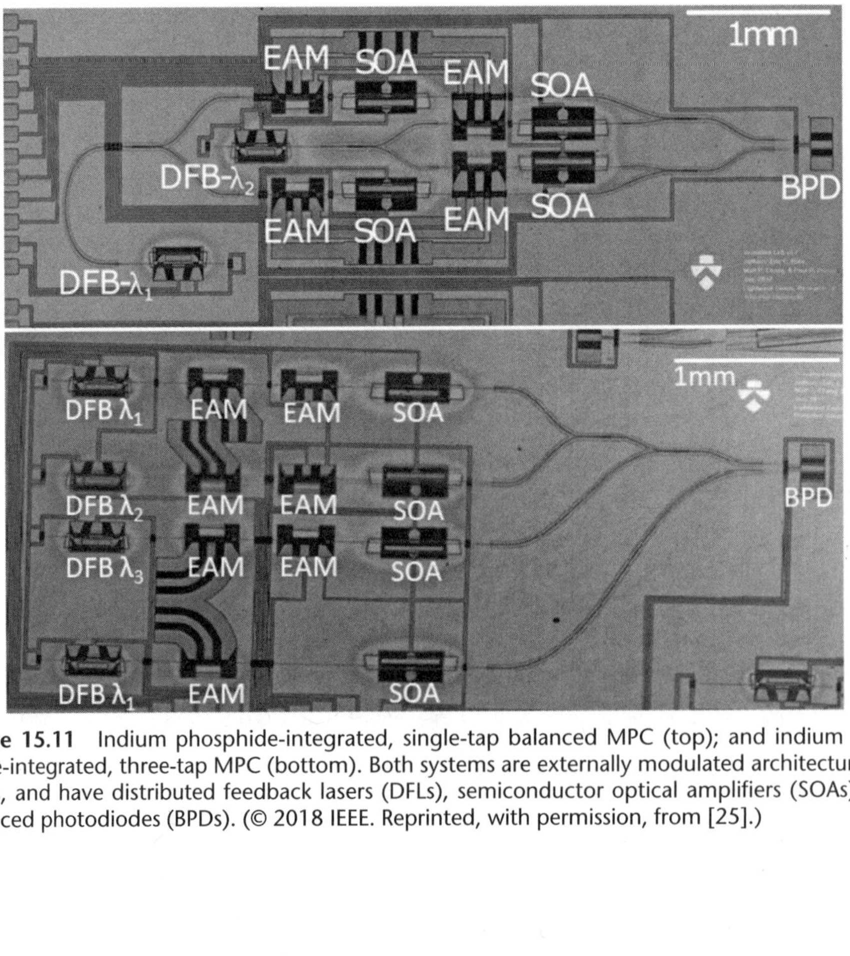

Figure 15.11 Indium phosphide-integrated, single-tap balanced MPC (top); and indium phosphide-integrated, three-tap MPC (bottom). Both systems are externally modulated architectures via EAMs, and have distributed feedback lasers (DFLs), semiconductor optical amplifiers (SOAs), and balanced photodiodes (BPDs). (© 2018 IEEE. Reprinted, with permission, from [25].)

lower-optical power systems such as a III-V integrated MPC, directly modulated architectures are ideal. For shot noise-dominated, medium-optical input power systems, externally modulated architectures become favorable. At high-optical-power systems such as discrete systems or passive integrated PICs with high-power, off-chip lasers, a balanced externally modulated architecture becomes optimal for RF performance metrics.

Another practical consideration is when to implement additional complementary cancellation and isolation technology. MPCs are excellent at cancelling high-power, line-of-sight SI. While multipath cancellation is achievable, each additional tap increases control complexity and limits performance. A more ideal approach is to use passive isolation, an analog canceller and digital cancellation working in harmony, as highlighted in Chapter 1 [47]. The analog canceller is responsible for the cancellation of the highest-power SI to return the remaining received signal into the range of the ADC. Once in the digital domain, digital cancellation systems can cancel many ISs with ease. In combination, the IBFD link becomes more robust to dynamically changing wireless environments.

Lastly, the photonic system designer should always consider the implementation of an LNA in front of the MPC to greatly reduce the overall noise figure of the system. The trade-off with including an LNA is the reduction of bandwidth to the operating frequency of the LNA as well as a reduction in maximum RF input power to the system. If the frequency requirements of the communication link match the frequency metrics of the LNA, the performance enhancement can be significant; the noise figure of the discrete balanced MPC at 16-dBm optical input power reduced from 32.2 dB to 4.3 dB with the addition of a 34-dB gain LNA with an LNA noise figure of 2.03 dB [30].

15.7 Conclusion

15.7.1 Summary

Rising internet traffic paired with finite spectral resources has created an ever-increasing spectral deficit. This effect results in congested networks and a limitation to the continued development of communication networks. IBFD communication links serve as an effective solution for the alleviation of the spectral deficit, but come at the cost of SI. The in-band SI requires active cancellation to protect the receiver and to effectively communicate. To enable this technology in 5G and future network standards, the transceiver must dynamically operate over a large frequency range in addition to use wide instantaneous bandwidths. While MPCs can achieve these requirements, traditional architectures can suffer from low RF performance. This caveat has motivated the development of novel architectures, such as balanced RIN suppression, for noise figure reduction and has made strides in the enhancement of overall performance. The parallel recent advancements in photonic integration offers an additional pathway forward for MPCs, not only with respect to SWaP, but also with continued maturity showing the potential to enable the highest-performing cancellers to date.

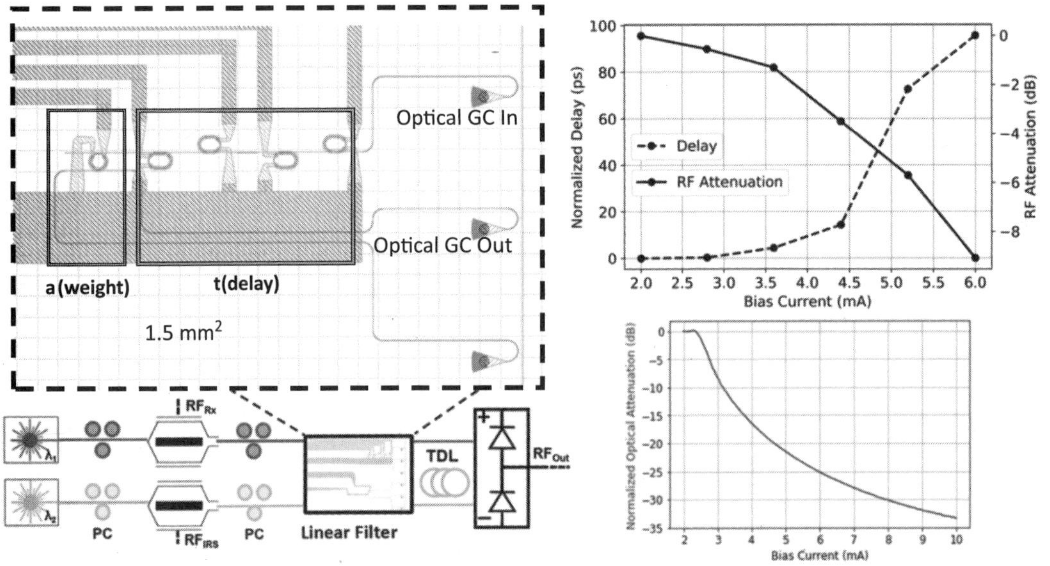

Figure 15.12 Silicon photonic MPC (left) with polarization controllers (PCs), tunable delay lines (TDLs) and grating couplers (GCs); the racetrack MRR tunable delay characterization (top right); and MRR weight characterization (bottom right). (© 2020 IEEE. Reprinted, with permission, from [31].)

15.7.2 Future Directions

Silicon photonics offers a promising future for integrated MPCs. This technology hits the sweet spot between the ultralow loss of the silicon nitride and the high functionality of the III-V semiconductor platform. The high fabrication precision and accuracy yield optical paths with extremely similar transfer functions, resulting in ultrawide instantaneous cancellation bandwidths. A passive silicon photonic linear filter is demonstrated in Figure 15.12 (left) and implemented into a discrete cancellation system [24, 31]. This system uses microring resonator (MRR)-based weights [42] and delays [49] to accomplish the amplitude and phase matching required by an MPC. The MRR can provide 34 dB of attenuation range and a resolution of 7.2 bits [42]. The MRR racetrack delays can achieve over 100 picoseconds of tunable delay with seven devices with a coupled RF loss of 8.4 dB [24, Figure 15.12 (right)]. Silicon photonics' combination of functionality, low loss, and high-power handling has the potential to yield the highest-performing MPC in cancellation and RF performance to date.

References

[1] Cisco, "Cisco Annual Internet Report (2018–2023) White Paper," 2020.

[2] Cisco, "Cisco Visual Networking Index: Global Mobile Data Traffic Forecast Update, 2015–2020," 2016.

[3] Shafi, M., et al., "5G: A Tutorial Overview of Standards, Trials, Challenges, Deployment, and Practice," *IEEE Journal on Selected Areas in Communications*, Vol. 35, No. 6, June 2017, pp. 1201–1221.

[4] Demestichas, P., et al., "5G on the Horizon: Key Challenges for the Radio-Access Network," *IEEE Veh. Technol. Mag.*, September 2013, pp. 47–53.

[5] Capmany, J., and D. Novak, "Microwave Photonics Combines Two Worlds," *Nat. Photon*, Vol. 1, 2007, pp. 319–330.

[6] Marpaung, D., et al., "Integrated Microwave Photonics," *Laser Photonics Rev.*, Vol. 7, 2013, pp. 506–538.

[7] Patel, D., et al., "High-Speed Compact Silicon Photonic Michelson Interferometric Modulator," *Optics Express*, Vol. 22, No. 22, 2014, pp. 26788–26802.

[8] Fard, M., et al., "Responsivity Optimization of a High-Speed Germanium-on-Silicon Photodetector," *Optics Express*, Vol. 24, No. 24, 2016, pp. 27738–27752.

[9] Marpaung, D., et al., "Integrated Microwave Photonics," *Nature Photonics*, Vol. 13, No. 2, 2019, pp. 80–90.

[10] Seeds, A. J., and K. J. Williams, "Microwave Photonics," *Journal of Lightwave Technology*, Vol. 24, No. 12, 2006, pp. 4628–4641.

[11] Yao, J., "Microwave Photonics," *Journal of Lightwave Technology*, Vol. 27, No. 3, 2009, pp. 314–335.

[12] Cox, C., et al., "Limits on the Performance of RF-over-Fiber Links and Their Impact on Device Design," *IEEE Transactions on Microwave Theory and Techniques*, Vol. 54, No. 2, 2006, pp. 906–920.

[13] Cox, C. H., "Gain and Noise Figure in Analogue Fibre-Optic Links," *IEE Proceedings J – Optoelectronics*, Vol. 139, No. 4, 1992, pp. 238–242.

[14] Cox, C. H., et al., "Techniques and Performance of Intensity Modulated Direct-Detection of Analog Optical Links," *IEEE Transactions on Microwave Theory and Techniques*, Vol. 45, No. 8, 1997, pp. 1375–1383.

[15] Ackerman, E. I., and C. H. Cox, "Optimization of Analog Optical Link Performance: First Minimize the Noise Figure," *2014 International Topical Meeting on Microwave Photonics (MWP) and the 2014 9th Asia-Pacific Microwave Photonics Conference (APMP)*, 2014.

[16] Suarez, J., et al., "Incoherent Method of Optical Interference Cancellation for Radio-Frequency Communications," *IEEE Journal of Quantum Electronics*, Vol. 45, No. 4, 2009, pp. 402–408.

[17] Chang, J., and P. R. Prucnal, "A Novel Analog Photonic Method for Broadband Multipath Interference Cancellation," *IEEE Microwave and Wireless Components Letters*, Vol. 23, 2013, pp. 377–379.

[18] Chang, M. P., et al., "Optical Analog Self-Interference Cancellation Using Electro-Absorption Modulators," *IEEE Microwave and Wireless Components Letters*, Vol. 23, No. 2, February 2013, pp. 99–101.

[19] Chang, M., et al., "An Integrated Optical Interference Cancellation System," *2014 23rd Wireless and Optical Communication Conference*, 2014, pp. 1–5.

[20] Lu, M., et al., "Performance Comparison of Optical Interference Cancellation System Architectures," *Appl. Opt.*, Vol. 52, 2013, pp. 2484–2493.

[21] Chang, M. P., et al., "A Microwave Photonic Canceller for Improved Interference Rejection in Full Duplex Radio," *2016 IEEE Photonics Conference (IPC)*, Waikoloa, HI, 2016, pp. 493-494.

[22] Blow, E. C., M. P. Chang, and P. R. Prucnal, "Microwave Photonic Interference Canceller: Noise Figure Reduction via Balanced Architecture," *2016 IEEE International Topical Meeting on Microwave Photonics (MWP)*, October 2016.

[23] Chang, M. P., et al., "Integrated Microwave Photonic Circuit for Self-Interference Cancellation," *IEEE Transactions on Microwave Theory and Techniques*, Vol. 65, No. 11, November 2017, pp. 4493–4501.

[24] Chang, M. P., et al., "RF Characterization of an Integrated Microwave Photonic Circuit for Self-Interference Cancellation," *IEEE Transactions on Microwave Theory and Techniques*, 2017.

[25] Blow, E. C., et al., "Silicon Photonic Weights for Microwave Photonic Canceller," *2020 Conference on Lasers and Electro-Optics (CLEO)*, 2020.

[26] Blow, E. C., P. Kaul, and P. R. Prucnal, "Integrated Balanced Microwave Photonic Canceller," *2018 IEEE Photonics Conference (IPC)*, Reston, VA, 2018, pp. 1–2.

[27] Kpodzo, E. B., et al., "Interference Cancellation System for Cancelling Interference in the Optical Domain," U.S. Patent No. 9,444,502, September 13, 2016.

[28] Kolodziej, K. E., et al., "Photonic-Enabled RF Canceller for In-Band Full-Duplex 5G Networks," *2018 IEEE/MTT-S International Microwave Symposium*, 2018.

[29] Kolodziej, K. E., et al., "Photonic-Enabled RF Canceller for Wideband In-Band Full-Duplex Wireless Systems," *IEEE Transactions on Microwave Theory and Techniques*, Vol. 67, No. 5, 2019, pp. 2076–2086.

[30] Chang, M. P., et al., "A Microwave Photonic Canceller for Improved Interference Rejection in Full Duplex Radio," *2016 IEEE Photonics Conference (IPC)*, 2016.

[31] Blow, E. C., et al., "Current Progress of Microwave Photonic Canceller: Balanced Architecture" (in preparation) 2021.

[32] Blow, E. C., et al., "Passive Microwave Silicon Photonic Canceller" (in preparation) 2020.

[33] Bharadia, D., et al., "Full Duplex Radios," *Proc. ACM SIGCOMM 2013 Conf. SIGCOMM – SIGCOMM '13*, Vol. 43, No. 4, August 2013, p. 375.

[34] Kolodziej, K. E., et al., "Multitap RF Canceller for In-Band Full-Duplex Wireless Communications," *IEEE Transactions on Wireless Communications*, Vol. 15, No. 6, 2016, pp. 4321–4334.

[35] Cox, C. H., *Analog Optical Links: Theory and Practice*, New York: Cambridge University Press, 2006.

[36] Chang, M. P., "A Microwave Photonic Interference Canceller: Architectures, Systems, and Integration," Dissertation, Princeton University, 2017.

[37] Madjar, A., and O. Malz, "A Balanced Fiberoptic Communication Link Featuring Laser RIN Cancellation," *IEEE MTT-S International Microwave Symposium Digest*, 1992, pp. 563–566.

[38] Ackerman, E., et al., "Balanced Receiver External Modulation Fiber-Optic Link Architecture with Reduced Noise Figure," *IEEE MTT-S International Microwave Symposium Digest*, 1993, pp. 723–726.

[39] Ackerman, E. I., et al., "Signal-to-Noise Performance of Two Analog Photonic Links Using Different Noise Reduction Techniques," *2007 IEEE/MTT-S International Microwave Symposium (IMS)*, 2007.

[40] Soref, R., "The Past, Present, and Future of Silicon Photonics," *IEEE Journal of Selected Topics in Quantum Electronics*, Vol. 12, 2006.

[41] Kish, F. A., et al., "Current Status of Large-Scale InP Photonic Integrated Circuits," *IEEE Journal of Quantum Electronics*, Vol. 17, No. 6, 2011, pp. 1470–1489.

[42] Leijtens, X., "JePPIX: The Platform for Indium Phosphide-Based Photonics," *IET Optoelectronics*, Vol. 5, No. 5, 2011, pp. 202–206.

[43] Smit, M., J. V. D. Tol, and M. Hill, "Moore's Law in Photonics," *Laser and Photon*, Vol. 13, No. 1, 2012, pp. 1–13.

[44] Roelkens, G., et al., "III-V/Silicon Photonics for On-Chip and Intra-Chip Optical Interconnects," *Laser and Photonics Reviews*, Vol. 4, 2010, pp. 751–779.

[45] Roeloffzen, C. G. H., et al., "Silicon Nitride Microwave Photonic Circuits," *Optics Express*, Vol. 21, No. 19, 2013, pp. 22937–22961.

[46] Chrostowski, L., and M. Hochberg, *Silicon Photonics Design: From Devices to Systems*, New York: Cambridge University Press, 2015.

[47] Kolodziej, K. E., et al., "In-Band Full-Duplex Technology: Techniques and Systems Survey," *IEEE Transactions on Microwave Theory and Techniques*, Vol. 67, July 2019, pp. 3025–3041.

[48] Huang, C., et al., "Demonstration of Scalable Microring Weight Bank Control for Large-Scale Photonic Integrated Circuits," *APL Photonics*, Vol. 5, No. 4, 2020, p. 040803.

[49] Cardenas, J., et al., "Wide-Bandwidth Continuously Tunable Optical Delay Line Using Silicon Microring Resonators," *Optics Express*, Vol. 18, No. 25, 2010, pp. 26525–26534.

CHAPTER 16

Backscatter Modulation Applications

Besma Smida and Seiran Khaledian

16.1 Introduction

In order to meet the rising demand for high-speed data services, wireless networks must exploit the available frequency spectrum more efficiently. One key limit to spectrum utilization efficiency involves the current practice of half-duplex communication. As introduced in Chapter 1, many recent studies have provided experimental evidence and approaches for IBFD communications. All methodologies regarding IBFD communications have been based upon the principle that the transmitter and receiver must generate their own radio-carrier waves separately, which results in a high level of SI, as discussed in the other chapters.

This chapter introduces a reliable IBFD wireless network architecture by means of reflected power, which utilizes the components highlighted within the IBFD transceiver diagram of Figure 16.1(a) and the SIC techniques tree of Figure 16.1(b). This approach offers a new paradigm for two-way wireless networks: rather than avoiding SI as in half-duplex, or combating SI as in full-duplex, nodes reuse the received interfering radio-carrier waves to transfer information. The power used to transfer data in one direction is reused to transfer data in the other. Our concept, named *ReflectFX*, is based on backscatter modulation (BM), where the same antenna that receives the electromagnetic (EM) waves modulates and reflects the waves back. Within this framework, transmission and reception are synchronized and the mobile receives a signal free of SI. The relevance of this concept follows from two facts:

1. *Communication is increasingly asymmetrical and becoming more so over time.* From the viewpoint of traffic symmetry between uplink (mobile to base station) and downlink (base station to mobile), data traffic started as highly symmetrical two-way communication with voice services. Over time, communication has become imbalanced, and the major growth of new multimedia services is characterized by asymmetric traffic (i.e., much

This research was funded by NSF-CAREER (award number CCF #1620902).

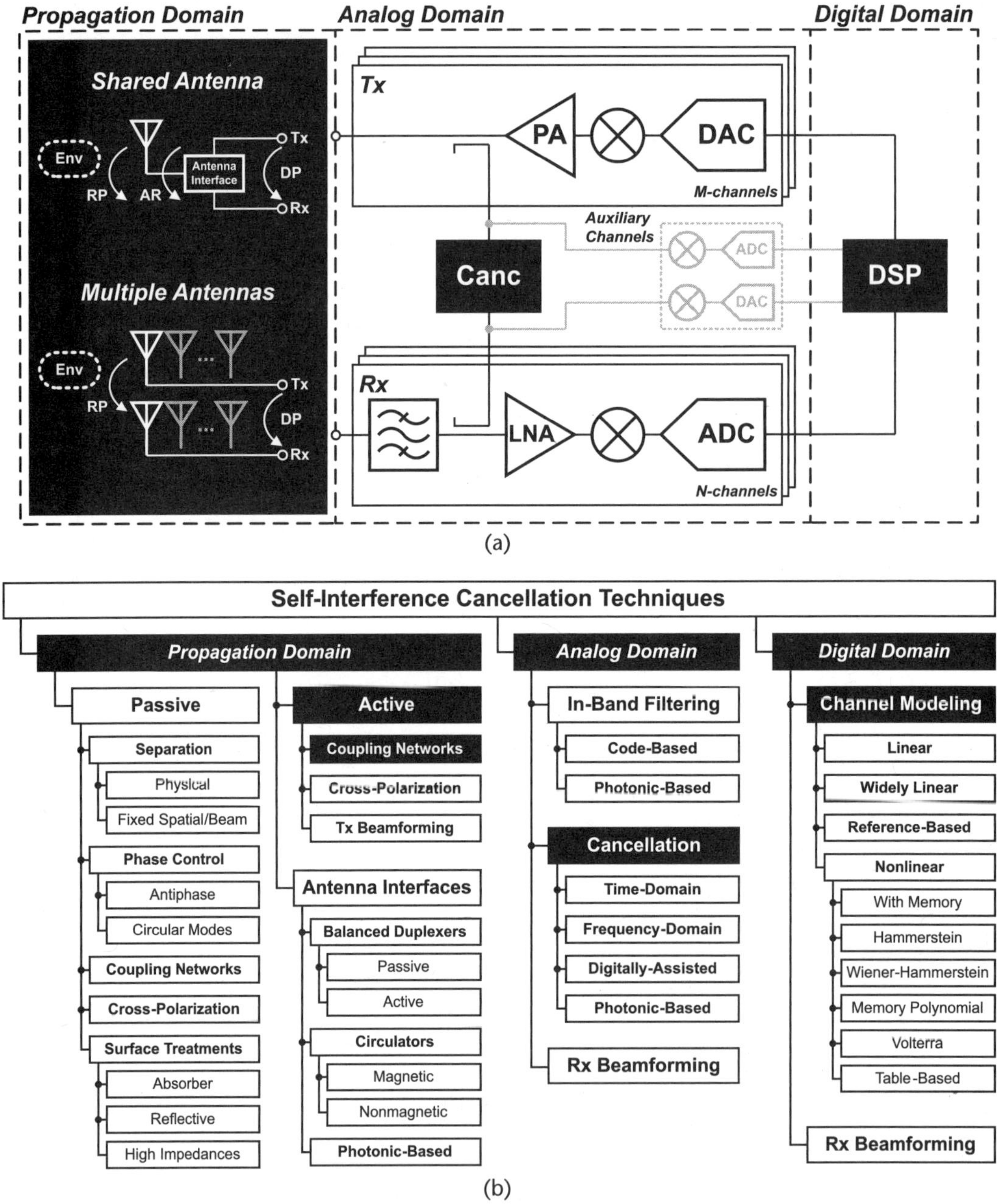

Figure 16.1 BM applications and specific chapter focus highlighted within (a) a generic IBFD transceiver and (b) a SIC techniques tree diagram.

higher downlink traffic). Overall traffic patterns are also becoming more asymmetrical, 3 to 5 times more in the downlink direction, because the largest bandwidth demand currently derives from downstream TV and internet video signals [1–3].

2. *BM for power-efficient communication.* The history of BM can be traced to Stockman's foundation article, "Communication by Means of Reflected Power," published in 1948 [4]. Stockman proposed a point-to-point

> communication, with the carrier power generated at the receiving end and the transmitter replaced by a modulated reflector. BM is based on the reflection of EM waves. It takes advantage of the presence of reflection coefficient variations at the interface between the transmit antenna and the receiver input load. By varying the load, Z_1 or Z_2, the reflection coefficient changes, as illustrated in Figure 16.2. This setup is capable of modulating data upon the original waveform, which can be received and decoded.

Much has happened in the more than 70 years since Stockman's work. The most prominent application of BM technology is radio frequency identification (RFID). As can be seen in [6–9], RFID tags are used as keys to open electronic locks, as a means of storing personal information in both credit cards and passports as well as tracking strategies in the shipping industry. BM is also implemented in a number of other areas, such as wireless sensor networks [10–14]. With BM, the power limitations normally associated with traditional wireless sensor systems can be circumvented [15–17]. Experimental evidence shows that BM can reach up to 112-Mbps data-rate and use narrowband, ultra-wideband or spread-spectrum signals [9, 18–23]. This leads to relatively low-complexity, low-power, and low-cost communication terminals.

Despite these benefits, BM has not, to date, been seriously considered in two-way communication. In this respect, BM-based approaches have been prevented by one primary, but relevant, drawback: it is essentially restricted to short-range communication. There are two facets to this challenge: first, the mobile does not generate its own radio wave; second, most of the research studies related to BM have added additional power constraints (e.g., many systems rely on the assumption that nodes are passive (battery-less)) [7, 8, 14, 22].

In contrast to the approach of the previously cited authors, we did not use BM to take advantage of its low-power requirements. Instead, we employed BM to allow the transmission of two-way flows of data over the same carrier at the same time. We therefore can significantly increase the range of BM by using negative

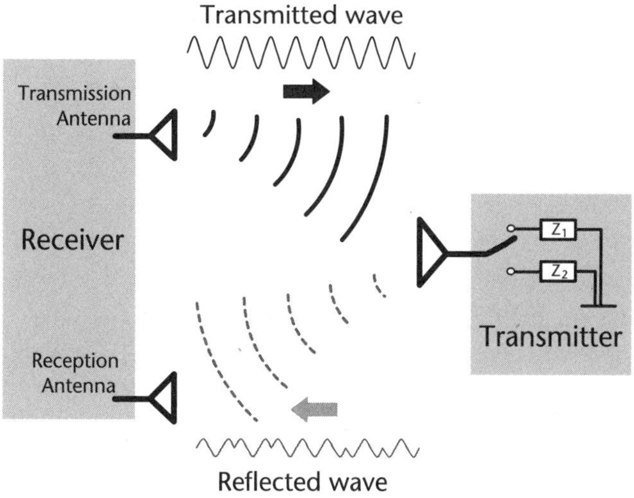

Figure 16.2 Illustrations of BM. (© 2017 IEEE. Reprinted, with permission, from [5].)

resistance at the mobile transceiver. Furthermore, because the main objective of this chapter is to illustrate the increased spectral efficiency provided by BM, many of the proposed RFID modulations to achieve two-way communication are not relevant to our work. Here, we propose a modulation based on analog-network coding that achieves IBFD, where the two flows of data simultaneously occupy the same bandwidth.

This chapter is based on our prior work on IBFD communication in [5, 24–26]. To the best of our knowledge, we were the first to show that it is beneficial to use BM for IBFD communication. In this chapter, the communication system introduced in [24, 25] is further analyzed and optimized with more accurate antenna-scattering models to provide a sense of the potential of IBFD BM applications.

16.2 Antenna Scattering

Because antenna scattering is of interest to both the wireless communication community (primarily concerned with antennas) and the radar and RFID communities (primarily concerned with scattering), two different terminologies have arisen from both groups of investigators. These terminologies make it difficult to correlate results from the two groups, evidenced by several articles and comments that have appeared recently concerning the question of how much power is scattered and absorbed by a receiving antenna [27–29]. We will therefore provide a concise review of the antenna-scattering theory before introducing a general framework that allows transmission of two-way data flows over the same carrier at the same time based on BM.

16.2.1 General Theory

An antenna illuminated by an incoming wave scatters EM energy as a result of current induced by the incident field [30]. In particular, the EM field scattered by an antenna is given by a relatively simple expression. For a given antenna in free space, the scattered electric field $E_{scat}(r, \theta, \phi|Z)$, when the antenna is loaded with impedance Z at a particular point in space (r, θ, ϕ), can be expressed as

$$E_{scat}(r,\theta,\phi|Z) = \overbrace{E_{scat}(r,\theta,\phi|Z^*_{ant})}^{\text{structure-depending}} - \overbrace{\Gamma(Z) I(Z^*_{ant}) E_r(r,\theta,\phi)}^{\text{load-depending}}, \tag{16.1}$$

where $E_{scat}(r,\theta,\phi|Z^*_{ant})$ is the field when the antenna is conjugate-matched $(Z = Z^*_{ant})$, Z_{ant} is the antenna impedance, $\Gamma(Z) = \frac{Z - Z^*_{ant}}{Z + Z_{ant}}$ is the reflection coefficient, $I(Z_{ant}^*)$ is the current flowing in the antenna when conjugate-matched, and $E_r(r, \theta, \phi)$ is the electric field generated by the antenna as a radiator when excited by a unit current source. Equation (16.1) is an old and fundamental result in antenna scattering [31–36]. It can be proven by using the compensation and superposition theorems of circuit theory [31]. An examination of (16.1) shows that the first term, called *structure scattering*, is independent of the load Z. The second term, called the antenna mode component of the scattering field, is load-dependent and vanishes when the antenna is conjugate-matched as the reflection coefficient becomes zero, $\Gamma(Z_{ant}^*) = 0$. Equation (16.1) may be rewritten as

$$E_{scat}(r,\theta,\phi|Z) = I(Z_{ant}^*)E_r(r,\theta,\phi)(A_{st} - \Gamma(Z)), \tag{16.2}$$

where $A_{st} = \dfrac{E_{scat}(r,\theta,\phi|Z_{ant}^*)}{I(Z_{ant}^*)E_r(r,\theta,\phi)}$. Note that A_{st} equals 1 for the special case of minimum scattering antennas (such as thin dipoles), but in general, $A_{st} \neq 1$.

16.2.1.1 Structure-Depending Scattering Field

In conventional communication systems, the antenna is matched to the load in order to increase the received power. The scattering field is therefore composed uniquely of the structure-depending scattering field. Contrary to the prevailing wisdom that "receiving antennas scatter as much as they absorb under matched load," most antennas will have absorption efficiencies below 50% (i.e., always scatter more power), as evidenced from the many examples of canonical structures cited in [29].

16.2.1.2 Load-Dependent Scattering Field

In radar and RFID applications, emphasis is placed on the parameter that characterizes the scattering from an object independently of the strength of the incident field and the distance from the scattered, and is named radar cross-section (RCS).

16.2.2 Thévenin Equivalent Circuit

To evaluate the power delivered to the receiver, we made use of the Thévenin equivalent model, as depicted in Figure 16.3. We propose two possible circuit configurations: series and parallel. The variable Z_{mod} represents the modulation circuit and is controlled by the end-user microcontroller. The input impedance of all of the mobile's integrated circuit (IC) components, except the modulator circuit, is modeled by Z_{ant}^*. Contrary to the prevailing wisdom, the Thévenin equivalent circuit can only be used to describe the current at the load and cannot be relied upon for calculating the scattered power [25, 27–29, 37]. In other words, we cannot rely on this circuit model to calculate the scattered EM field. Therefore, in contrast to common practice [22, 36], we used (16.2) rather than the Thévenin circuit model to evaluate more accurately the transmitted backscatter power as a function of the load Z_{mod}.

16.3 Backscatter Modulation Communication Framework

In this section, we describe how the base station and mobile exchange data at the same time and over the same frequency band using ReflectFX. For the sake of simplicity, we will intentionally ignore some details that are not needed for understanding the basic conceptual idea proposed in this work (e.g., error correction coding, upconversion, and downconversion). We will also assume for this discussion that the base station has one transmitting antenna and one receiving antenna, while the mobile has just one antenna, all polarization-matched.

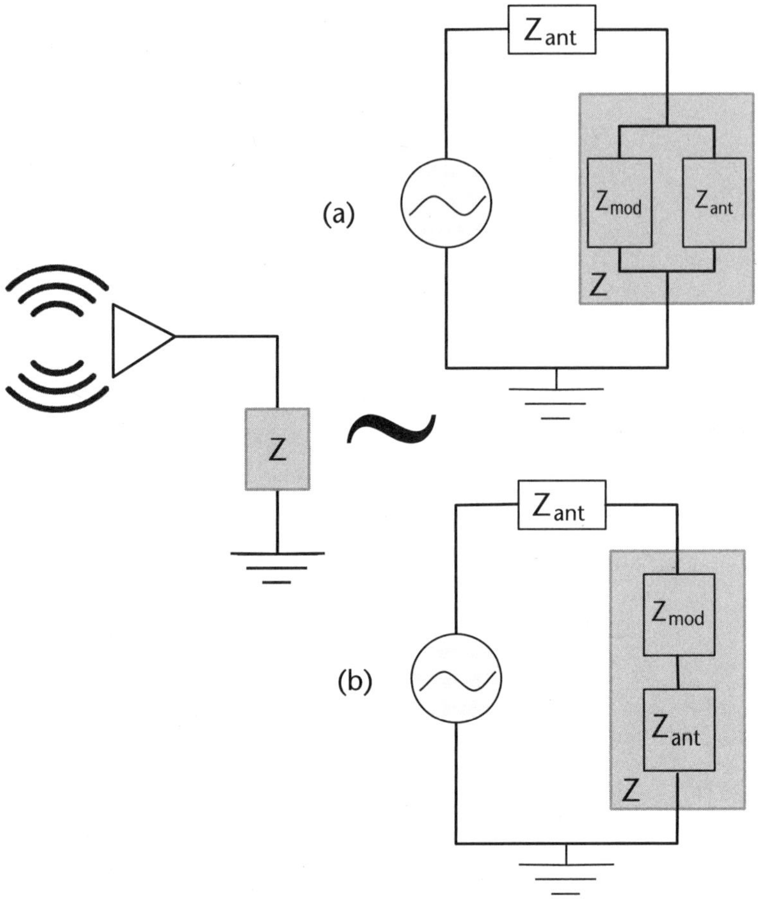

Figure 16.3 Receiving-antenna equivalent circuit: (a) parallel circuit and (b) series circuit configurations.

16.3.1 Channel Model for ReflectFX Communication

In this section, we assume that two nodes intend to exchange messages over a wireless channel, as depicted in Figure 16.4. This model system, in which a mobile m wishes to exchange messages with a central node, or base station B, is a model that captures the behavior of cellular systems, including indoor and Internet of Things (IoT) networks. The mobile m has an independent message $w_{m,B}$ destined for base station B, and the base station has an independent message $w_{B,m}$ destined for the mobile m. We assume that the channel outputs are linear combinations of the transmitted signals scaled by random channel gains plus independent additive white Gaussian noise (AWGN). The Rayleigh fading channels are named $h_{B\rightarrow m}$ (base station to mobile), $h_{m\rightarrow B}$ (mobile to base station), and $h_{B\rightarrow B}$ (base station transmitter to base station receiver).

16.3.2 Communication Forward Link

The base station transmitter sends the symbol sequence to the transmission radio. In Figure 16.4, signal $x_B(i)$ denotes the ith symbol transmitted from the base station. We define $x_B(i) = \sqrt{E}s_B(i)$ where $s_B(i)$ is the transmitted constellation symbol

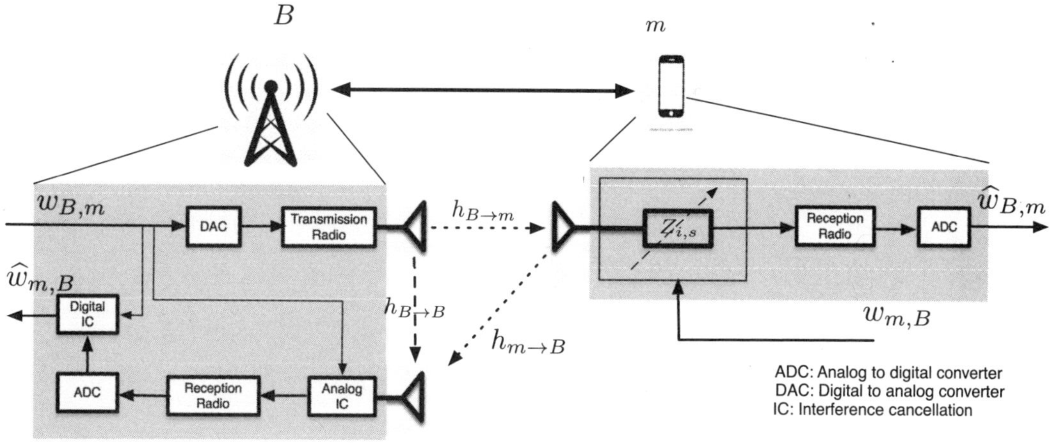

Figure 16.4 Framework for IBFD communication with binary BM. (© 2017 IEEE. Reprinted, with permission, from [5].)

normalized to unit energy and E denotes the average symbol energy. The received ith symbol by the mobile is

$$y_m(i) = \sqrt{\xi_{B\to m}(Z_i)}h_{B\to m}(i)x_B(i) + n_m(i), \tag{16.3}$$

where $\xi_{B\to m}(Z_i) = \left(\frac{\lambda}{4\pi d}\right)^2 G_{Bt}G_m\left(1-|\Gamma(Z_i)|^2\right)$ is the power reduction (average large-scale attenuation) from the base station transmitter to the mobile receiver, d is the distance between the base station antenna and mobile, λ is the wavelength, G_{Bt} and G_m are the base station transmitter and the mobile antenna gains, $n_m(i)$ is the AWGN at the mobile, and $\Gamma(Z_i) = \frac{Z_i - Z_{ant}^*}{Z_i + Z_{ant}}$. Note that the received symbol $y_m(i)$ varies with the receiver load Z_i at the ith symbol. Because the load is changing with the data, the receiving antenna cannot be conjugate-matched to the load; hence, the received power may be diminished when compared with conventional communication (see Figure 16.3). Therefore, we used negative resistance in the load to increase the received power.

16.3.3 Communication Backscatter Link

At the mobile, the antenna receives and scatters back at the same time. By choosing from a set of impedance loadings, the mobile can map a sequence of digital symbols onto the backscattered waveforms. More specifically, we consider M-ary phase-shift keying (M-PSK) modulation, so the backscattered signal becomes a network-coded combination of the messages sent by the mobile and the base station.

Our over-the-air experiments, simulations, and related analyses demonstrate the following result. If we use M-PSK modulation at both the base station and the mobile, then we set $x_B(i) = \sqrt{E}e^{j(l-1)/M}$ and represent the M symbols, at the mobile, on a circle of radius $\sqrt{E'}$ centered at $c_{st} \in C$ as $(A_{st} - \Gamma(Z_i)) = \sqrt{E'}e(j(n-1)\pi/M + j\psi) + c_{st}$, where l and $n \in \{1, ..., M\}$ and M is the modulation constellation size. The backscattered signal $x_m(i) = x_B(i-t)(A_{st} - \Gamma(Z_i)) = \sqrt{EE'}e(j(l-1)\pi/M + j(n-1)\pi/M + j\psi) + c_{st}\sqrt{E}e(j(l-1)\pi/M)$ becomes an analog-network-coded combination of the messages sent by the base station and the mobile (modulated by Z_i), and t accounts for the propagation delay between the base station and the mobile.

Recently, many researchers [19, 36, 37] have demonstrated the feasibility of building backscatter systems that support higher-order constellations. This introduces the possibility of coded modulation and other advanced modulation and coding combinations [38].

Due to leakage from the base station transmitting antenna, the signal received at the base station receiver is given by

$$y_B(i) = \sqrt{\xi_{m\to B}}h_{B\to m}(i)h_{m\to B}(i)x_B(i-t)\cdot \\ \left(A_{st} - \Gamma(Z_i)\right)\sqrt{\xi_{B\to B}}h_{B\to B}(i)x_B(i) + n_B(i)' \tag{16.4}$$

where $\xi_{m\to B} = \left(\frac{\lambda}{4\pi d}\right)^4 G_{Bt}G_m^2G_{Br}p_{Bm}p_{mB}$ is the average power attenuation for the two-way link, G_{Br} is the base station receiver antenna gain, p_{mB} accounts for polarization mismatch at the base station receiver antenna, $\xi_{B\to B} = \left(\frac{\lambda}{4\pi d_{B\to B}}\right)^2 G_{Bt}G_{Br}p_{BB}$, $d_{B\to B}$ is the distance between the transmitter and receiver antennas at the base station, p_{BB} accounts for the polarization mismatch between base station antennas, and $n_B(i)$ is the AWGN noise at the base station.

The base station will use multiple analog and digital SIC techniques to extract the backscattered signal, as discussed in Chapters 2 through 12. The base station is left with the backscattered signal $x_m(i)$ that is a function of the messages sent by both the mobile station and the base station. Knowing the message that was initially sent, the base station can extract its own message after removing the structure-dependent term of $x_m(i)$.

16.4 ReflectFX Load Optimization

The ReflectFX setting is different from the conventional setting, where a total power is imposed on the two nodes. Indeed, the mobile has no radio power available for transmission besides the received radio power. This guarantees no SI at the mobile, but it limits the backscatter link range. In addition, because the load is allowed to change, the receiving antenna is not conjugate-matched to that of the load. The performance of ReflectFX is mainly determined by how we manage the SI at the base station and how we optimize the power between demodulation and modulation at the mobile. A modulation-impedance design should provide sufficient power for the mobile demodulator while exhibiting maximal difference in the backscattered field (i.e., maximum distance between BM constellation points). To analyze these trade-offs, we combined the scattering theory with the Thévenin equivalent model to express the received and backscattered powers at the mobile as a function of the modulation load Z_{mod}. Next, we added negative resistance to the mobile load to improve the transmission range. Finally, we derived our optimization method, which we evaluated numerically.

16.4.1 Parallel and Series Circuit Implementations

Most of the current RFID systems restrict the modulation impedance Z_{mod} to non-negative resistive, that is, $Re(Z_{mod}) \geq 0$ because they rely on the assumption of low-complexity nodes. In this chapter, we propose the usage of negative resistance to increase the backscatter power. We proposed two active BM mobile implementations, named parallel and series (Figure 16.3). For both circuit configurations, we used active loads, which include negative resistance, in the mobile modulators to maximize the uplink range and implement the desired backscattered constellation, subject to no data loss in the downlink path [29]. Note that any active circuit has the potential to become unstable and start to oscillate. This potential increases as the gain increases [30]. While taking into account the stability of both circuits, we compared the performance of both implementations [21]. The simulation results proved that parallel configurations outperform series configurations. Therefore, we will only consider parallel implementation for the rest of the chapter. To increase the reflection coefficient $\Gamma\left(\frac{\beta_n Z_{ant} Z^*_{ant}}{\beta_n Z_{ant} + Z^*_{ant}}\right) = \frac{Z^*_{ant}}{\beta_n\left(Z_{ant} + Z^*_{ant}\right) + Z^*_{ant}}$, the real part of $\beta_n(Z_{ant} + Z^*_{ant}) + Z^*_{ant}$ should have a small positive value, while the imaginary part should be kept different from zero in order to reduce any undesired self-oscillations [39].

16.4.2 Mobile Received Power

To evaluate the received power at the mobile, we made use of the Thévenin equivalent model, as depicted in Figure 16.3. The variable Z_{mod} represents the modulation circuit and is controlled by the mobile microcontroller. The input impedance of all the mobile's IC components, except the modulator circuit, is modeled by Z^*_{ant}. The power delivered to Z^*_{ant} is the power available for the demodulation at the mobile. As stated earlier, an optimum modulation-impedance design is needed to establish ReflectFX communication. We assume $Z_{mod} = \beta_n Z_{ant}$ where $\beta_n \in C$, $n \in \{1, 2, ..., M\}$ and M is the modulation constellation size. Under this model, the power available to the mobile demodulation is

$$P_n = 4P_{match}\left|\frac{\beta_n\left(Z^*_{ant} + Z_{ant}\right)}{2Z^*_{ant} + 2\beta_n\left(Z^*_{ant} + Z_{ant}\right)}\right|^2. \tag{16.5}$$

The average available power for the downlink is then evaluated as $P_{avg} = \frac{1}{M}\sum_n^M P_n$.

16.4.3 Backscatter Transmission

With ReflectFX, the antenna load $Z_i = \frac{\beta_n Z_{ant} Z^*_{ant}}{\beta_n Z_{ant} + Z^*_{ant}}$ varies with the mobile data. We represent the M-PSK modulation constellation as M symbols on a circle of radius $\sqrt{E'}$ centered at $c_{st} \in C$, that is,

$$A_{st} - \Gamma\left(\frac{\beta_n Z_{ant} Z^*_{ant}}{\beta_n Z_{ant} + Z^*_{ant}}\right) := \sqrt{E'}\exp\left(j\theta_n + j\psi\right) + c_{st} \text{ for } n \in [1, 2, \ldots, M], \tag{16.6}$$

where $\Gamma\left(\frac{\beta_n Z_{ant} Z_{ant}^*}{\beta_n Z_{ant} + Z_{ant}^*}\right) = \frac{Z_{ant}^*}{\beta_n\left(Z_{ant} + Z_{ant}^*\right) + Z_{ant}^*}$, $\theta_n = (n-1)\pi/M$ and ψ is a rotation for all symbols. E' represents the energy-gain generated by the antenna mode component of the antenna-scattering field. We rewrite (16.6) as

$$\beta_n = \frac{Z_{ant}^*\left(1 + A_{st} - \sqrt{E'}\exp(j\theta_n + j\psi) - c_{st}\right)}{\left(\sqrt{E'}\exp(j\theta_n + j\psi) + c_{st} - A_{st}\right)\left(Z_{ant} + Z_{ant}^*\right)}. \tag{16.7}$$

16.4.4 Optimum Modulation Impedance

In this section, we propose an optimum ReflectFX modulation-impedance design that captures the trade-offs between received and backscatter power. To improve the performance of the uplink, we must maximize the distance between the constellation points of the BM by increasing E' subject to $\mathrm{Re}(Z_{mod}) = \mathrm{Re}(\beta_n Z_{ant}) \geq R_N$, where R_N is the minimum negative resistance that we can implement. However, to guarantee no loss in the downlink data rate, the P_{avg} has to be at least equal to P_{match}.[2] More specifically, our modulation-load optimization problem seeks to:

Maximize c_{st}, ψ, E'

Subject to

$$\forall n \in \{1, 2, \ldots, M\}, \mathrm{Re}\left(\frac{Z_{ant} Z_{ant}^*\left(1 + A_{st} - \sqrt{E'}\exp(j\theta_n + j\psi) - c_{st}\right)}{\left(\sqrt{E'}\exp(j\theta_n + j\psi) + c_{st} - A_{st}\right)\left(Z_{ant} + Z_{ant}^*\right)}\right) \geq R_N$$

and

$$\frac{1}{M}\sum_{n=1}^{M}\left|A_{st} + 1 - \sqrt{E'}\exp(j\theta_n + j\psi) - c_{st}\right|^2 = \left|A_{st} + 1 - c_{st}\right|^2 + E' \geq 1, \tag{16.8}$$

where we derived (16.8) by combining (16.7) and (16.5) and using the definition of $\theta_n = (n-1)\pi/M$. We have optimized the modulation-impedance design through an exhaustive search.

16.5 ReflectFX Performance Analysis

In this section, we numerically evaluate the ergodic capacity of ReflectFX. Because ReflectFX utilizes load change to modulate the signal, the input messages are constrained to a discrete and finite set. Hence, we used the constellation-constrained capacity expression for channel with a finite alphabet input [22]. We also account for the impact of SI at the base station receiver.

2. Note that in the case of conventional half-duplex and conventional IBFD communication, the received power equals P_{match}.

16.5.1 Transmit Power Normalization

For a fair comparison between ReflectFX, conventional IBFD, and half-duplex systems, the total energy transmitted by the ReflectFX base station must be the same as the total energy transmitted by the two in half-duplex because the ReflectFX mobile does not generate its own radio wave. Because the energy is power multiplied by time, then the power allocated for each half-duplex node is identical to the power allocated to the ReflectFX base station. Analogously, when comparing the half-duplex and conventional IBFD systems, the power allocated to the IBFD node is half the power allocated to the half-duplex node [36].

16.5.2 Modulation-Impedance Design

To guarantee that the ReflectFX downlink data rate equals the half-duplex data rate, the modulation impedances are designed to meet the downlink requirement of $P_{avg} \geq P_{match}$ while also being feasible to implement. To solve for optimum E', named E'_{opt}, in (16.8), we exhaustively searched over all possible $\{c_{st}, \psi\}$. Table 16.1 displays the optimal impedances and the optimal E'_{opt} for QPSK modulation. Our simulation results show that with negative resistance of –22.5Ω, –24.5Ω, and –24.95Ω, we obtained a backscatter power gain of 8 dB, 22 dB, and 42 dB, respectively. Note that the transmitted power at the mobile varies with the transmitted power from the base station, the distance between the base station and the mobile, and the backscatter gain at the mobile. For instance, if the base station transmit power is set to 15 dBm, the distance between the base station and the mobile node is d = 10m, the wavelength is λ = 0.12m (i.e., the path loss is –60 dB), the antenna gains are 0 dBi, and the backscatter gain is 42 dB, then the power transmitted at the mobile is –3 dBm.

16.5.3 Ergodic Capacity Evaluation

We use the constellation-constrained capacity expression for a channel with a finite alphabet input derived in [5]. We assume $\lambda = 0.12$m, $d_{B \to B} = 1$m, $G_{BT} = G_{BR} = G_m$ = 1.6 (dipole antenna). We also set the path-loss exponent to 2, the polarization

Table 16.1 Optimum Modulation Impedance for QPSK, $Z_{ant} = 50\Omega$, and $A_{st} = 1$

Optimum Modulation Loads	R_N (Ω)	$\sqrt{E'_{opt}}$	*Optimum* c_{st}	*Optimum* ψ
$\beta_1 Z_{ant} = -18.89 + 11.48i, \beta_2 Z_{ant} = -18.89 - 11.48i,$ $\beta_3 Z_{ant} = -19.96 + 1.99i, \beta_4 Z_{ant} = -19.96 - 1.99i$	–20	1.2	1.8	$\frac{\pi}{4}$
$\beta_1 Z_{ant} = -21.94 + 5.74i, \beta_2 Z_{ant} = -21.94 - 5.74i,$ $\beta_3 Z_{ant} = -22.48 + 0.99i, \beta_4 Z_{ant} = -22.48 - 0.99i$	–22.5	2.4	3.1	$\frac{\pi}{4}$
$\beta_1 Z_{ant} = -24.47 + 1.2i, \beta_2 Z_{ant} = -24.47 - 1.2i,$ $\beta_3 Z_{ant} = -24.5 + 0.2i, \beta_4 Z_{ant} = -24.5 - 0.2i$	–24.5	12.4	13.1	$\frac{\pi}{4}$
$\beta_1 Z_{ant} = -24.95 - 0.12i, \beta_2 Z_{ant} = -24.95 + 0.12i,$ $\beta_3 Z_{ant} = -24.95 - 0.02i, \beta_4 Z_{ant} = -24.95 + 0.02i$	–24.95	124.5	126	$\frac{\pi}{4}$

Source: [26]

mismatch between antennas to 1, and $A_{st} = 1$. For simplicity, we grouped the SIC techniques into two classes: (1) passive cancellation (as discussed in Chapter 2), and (2) analog-and-digital cancellation (as discussed in Chapters 6 through 12). We assume that both the mobile and the base station are using the same analog-and-digital SIC techniques denoted by σ_{IC}^2 (e.g., –20 and –40). Combining the analog-and-digital cancellation with the passive cancellation generated by the propagation loss between the transmitting and receiving antennas, the total SIC values are 60 and 80 dB at the base station. In the case of the conventional IBFD, we assumed the distance between the transmitting and the receiving antennas at the mobile to be 0.1m. The total SIC values are hence 40 and 60 dB at the mobile.

16.5.4 ReflectFX Versus Half-Duplex

We first compared the ergodic capacity of the ReflectFX and half-duplex systems as a function of the distance separating the mobile from the base station. We set the maximum transmit power to 15 dBm and the noise level to –174 dBm. Figure 16.5 shows the ergodic capacity for QPSK input constellation. ReflectFX always outperforms the half-duplex system. The gain provided by ReflectFX compared to the half-duplex increases with the SIC and the backscatter gain, as expected. Note that the performance of half-duplex does not vary because the transmit power is very high. In Figure 16.6, we reduced the transmit power to –15 dBm and set the noise level to the practical level of –90 dBm. Note that in the low SNR regime (i.e., large distance *d*), the mobile does not receive enough power to transmit back to the base station, and so ReflectFX performs as well as the half-duplex.

16.5.5 ReflectFX Versus Conventional Full-Duplex

As expected, the performance of the conventional IBFD depends largely on the amount of SIC. In the high SNR regime (i.e., small distance *d*) the ergodic capacity and throughput of ReflectFX are higher than the conventional IBFD. The gain

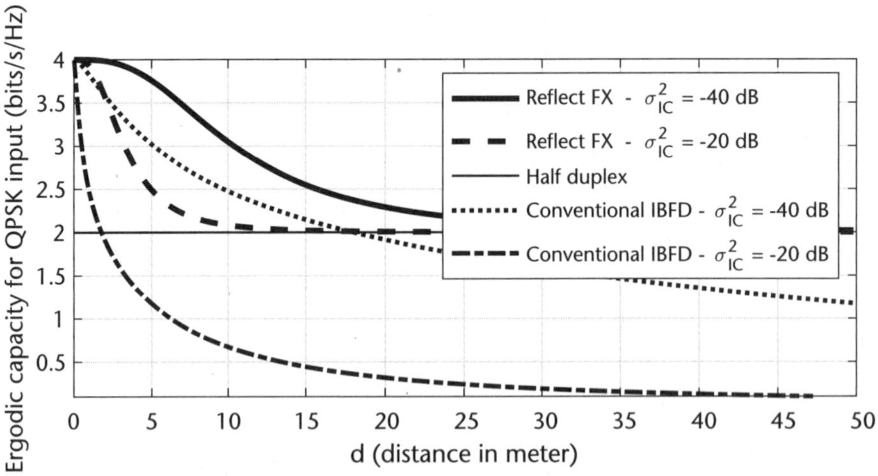

Figure 16.5 The ergodic capacity versus the distance separating the base station from the mobile, $E'_{opt} = 42$ dB.

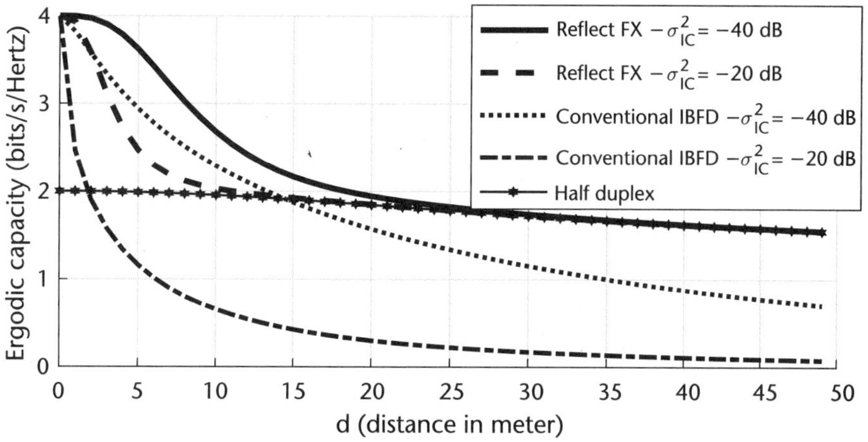

Figure 16.6 The ergodic capacity versus the distance separating the base station from the mobile, $E'_{opt} = 42$ dB and low power.

provided by ReflectFX compared to the IBFD increases with backscatter gain. In the low SNR regime (i.e., large distance d), ReflectFX performs as well as half-duplex but better than the conventional IBFD, with the worst performance because both nodes suffer from SI. The conventional IBFD outperforms ReflectFX when the amount of SIC at the mobile is high and the backscatter gain is low.

To further analyze the impact of the SIC at the mobile on the competitiveness of ReflectFX, we plot the throughput and ergodic capacity versus the total SIC at the mobile node. Here we set the maximum transmit power to 15 dBm and the noise level to –174 dBm. In Figure 16.7, we analyzed the impact of not having the same analog-and-digital SIC at the mobile and the base station. We fixed the analog-and-digital SIC at the base station to 50 dB, generating a total SIC of 90 dB at the base station. We set the distance between the mobile and base station to $d =$ 10m. We varied the level of the analog-and-digital interference cancellation at the mobile from 20 to 70 dB, generating 35 to 85-dB total SIC. As shown in Figure 16.7, the conventional IBFD performance greatly depends on the value of the mobile interference cancellation. Figure 16.7 illustrates that ReflectFX outperforms

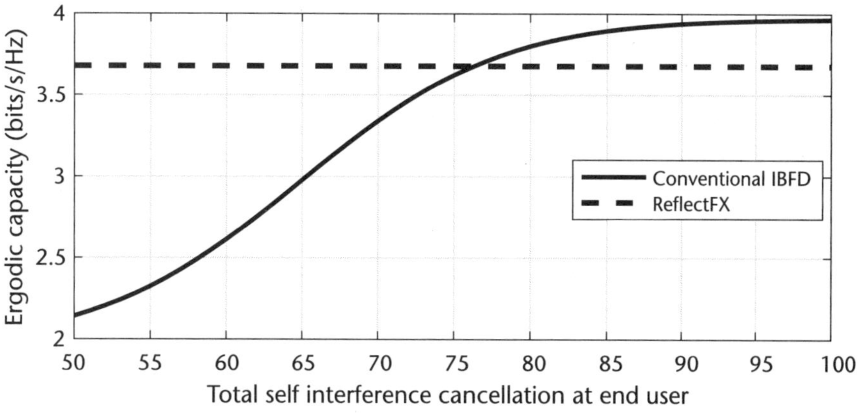

Figure 16.7 ReflectFX and conventional IBFD performance comparison versus the total SIC at mobile (passive cancellation = 15 dB).

conventional IBFD in the case where the mobile cannot achieve a high SIC because of space or processing limitations. Indeed, conventional IBFD outperforms Reflect-FX when the analog-and-digital cancellation techniques implemented at the mobile are better than those implemented at the base station.

16.6 Practical Considerations

Based on the proposed ReflectFX framework, the mobile sends a sequence of symbols onto the backscattered waveforms by switching the impedance loads. An efficient fast switch can be implemented using a switched pulse-injected oscillator. Note that this framework does not require the mobile and base station to transmit at the same data rate. We can modify the mobile data rate by reducing the impedance load change (i.e., factor of the symbol rate), changing the mobile modulation constellation size and/or changing the channel encoder rate.

Note that negative resistance is a characteristic of certain devices where the current flowing through the device is opposite to the applied voltage [40]. In a previous publication, two types of approaches have been proposed to implement negative resistances. One possible implementation of negative resistance can use a gallium arsenide (GaAs) heterojunction field-effect transistor (HFET) for the 2.45-GHz carrier frequency.

16.7 Conclusion

16.7.1 Summary

In this chapter, we proposed an IBFD wireless communication system that uses reflected power, named ReflectFX. Our approach is based on BM, where the same antenna (that receives the EM waves) modulates and reflects the waves back. We considered two nodes (a base station and a mobile) that intended to exchange data over a wireless Rayleigh fading channel with AWGN. We derived an optimum and accurate design that provides sufficient power for the mobile demodulator while exhibiting maximal difference in the backscattered field. Ergodic capacity expressions were numerically evaluated. The simulation results allowed us to conclude there are potentially significant benefits to be gained from including ReflectFX in indoor communication systems.

16.7.2 Future Directions

Future efforts will examine the impact of external interferences, such as intermobile or intercell interferences, in RefectFX. The impact of the external interferences on the ReflectFX downlink is identical to the half-duplex case. This link is more robust to external interferences than conventional full-duplex systems. However, the external interferences may have a combined impact on the ReflectFX uplink. The external interference effects on the downlink can propagate to the uplink because of the BM. The external interference is different from SI. Indeed, the power levels are much lower, and we do not have knowledge of what interference signal is transmitted. We can cancel them using DSP. We also assumed a perfect symbol-level synchronization. Therefore, we also plan to evaluate the use of pilot-aided synchronization and/or global time synchronization. Note that the channel estimation techniques used in half-duplex systems can also be employed in ReflectFX.

References

[1] Pesovic, A., and R. Sharpe, "Is Symmetrical Bandwidth a Myth or a Must?" http://www2.alcatel-lucent.com/techzine/issymmetrical-bandwidth-a-myth-or-a-must/#is-symmetrical-bandwidth-a-myth-or-a-must-n-5.

[2] Shaw, T. D., "Traffic Flow on Broadband Access Networks," http://www.ofcnfoec.org/osa.ofc/media/Default/2012/Presentations/Harstead-Panel-Shaw.pdf.

[3] CISCO, "Cisco Annual Internet Report (2018–2023) White Paper," March 2020, https://www.cisco.com/c/en/us/solutions/collateral/executive-perspectives/annual-internet-report/white-paper-c11-741490.html#Executivesummary.

[4] Stockman, H., "Communication by Means of Reflected Power," *Proc. IRE*, Vol. 36, No. 10, 1948, pp. 1196–1204.

[5] Smida, B., and S. Khaledian, "ReflectFX: In-Band Full-Duplex Wireless Communication by Means of Reflected Power," *IEEE Transactions on Communications*, No. 2, February 2017.

[6] Greeff, R., and D. K. Ovard, "Backscatter Communication Systems, Interrogators, Methods of Communicating in a Backscatter System, and Backscatter Communication Methods," U.S. Patent US6192222, Micron Technology Inc., 2001.

[7] Finkenzeller, K., and D. Muller, *RFID Handbook: Fundamentals and Applications in Contactless Smart Cards, Radio Frequency Identification and Near-Field Communication*, New York: John Wiley & Sons, 1999.

[8] Dobkin, D. M., *The RF in RFID Passive UHF RFID in Practice*, Boston, MA: Elsevier, 2008.

[9] Dardari, D., et al., "Ultrawide Bandwidth RFID: The Next Generation?" *Proc. IEEE*, Vol. 98, No. 9, September 2010, pp. 1570–1582.

[10] Vannucci, G., A. Bletsas, and D. Leigh, "Implementing Backscatter Radio for Wireless Sensor Networks," *Proc. IEEE Int. Symp. Personal, Indoor, Mobile Radio Commun.*, Athens, Greece, September 2007.

[11] Bletsas, A., S. Siachalou, and J. Sahalos, "Anti-Collision Backscatter Sensor Networks," *IEEE Transactions on Wireless Communications*, Vol. 8, No. 10, October 2009, pp. 5018–5029.

[12] Bilstrup, U., and P. Wiberg, "An Architecture Comparison Between a Wireless Sensor Network and an Active RFID System," *Proc. IEEE Int. Conf. on Local Computer Networks*, Tampa, FL, November 2004, pp. 583–584.

[13] Vannucci, G., D. Leigh, and J. Katz, "Wide-Area Long-Range Unidirectional Sensor (WALRUS) Network," *Proc. IEEE Int. Conf. Mobile Ad Hoc and Sensor Systems*, Washington, D.C., November 2005.

[14] Bereketli, A., and O. Akan, "Communication Coverage in Wireless Passive Sensor Networks," *IEEE Commun. Lett.*, Vol. 13, No. 2, February 2009, pp. 133–135.

[15] Vannucci, G., A. Bletsas, and D. Leigh, "A Software-Defined Radio System for Backscatter Sensor Networks," *IEEE Transactions on Wireless Communications*, Vol. 7, No. 6, June 2008, pp. 2170–2179.

[16] Vannucci, G., A. Bletsas, and D. Leigh, "Implementing Backscatter Radio for Wireless Sensor Networks," *Proc. IEEE Int. Symp. Personal, Indoor, Mobile Radio Commun.*, Athens, Greece, September 2007.

[17] Bolomey, J., "Review of Load-Modulated Transponders for Wireless Sensing Applications," *Proc. IEEE Int. Conf. RFID-Tech. and Applications*, Grenoble, France, November 2012, pp. 62–67.

[18] Pelissier, M., et al., "A 112 mb/s Full Duplex Remotely-Powered Impulse-UWB RFID Transceiver for Wireless NV-Memory Applications," *IEEE Journal of Solid-State Circuits*, Vol. 46, No. 4, April 2011, pp. 916–927.

[19] Thomas, S., and M. Reynolds, "A 96 mbit/sec, 15.5 pj/bit 16-QAM Modulator for UHF Backscatter Communication," *IEEE International Conference on RFID*, 2012, pp. 185–190.

[20] Lee, M. -H., C. -Y. Yao, and H. -C. Liu, "Passive Tag for Multi-Carrier RFID Systems," *2011 IEEE 17th International Conference on Parallel and Distributed Systems (ICPADS)*, December 2011, pp. 872–876.

[21] Carlowitz, C., et al., "A MM-Wave RFID System with Locatable Active Backscatter Tag," *2012 IEEE International Conference on Wireless Information Technology and Systems (ICWITS)*, November 2012, pp.1–4.

[22] Yu, P., P. Schaumont, and D. Ha, "Securing RFID with Ultra-Wideband Modulation," *Proceedings of Workshop on RFID Security (RFIDSec)*, 2006, pp. 27–39.

[23] Nikitin, P., et al., "Passive Tag-to-Tag Communication," *Proc. IEEE Int. Conf. RFID-Tech. and Applications*, Orlando, FL, April 2012, pp. 177–184.

[24] Smida, B., and S. Islam, "Full-Duplex Wireless Communication Based on Backscatter Amplification," *Proc. IEEE Int. Conf. Commun.*, Sydney, Australia, 2014.

[25] Smida, B., "Communicating Data Using Backscatter Modulation," U.S. Patent 2013-SMID-66564-01, filed, 2013.

[26] Khaledian, S., et al., "Active Two-Way Backscatter Modulation: An Analytical Study," *IEEE Transactions on Wireless Communications*, Vol. 18, No. 3, March 2019, pp. 1874–1886.

[27] Best, S., and B. Kaanta, "A Tutorial on the Receiving and Scattering Properties of Antennas," *IEEE Transactions on Antennas and Propagation*, Vol. 51, No. 5, May 2009, pp. 26–37.

[28] Pozar, D., "Scattered and Absorbed Powers in Receiving Antennas," *IEEE Antennas and Propagation Magazine*, Vol. 46, No. 1, January 2004, pp. 144–145.

[29] Andersen, J. B., and R. G. Vaughan, "Transmitting, Receiving, and Scattering Properties of Antennas," *IEEE Antennas and Propagation Magazine*, Vol. 45, No. 4, April 2003, pp. 93–98.

[30] Green, R., "The General Theory of Antenna Scattering," Ph.D. Dissertation, Ohio State University, 1963.

[31] King, R., and C. Harrison, "The Receiving Antenna," *Proc. IRE*, January 1944, p. 35.

[32] Stevenson, A., "Relations Between the Transmitting and Receiving Properties of Antennas," *Quarterly of Applied Mathematics*, January 1948, pp. 140–148.

[33] Hu, Y., "Back-Scattering Cross Section of a Center-Loaded Antenna," *IEEE Transactions on Antennas and Propagation*, January 1958, pp. 140–148.

[34] Collin, R., *Antenna and Radio Wave Propagation*, New York: McGraw-Hill, 1985.

[35] Silver, S., *Microwave Antenna Theory and Design*, New York, McGraw-Hill, 1949.

[36] Boyer, C., and S. Roy, "Coded QAM Backscatter Modulation for RFID," *IEEE Transactions on Communications*, Vol. 60, No. 7, July 2012, pp. 1925–1934.

[37] Thomas, S., et al., "Quadrature Amplitude Modulated Backscatter in Passive and Semi-passive UHF RFID Systems," *IEEE Transactions on Microwave Theory and Techniques*, Vol. 60, No. 4, April 2012, pp. 1175–1182.

[38] Smida, B., et al., "Backscatter-Modulation Constellation for Full-Duplex Wireless Communication," *2015 IEEE 12th International Conference on Mobile Ad Hoc and Sensor Systems (MASS)*, October 2015, pp. 561–566.

[39] Marshak, A. H., "A Unique Current-Controlled Negative Resistance Generator," *IEEE Transactions on Communication and Electronics*, 1964, pp. 182–184.

[40] Chan, P., "Bi-Static 5.8ghz RFID Range Enhancement Using Retrodirective Techniques," *41st European Microwave Conference (EuMC)*, 2011.

CHAPTER 17

Military Applications

Taneli Riihonen

17.1 Introduction

As mentioned in Chapter 1, the scientific research on nonmilitary (e.g., civilian and commercial), full-duplex systems has been progressing on a broad front since 2010, and I (among many other scientists) have had the honor to witness its beginnings on the frontline, even initiating some developments. The full coverage of the history and overviews of the latest advances can be found in [1, 2], as well as the preceding chapters of this book. This chapter concerns an emerging, more specialized research area within the general development of IBFD and/or STAR[1] techniques and systems by providing a dissection of the scientific research on their military applications since 2017, which utilized the elements highlighted within the IBFD transceiver diagram of Figure 17.1(a) and the SIC techniques tree of Figure 17.1(b).

Several years ago, it became evident that, from an academic perspective, IBFD radio technology had already reached a rather mature state in the context of wireless communication applications, to the point where research and development (R&D) activities in industry are today overtaking basic scientific research. Our team started looking beyond plain data transmission and envisioning prospective emerging defense and security applications facilitated by the general STAR capability of IBFD radios. In early 2016, this thinking of renewed research lines led to a successful project proposal to the Ministry of Defence of Finland (namely, the Finnish Scientific Advisory Board for Defence (MATINE)) [3, 4] and, by the end of 2017, escalated into an international collaboration activity within the NATO Science and Technology Organization (STO) [5–7]. In general, these developments represent technology transfer from civilian and commercial communications to the military and defense and the security domain [8, 9].

The research started as a feasibility study to survey the prospects and challenges of exploiting IBFD radio technology in novel military communication applications.

1. In this chapter, it would be often appropriate to refer to STAR operation rather than IBFD operation because the word "duplexing" is surrounded by the connotation of two-way communications, which does not pertain to many of the considered military applications. Thus, the terms are here combined as IBFD/STAR.

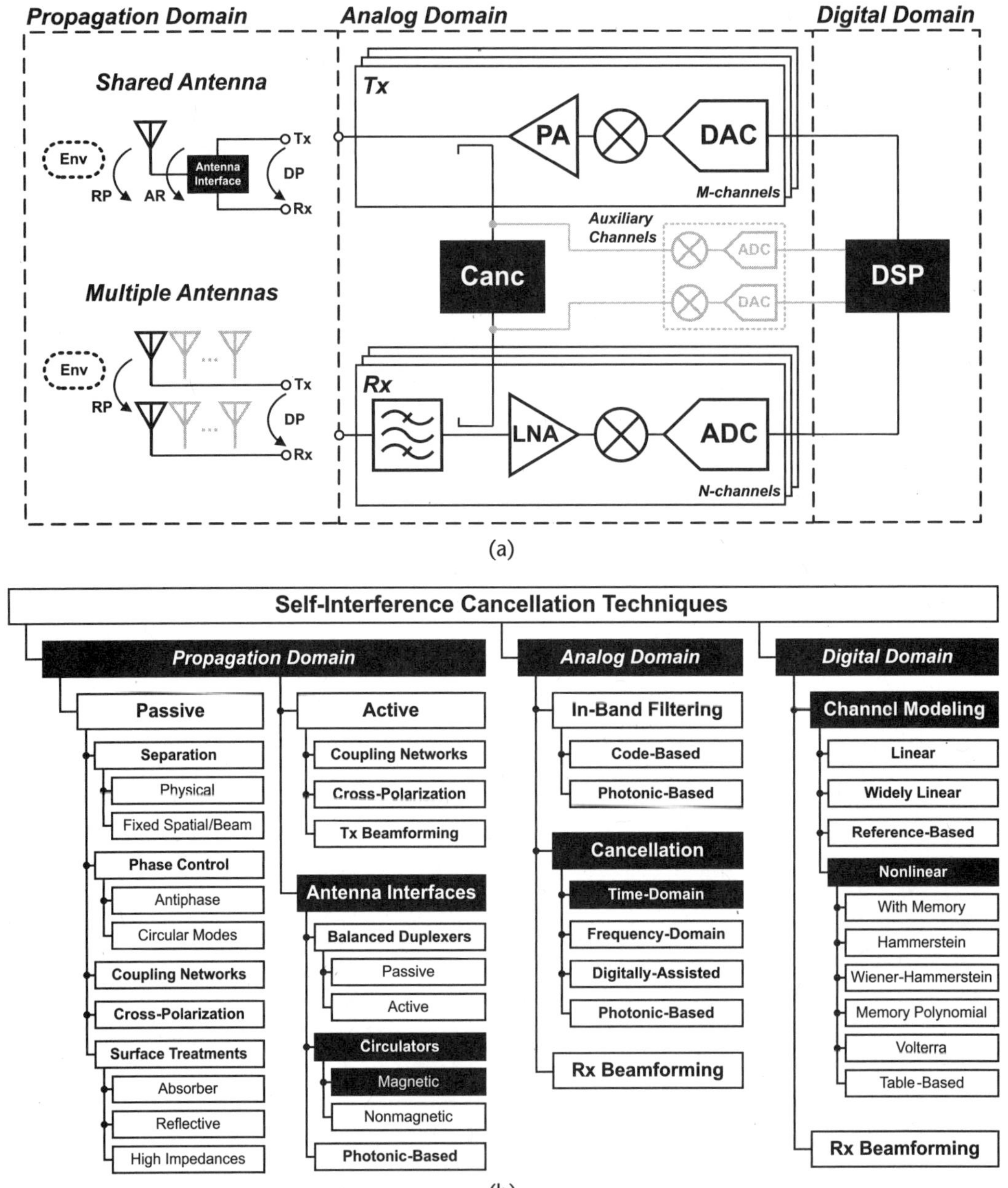

Figure 17.1 Military applications and specific chapter focus highlighted within (a) a generic IBFD transceiver and (b) a SIC techniques tree diagram.

In the technology transfer, the first step has been to understand how the requirements for IBFD/STAR transceivers differ in military scenarios with respect to nonmilitary scenarios, and what new developments are needed for repurposing the IBFD radio communication technology to military IBFD/STAR applications. Military spectra are also scarce and congested, whereby the defense industry shares telecom's motivation to improve spectral efficiency in tactical communications [5], but the main originality in the emerging research area actually pertains to envisioning and characterizing other, novel applications in the battlefield [10]. A general overview of different potential military usage scenarios is illustrated in Figure 17.2.

When extending beyond the communications context, prospective military full-duplex radios will have the progressive capability for STAR operation where they can conduct electronic warfare (EW) while concurrently using the same frequency band for communication or performing an electronic attack with simultaneous signals intelligence [11]. In particular, we have suggested that IBFD/STAR transceivers could create a paradigm shift in defense applications by way of a multifunction military full-duplex radio (MFDR) integrating tactical communications, signals intelligence, and EW [9, 10] as well as possibly introducing the concept to civilian/commercial security and privacy applications in the form of a radio shield [8, 9].

Because certain sensitivity is always associated with military-related topics, the scope of this chapter is strictly on scientific research involving some explicit application of IBFD/STAR technology in some military context. It is true that scientific research and military technology development do not fit perfectly together, at least not classically. This somewhat restrains the evolution of disruptive defense concepts, but the research advances reported here also demonstrate that significant developments are possible, especially in basic research, without concealing true objectives and outcomes. The critical point for success in scientific research on military/defense technology seems to be starting transparently with the principle of openness characteristic of MATINE, where all funded projects are public unless otherwise specified, and keep communicating this attitude clearly in all interaction with stakeholders. Similarly, we must also recognize that much of the general basic scientific research on STAR concepts could apply to military systems as well, and much of it is even funded by the defense sector. However, this chapter does not

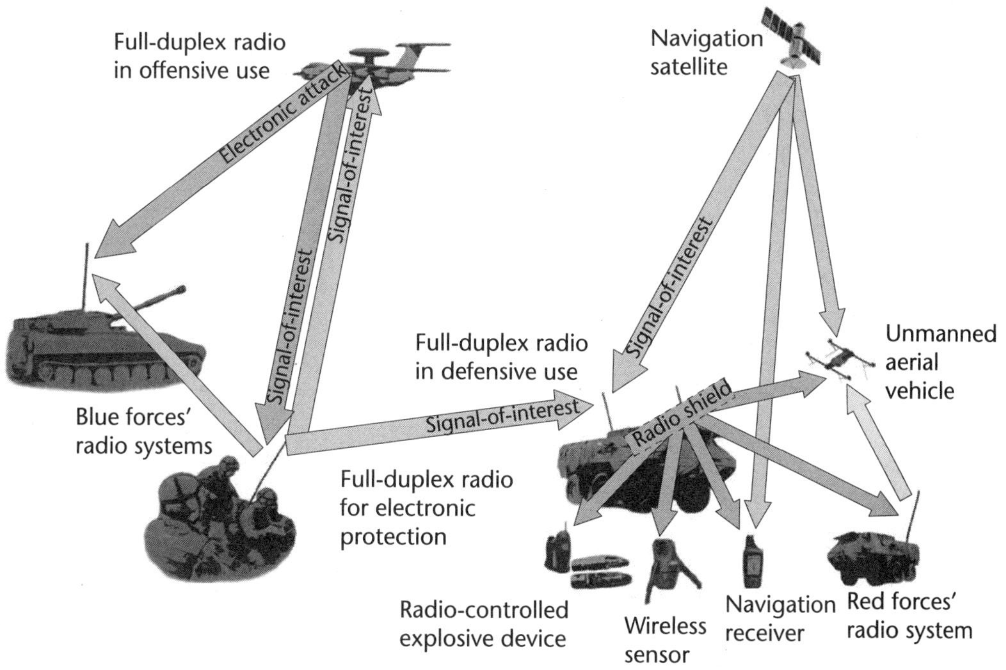

Figure 17.2 Prospective military applications for IBFD and/or STAR radios beyond communications.

attempt to cover the vast body of important research that has implicit military applications, but has not disclosed explicit ones.

The classical application for full-duplex operation in a military context is certain radar transceivers whose technology gradually became public information in the years following World War II, namely, that an IBFD/STAR radio is analogous to a nonpulsed monostatic radar. The main conceptual difference between a monostatic radar and an IBFD/STAR transceiver is that the signal reflecting back to the IBFD/STAR transceiver is completely undesired, while the very same signal is actually the signal-of-interest (SOI) for a radar receiver, except for direct path (DP) leakage and near-field clutter. Moreover, compared to IBFD/STAR systems, conventional nonpulsed radars employ rather primitive transmitted waveforms such as unmodulated CW and variations of frequency-modulated CW (FMCW), as mentioned in Chapter 1. The mixer structure that is typical for FMCW radars can be effectively used to implement an IBFD/STAR transceiver if considering such military applications (e.g., some form of tactical communications and jamming), where the restriction to constant-envelope transmission is feasible [12]. Conversely, IBFD/STAR transceiver techniques could allow for implementing more advanced continuous radar waveforms with arbitrary I/Q modulation, which may lead to significant advances in noise radar technology and manifest itself as a key component for the RF convergence of MFDRs [13] that implement radar functionalities among those for tactical communications, radio-based sensing, and EW.

17.2 Overview of Global Research

17.2.1 Prior Art Before 2017

Scientific research on the military/defense applications of IBFD/STAR radios was virtually nonexistent until early 2017, when the first study on the subject [11] was completed by my associates and me. It should be again emphasized that herein scientific implies public in nature. The concept of IBFD/STAR operation is so strikingly advantageous that it would be foolish to assume that armed forces and defense industries are not developing it and related applications in a confidential/classified manner. In addition, as already mentioned above, the prior art should be considered to include many research works on generic STAR transceivers and SIC techniques without explicit military applications, although they are not discussed herein. These remarks apply especially to East Asia, from where a massive body of scientific knowledge on IBFD techniques and systems currently emerge, but explicitly military/defense-oriented research is conspicuously absent unlike studies into authoritarian security applications.

One military application is concretely suggested by a passing sentence in [14, p. 120], which is perhaps the very first, although tiny, record of a precursor for greater scientific discourse. In particular, IBFD/STAR radios' self-interference cancellation capability could allow them, when operating as jammers, to "flood the airwaves with strong transmission ... to disrupt enemy communications and at the same time listen to its own troop communications" [14]. To my knowledge, the first scientifically published evidence of defense industry's R&D toward military IBFD/STAR products are [15–17], where the authors present their company's development of a joint analog and digital interference cancellation (JADIC) system. The two versions of the same paper [15, 17] reveal that the JADIC system is developed for blue force communications signal reception with simultaneous jamming transmission by a colocated EW system or a high-powered radar system (e.g., in vehicles).

However, although the military applications of interference cancellation are emphasized by the title of [16], they are actually not specified in more detail than mentioning that the "reduction of the level of unknown interference is an important goal of … electronic warfare" [16]. Thus, the publication is at the borderline of scientific discourse on the applications of the technology in the scope of this chapter. It is also a standard example of how scientific research on STAR techniques conceals the underlying true objectives and target applications while still appropriately acknowledging affiliations to or funding from the defense sector. The above broadly represents the very limited state of the art before the greater scientific discourse commenced at the International Conference on Military Communications and Information Systems (ICMCIS), Oulu, Finland, in May 2017.

17.2.2 Seminal Research in Finland

17.2.2.1 Finnish Scientific Advisory Board for Defence

As mentioned earlier, my team proposed a two-part feasibility study "Full-Duplex Radio Technology in Military Applications" in 2016 to MATINE. The first part of the project in 2017 laid out theoretical foundations for the new research area by reconsidering the architectures and requirements of existing IBFD transceivers in the military domain as well as envisioning and analyzing some of the highest potential military applications of the IBFD/STAR capability [3, 10, 11, 18]. The second part of the project in 2018 contributed the world's first practical experiments that demonstrated the viability of some of the proposed applications for MFDRs in a laboratory and outdoors [4, 19–22].

How do we even start implementing IBFD/STAR capability for military applications? This was our baseline research question when investigating how to transfer the maturing nonmilitary techniques to the new domain. Beyond requiring both extreme sensitivity for desired signals and extreme robustness against strong hostile signals, many military radios would actually operate at very high frequency (VHF) or, even lower, at high frequency (HF), unlike almost all academic prototypes that demonstrate the feasibility of SIC at upper ultrahigh-frequency (UHF) bands only. Moreover, the previous experimental research has not focused on supporting any larger transmission powers than what mobile phones and local area networks can use in civilian/commercial communications.

Figure 17.3 surveys existing research prototypes in terms of their operation frequencies and maximum transmission powers. Our feasibility study concluded that nothing in principle should prevent redeveloping their state-of-the-art nonmilitary techniques to military use, and we could not find evidence suggesting a technical impossibility to employing MFDRs for successful battlefield operations. It is rather obvious that military IBFD/STAR techniques will need to reach much higher transmission powers (especially in jamming), while still achieving much higher SIC capability, and different frequencies than the state-of-the-art nonmilitary technology,

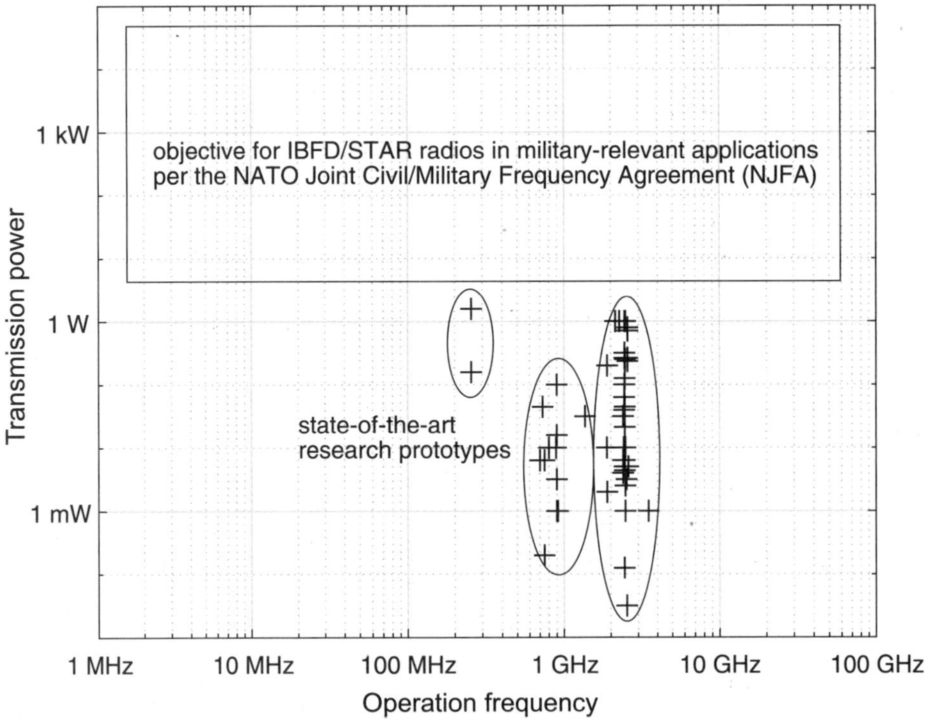

Figure 17.3 Characteristics of existing prototypes. (Scatter data courtesy of K. E. Kolodziej et al. [2].)

which will be a tough research challenge. However, bandwidth requirements are a two-sided issue: while tactical communication bandwidths are in general very small compared to modern wireless networks, the usage of frequency hopping may require supporting wider bandwidth than typical civilian/commercial mobile communications, the networks' latest 5G notwithstanding. Moreover, radar systems benefit from a wider bandwidth, as their resolution is inversely proportional to it and the bandwidth requirements grow to the extreme with noise waveforms.

What are, then, the best ways to exploit MFDRs in cyber-electromagnetic battles? We envisioned that armed forces could gain a major technical advantage from radio transceivers that conduct EW (e.g., signals intelligence or jamming), both when they are receiving or transmitting other signals (e.g., tactical communications) on the same band. Thus, our project characterized all and, especially, the best uses for STAR capability in battlefield scenarios [10]. Beyond compelling theoretical evidence, it still remained somewhat unknown whether IBFD/STAR capability could be useful for military radio systems in practice. Thus, it became necessary to also consider the following question: is it reasonable to expect that the blue-sky concepts could be viable in real life, too? In particular, the second part of the MATINE project took the feasibility study to a practical level by experimenting with selected example applications in indoor and outdoor laboratory settings and provided general proofs-of-concept, as detailed in Section 17.3.2.

The question of ultimate viability still remains unanswered because any field experiments with military-grade transceivers have not been reported to date. The prototype MFDR transceiver used in our experiments (Figure 17.4) was also

Figure 17.4 Photo of the prototype full-duplex (FD) transceiver used in the experiments.

repurposed from earlier research on nonmilitary wireless communications, which operates at the unlicensed 2.4-GHz ISM radio band and is limited to low transmission powers below 0.2W. The prototype is built atop a high-quality vector signal transceiver (VST) that acts as the basic digital transceiver. The necessary SI suppression is implemented in three stages: first, a circulator allows the transmit and receive chains to share a single antenna, while also providing some passive isolation (as discussed in Chapter 2); second, the SI is suppressed by custom-made active analog cancellation electronics (similar to those in Chapter 8) to ensure that SI power entering the ADC does not saturate it; finally, the remaining SI is suppressed by nonlinear digital cancellation in software, as mentioned in Chapter 11. In the most extreme case, we only took the prototype setup to our campus yard for outdoor experiments, which, while pioneering work at this early development stage, is still very far from convincing field trials.

17.2.2.2 Recent Work

We are currently implementing another MATINE project "Electronic Interception of Unmanned Aerial Vehicles," where one task must develop further IBFD/STAR radios for simultaneous consumer drone jamming and radio control (RC) signal detection. The previous feasibility study has provided laboratory proofs of concept and now the research continues towards more realistic outdoor testing and initiating R&D for eventual productization. The research is therefore done in collaboration with industry, which has resulted in scientific publications too [23–26].

Our parallel project, "Radio Shield Against Malign Wireless Communication," is funded by the Academy of Finland, where some applications for IBFD/STAR techniques have similarities with those considered in the military domain, despite the scope being purely on civilian/commercial security and safety applications. Three related articles [26–28] were submitted to ICMCIS 2020.

17.2.3 NATO Science and Technology Organization

17.2.3.1 Exploratory Team

Inspired by our conference presentation of [11], the NATO STO's Information Systems Technology (IST) panel established an exploratory team (ET) under the activity reference IST-ET-101 (COM) and the title "Full Duplex Radio Technology – Increasing the Spectral Efficiency for Military Applications" [5]. The ET was to be active through 2018 only, although it started a month early. The main scientific outcomes were a conference article [29] with presentation at ICMCIS 2019 and the final report [6], published with open access on the NATO STO's website (www.sto.nato.int).

The key objective of the ET was to introduce full-duplex technology to the military domain. To that end, the ET conducted a feasibility study with similar objectives as the Finnish MATINE project in 2017, but added greater international perspective to corroborate the findings. As the title implies, the ET commenced with the main objective to increase the spectral efficiency of military communications, although it also considered IBFD/STAR techniques as a promising approach to efficiently combine tactical communication and EW applications. The military applications can be categorized as shown in Figure 17.5, for which the block for jamming detection during information transmission is recently added [3, 6, 9, 10] following the research for [27]. As for the requirements and challenges to implement IBFD/STAR techniques in military environments, the ET reaffirmed the conclusion that they must reach much higher transmission powers and lower frequencies than the existing nonmilitary technology. Unlike the MATINE project, which did not experiment with plain communication applications, the ET contributed the world's first practical experiments on two-way IBFD military communications [6, 29], as described in Section 17.3.1.

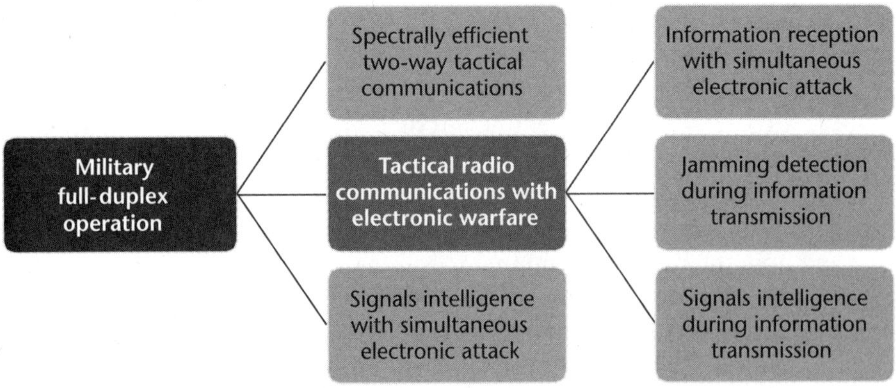

Figure 17.5 Classification of military applications for full-duplex (FD) radios.

17.2.3.2 Research Task Group

The exploratory team's work continues as a research task group (RTG) under the activity reference IST-175 (COM) and the title "Full Duplex Radio Technology for Military Applications" until 2022 [7]. The objectives and topics for the RTG can be categorized under three themes, as paraphrased from the public activity description [7].

First, the RTG is working to identify further candidate military applications and areas thereof, where they also consider other, primarily noncommunications, systems that can benefit from IBFD/STAR techniques such as jamming detection. Second, the RTG further assesses state-of-the-art SIC solutions and the efforts needed to integrate IBFD/STAR techniques into existing communications systems, where the specific objective is to move from civilian/commercial frequency bands to military-relevant spectrum (or make techniques frequency-agnostic) and from rather restricted transmit powers to high military-relevant levels. Third, the RTG is paving the way for a common multinational demonstrator based on COTS software-defined radios by identifying and describing internal and external interfaces thereof. Finally, the activity description concludes by asking: "what are the short-, middle-, and long-term challenges and opportunities of the emerging full-duplex technology?" [7]. This suggests that the RTG is also outlining a potential roadmap for the long way towards field trials and operational use.

17.2.4 Related Activities

Within the NATO STO, there are also other activities related to full-duplex techniques' and systems' research, at least tangentially, two of which can be found under the Sensors and Electronics Technology (SET) panel. The RTG for SET-285 "Multifunction RF Systems" [30] is developing technology for radio systems that jointly provide radar, communications, electronic attack, and electronic support functions. They have also recognized that IBFD/STAR techniques will be a key component for such systems by allowing flexibility in simultaneous usage of transmit- and receive-side functionalities, as one of the scientific objectives is indeed to "evaluate the benefits and challenges of STAR operation" [30]. In addition, the RTG for SET-287, "Characterization of Noise Radar," is among other topics studying the "control and characterization of direct leakage between transmitter and receiver" [31], which is essentially the same as the SI challenge addressed by IBFD/STAR techniques. We believe that significant synergy would emerge by fostering interaction between all the aforementioned NATO activities, thus bypassing potential silo mentality between the STO's panels.

There is also private research on the military applications of IBFD/STAR techniques using public funding. These are hardly scientific activities, but are briefly mentioned for completeness because their existence and general objectives, although not yet the outcomes, have been communicated openly to the public. The biggest of them ($15,700,000), to our knowledge, is the project "Military Full Duplex Radio" [32], funded by the National Spectrum Consortium (NSC). According to the press release [32], the research focused on applying STAR techniques for improving the spectral efficiency of tactical radio communications. The work was likely a continuation of the development of SIC techniques [14] and the JADIC system [15,

17] by the project partner companies. There have also been several solicitations for applications and awards for projects in the Small Business Innovation Research (SBIR) and Small Business Technology Transfer (STTR) programs authorized by the U.S. government. We can assume the IBFD/STAR techniques developed in the projects concern military applications in some way whenever the funding agency is the Department of Defense. The Swedish Defense Research Agency also recently published results from its research [33, 34].

17.3 System Concepts and Experimentation

17.3.1 Tactical Communications

The IBFD radios' applications in plain military communication systems are not fundamentally different from their civilian/commercial counterparts, especially when primarily aiming at improving spectral efficiency. As illustrated in Figure 17.6, two principal communication scenarios are two-hop relaying from a half-duplex (HD) source transmitter via a full-duplex relay transceiver to an HD destination receiver and two-way communications between two full-duplex transceivers. In addition, the STAR capability may indirectly facilitate spectrum use by allowing novel medium access control schemes to be implemented (e.g., for alleviating the hidden node problem) [6, 9]. This is illustrated by the half-duplex receiver in the top-right corner of Figure 17.6, which is able to detect that the channel is already allocated by other tactical transmissions; the STAR capability can be used also for transmitting a busy tone from receivers to mark channel allocation, even if the system is not performing actual IBFD communications.

Studying the feasibility of IBFD communications at large would not (anymore) be very novel per se, but special novelty in the experiments reported by the NATO

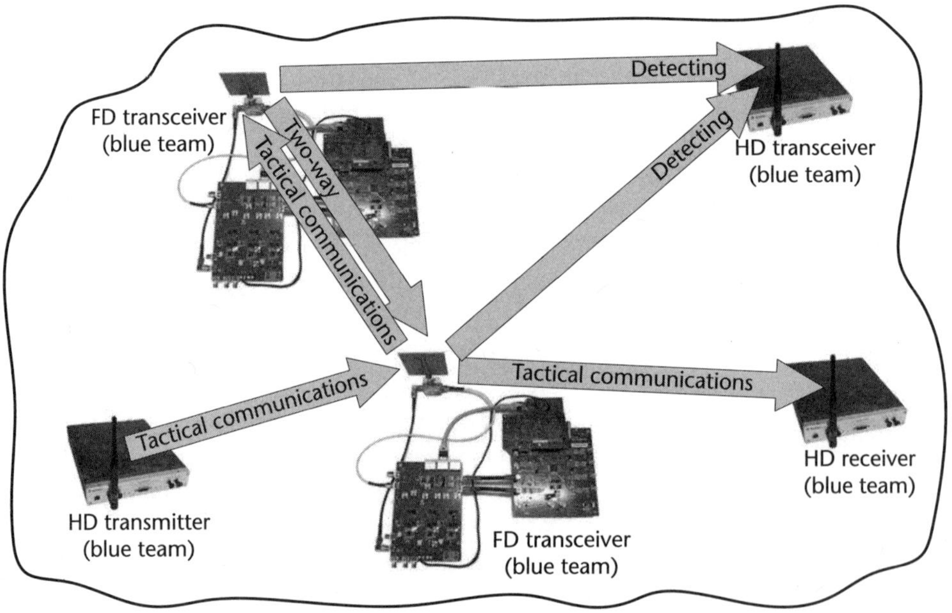

Figure 17.6 FD operation in tactical communications.

STO's ET [6, 29] stem from the fact that they used a real tactical communications signal, namely, the NATO Narrowband Waveform (NBWF), for the first time ever alongside the world's first SIC prototype developed specifically for tactical communication applications. In particular, the ET partners from Germany and Belgium demonstrated the feasibility of two-way IBFD tactical communications in an indoor laboratory environment.

17.3.2 Reception While Transmitting Jamming

The experiments conducted in the MATINE project focused on the following applications, where the MFDR transmits a jamming signal while simultaneously receiving some SOI. All the considered variations of the concept are illustrated collectively in Figure 17.7. The jamming can represent an electronic attack against a remote RC system or an electronic protective measure against interception (i.e., eavesdropping in a nonmilitary context), while the SOI can pertain to detecting the RC system, intercepting enemy transmissions or just receiving own tactical communications.

17.3.2.1 Simultaneously Detecting and Preventing Radio Control Systems

In our first (perhaps in the whole world too) experiment [19] in January 2018, we considered a battlefield scenario, where the red team attempts to use an improvised RC system to activate an improvised explosive device (IED) (e.g., a roadside bomb) (but the results are applicable also with drones to some extent [28] due to the RC components in use). The blue team's MFDR attempts to prevent that by broadcasting jamming, which is conventional, but its STAR capability now enables

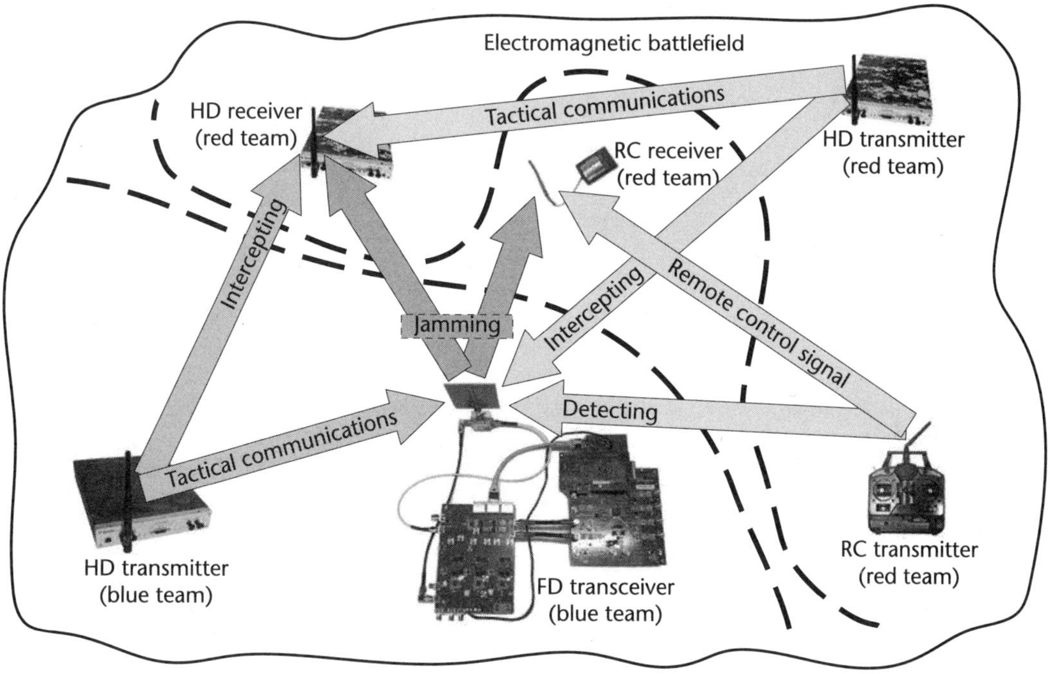

Figure 17.7 Applications for FD reception while transmitting jamming.

simultaneous spectrum monitoring and signal surveillance for detecting the red team's RC transmitter. In contrast, a conventional HD radio would alternate between the transmit and receive functions. The experiments proved that an MFDR can successfully detect attempts to trigger RC IEDs or control RC multicopters while simultaneously also preempting them, because the RC transmission is visible in the measured signals at the same time when the RC receivers are jammed to become inoperable. In the following summer, we continued the experiments outdoors [22].

17.3.2.2 Simultaneously Jamming and Intercepting Enemy Tactical Communications

Our second indoor experiment [20] in April 2018 demonstrated that, by employing an IBFD/STAR transceiver, the blue team can simultaneously transmit jamming against the red team while intercepting tactical communications. A real tactical communications signal, namely the Soldier Radio Waveform (SRW), was used in the experiments. The simultaneous jamming and interception results in technical advantage, because the red team's half-duplex transmitter may increase its transmission power as a countermeasure to jamming, which causes easier interception. The advantage is because jamming represents unavoidable interference to the red team's half-duplex receiver in the top-left corner of Figure 17.7, while the blue team's full-duplex transceiver can cancel the corresponding SI effectively. Continuing these experiments outdoors is a potential direction for future research.

17.3.2.3 Radio Shield for Protection Against Enemy Receivers

Our third indoor experiment [21] in April 2018 studied how well an IBFD/STAR transceiver could receive tactical communication signals from the same team's half-duplex transmitter while simultaneously transmitting jamming that would prevent an enemy from intercepting the communications or using an improvised RC system on the same frequency band. A real tactical communications signal, namely, a multicarrier version of the SRW, was used in the experiments. The measurements showed that, with the prototype MFDR used in all the experiments, the quality of the tactical communications signal remained almost constant, regardless of jamming power. Thus, IBFD/STAR operation allows one to create a protective radio shield of jamming at a reasonably low cost of reduced communications capability. These experiments were also continued outdoors in the following summer [22].

17.3.3 Reception While Transmitting Tactical Communications

In January 2020, we performed a measurement-based evaluation of the novel scenario depicted in Figure 17.8, where the IBFD/STAR capability is used for defensive purposes by detecting the enemy's electronic attack during the MFDR's own transmission [27]. In particular, the blue team's MFDR transmits tactical communications while simultaneously detecting if the red team is attempting to jam the communications, thereby detecting the presence of the enemy, enabling taking countermeasures against the jamming, and informing team members about the attack. The experiments evaluated the detection threshold at how low the MFDR can

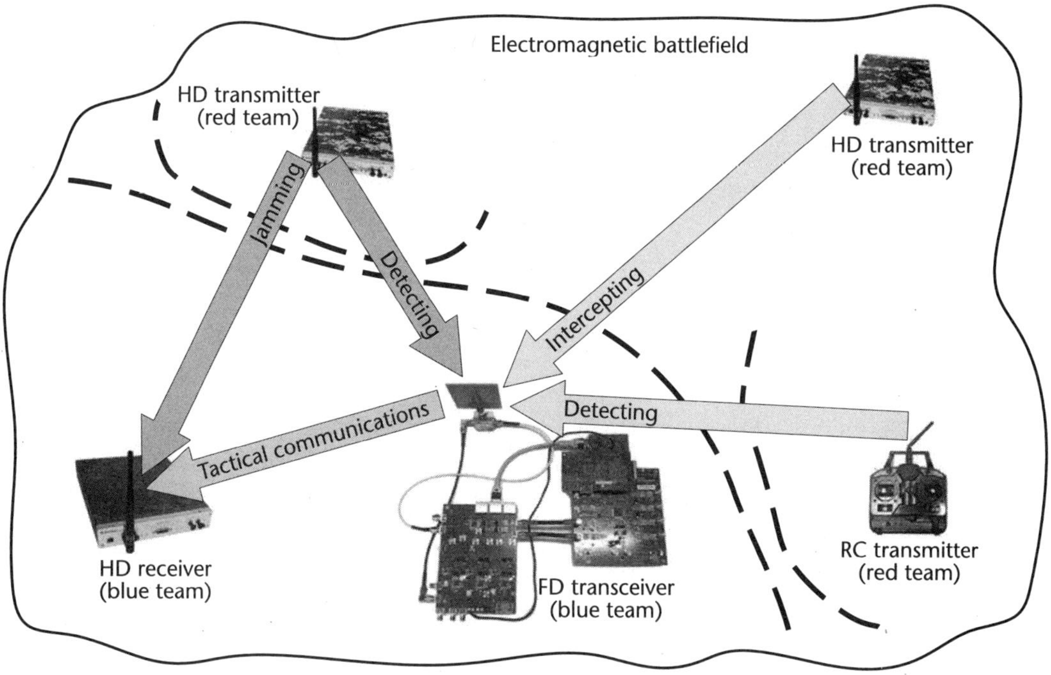

Figure 17.8 Applications for FD reception while transmitting tactical communications.

detect the presence of the electronic attack and thus demonstrated that the concept is indeed viable. The special benefit of the concept is that it allows the detection of the jamming even if a smart stealth opponent stops the jamming signal whenever the MFDR itself stops transmitting. We envision that the MFDR could similarly perform signals intelligence (e.g., for detecting RC transmitters or intercepting tactical communications) during information transmission (at no cost to it), as a future area of research.

17.4 Practical Considerations

The full-duplex technologies are located around a sweet spot between multiple branches of electrical engineering. As mentioned in Chapter 1, implementing an IBFD/STAR radio requires mastering antennas and propagation, microwave electronics, and analog and digital signal processing, while developing it for some specific application requires mastering systems design and the fundamentals of all the various functionalities implemented. The diverse skill requirements can soar to the extreme when considering MFDRs for total RF convergence that combines tactical communications, EW, and radar functionalities. Moreover, although theoretical developments continue to be valuable, the topic is becoming mature enough whereby it is necessary to corroborate findings with measurements and prototype experiments. As a result, scientific research on military full-duplex techniques inherently calls for teamwork and global collaboration between all teams involved.

It can also be recognized that the art of full-duplex research is that of addressing a complete system design problem. This is especially true with military radio transceivers with extreme requirements for both sensitivity and robustness. It may not be possible to develop general-purpose IBFD/STAR radios that would be effective in many military applications at large, but the overall system configuration starting from antenna and transceiver design must be tailored for the specific application in question. However, general principles and technical solutions can be concluded through scientific research so they are applicable in R&D towards specific products.

Finally, one should also consider the practicalities of how cyber-electromagnetic warfare affects physical warfare techniques and tactics or, conversely, how the real operation environment at the battlefield must be recognized in the design of IBFD/STAR systems. For example, I was intuitively expecting that jammers in vehicles against RC IEDs would benefit from using directive transmit antennas before learning that it is not good at all for the soldiers in practice. This is because, despite antenna directivity allowing for longer jamming range, adversaries would likely learn to identify them on vehicles and place the RC receiver farther behind the main IED so that jamming becomes less effective than with omnidirectional jamming and the risk of casualties increases. Although this advice may be difficult to follow by just anyone, one must participate in military exercises to gain hands-on experience of the end user's everyday life in the field as well as to receive active officers' feedback (preferably lower-ranked ones, as higher-ranked ones are strategic thinkers and detached from physical action). This will help broaden the academic bubble, avoid pitfalls, and develop better technology.

17.5 Conclusion

17.5.1 Summary

This chapter presented a comprehensive overview to the nascent scientific research on the applications of IBFD/STAR radios in military systems. Extrapolating from the rapid advances in their civilian and commercial counterparts, the disruptive and unprejudiced idea of in-band STAR operation is expected to find its way in many forms to military radio systems before long. We may even be witnessing now the beginning of a paradigm shift: the RF convergence. Thus, this chapter explored the prospects of IBFD/STAR techniques in cyber-electromagnetic battles to encourage more scientific research on this emerging topic and to disseminate the ideas further in the scientific community and beyond. It is not unreasonable that armed forces could gain a major technical advantage over an opponent that does not possess the IBFD/STAR technology, or that they need new procedures and tactics to counteract opponents' IBFD/STAR capability.

The current stage of development can be reflected with Everett M. Rogers's famous theory for the diffusion of innovations that has four main elements influencing the spread of a new idea: the innovation itself, a social system, communication channels, and time. The innovation in question is obviously the application of IBFD/STAR radios in explicit military contexts, while the social system is understood herein as that formed by the global scientific community of researchers, educators, practitioners, or leaders, which nurtures knowledge and civilization for humankind. The communication channels are scientific publications (namely, this chapter's references), conference and workshop presentations, and interaction in task groups. As justified in Section 17.2.1, the scientific community has not had open discourse on the subject before 2017 so it is fair to say the innovation has been spreading only among the innovators (per Rogers) defined in this section thus far. In contributing to this book, we aim to spread this knowledge to early adopters and early majority (per Rogers) over time.

Another way to assess the maturity of the IBFD/STAR technology is through technology readiness levels (TRLs). We estimate based on unclassified information that the technical readiness of military IBFD/STAR systems is currently somewhere at the third and fourth levels, where experimental proofs of concept have been successful, and the technology has been somewhat validated in laboratory environments. In particular, the Finnish MATINE project contributed toward achieving TRL 1 and TRL 2 in 2017 as well as TRL 3 and TRL 4 in 2018, while the NATO STO's international exploratory team attained TRL 1 and TRL 2 with some contributions toward TRL 3 in 2018. However, the technology readiness varies considerably between different specific concepts that are below the generic umbrella of military IBFD/STAR systems, and even the two lowest levels may not have been fully achieved yet, because there are still further basic principles to be observed and new technology concepts to be formulated. The higher TRLs have not yet even been approached, because no techniques have been validated or demonstrated with prototypes in relevant military environments, yet alone with pilot products in operational situations. In the strictest sense, it remains somewhat unknown how and why the IBFD/STAR capability should be implemented for real military radios. This is only a positive aspect for scientists, as it implies that there is much room for original, challenging research.

17.5.2 Future Directions

Apart from the research overviewed here, the scientific literature still remains rather limited; for example, the authors of [34–36] are examples of the first early adopters (per Rogers). Thus, one future direction is to keep communicating the innovation to a wider audience, because this is necessary to self-sustain. Forward-looking visionaries in the electrical engineering community would right now have a window of opportunity to engage in original, potentially high-impact scientific research on IBFD/STAR military radio systems. The work done now to increase the technology readiness will be seminal and extremely valuable when and if the technology is adopted in operational use. The short-term objectives would be to fully establish TRL 1 and TRL 2, where there is still room for innovators (per Rogers), and to validate or demonstrate the technology in relevant environments.

In conclusion, we scientists will be challenged by not only tough research problems to solve, but also the fact that defense sector and armed forces are notorious for conservativeness in adopting new technologies. The prospects are great, but we have yet to see whether military IBFD/STAR radios ever become more than an academic exercise. The development needs global collaboration between all scientists working on the subject. We need you, dear reader, to join the movement.

References

[1] Zhang, Z., et al., "Full-Duplex Wireless Communications: Challenges, Solutions, and Future Research Directions," *Proceedings of the IEEE*, Vol. 104, No. 7, July 2016, pp. 1369–1409.

[2] Kolodziej, K. E., B. T. Perry, and J. S. Herd, "In-Band Full-Duplex Technology: Techniques and Systems Survey," *IEEE Transactions on Microwave Theory and Techniques*, Vol. 67, No. 7, July 2019, pp. 3025–3041.

[3] Riihonen, T., et al., "Full-Duplex Radio Technology in Military Applications: Theoretical Foundations," summary report, Finnish Scientific Advisory Board for Defence, December 2017.

[4] Riihonen, T., et al., "Full-Duplex Radio Technology in Military Applications: Practical Experiments," summary report, Finnish Scientific Advisory Board for Defence, December 2018.

[5] Exploratory Team IST-ET-101 (COM), "Full Duplex Radio Technology – Increasing the Spectral Efficiency for Military Applications," activity details, NATO Science and Technology Organization, December 2019, https://www.sto.nato.int/Lists/test1/activitydetails.aspx?ID=16705.

[6] Exploratory Team IST-ET-101, "Full-Duplex Radio: Increasing the Spectral Efficiency for Military Applications," technical report, NATO Science and Technology Organization, March 2019.

[7] Research Task Group IST-175 (COM), "Full Duplex Radio Technology for Military Applications," activity details, NATO Science and Technology Organization, April 2020, https://www.sto.nato.int/Lists/test1/activitydetails.aspx?ID=16649.

[8] Pärlin, K., et al., "Transferring the Full-Duplex Radio Technology from Wireless Networking to Defense and Security," *Proc. 52nd Annual Asilomar Conference on Signals, Systems, and Computers*, October 2018.

[9] Pärlin, K., and T. Riihonen, "Full-Duplex Transceivers for Defense and Security Applications," in Alves, H., Riihonen, T., and Suraweera, H. A. (eds.), *Full-Duplex Communications for Future Wireless Networks*, Ch. 9, New York: Springer Nature, 2020, pp. 249–274.

[10] Riihonen, T., et al., "Inband Full-Duplex Radio Transceivers: A Paradigm Shift in Tactical Communications and Electronic Warfare?" *IEEE Communications Magazine*, Vol. 55, No. 10, October 2017, pp. 30–36.

[11] Riihonen, T., et al., "On the Prospects of Full-Duplex Military Radios," *Proc. International Conference on Military Communications and Information Systems*, May 2017.

[12] Marin, J., "Full-Duplex Transceiver Inspired by Frequency-Modulated Continuous-Wave Radar," M.Sc. thesis, Tampere University, Finland, April 2020.

[13] Saikanmäki, J., "Inband Full-Duplex Technology for Multi-Functional Multi-Mode Radios: Applications and System Experiments," M.Sc. thesis, Tampere University of Technology, Finland, November 2018.

[14] Hong, S., et al., "Applications of Self-Interference Cancellation in 5G and Beyond," *IEEE Communications Magazine*, Vol. 52, No. 2, February 2014, pp. 114–121.

[15] Enserink, S., et al., "Joint Analog and Digital Interference Cancellation," *Proc. IEEE International Symposium on Dynamic Spectrum Access Networks*, April 2014.

[16] Polydoros, A., et al., "Tethered Interference Cancellation: Military Versus Full-Duplex Applications and Comparisons," *Proc. 20th European Wireless Conference*, May 2014.

[17] Enserink, S., et al., "Joint Analog and Digital Interference Cancellation," *Proc. IEEE MTT-S International Microwave Symposium*, June 2014.

[18] Korpi, D., et al., "In-Band Full-Duplex Radios for Military Communications," *Proc. XXXIV Finnish URSI Convention on Radio Science*, September 2017.

[19] Riihonen, T., et al., "Full-Duplex Radio Technology for Simultaneously Detecting and Preventing Improvised Explosive Device Activation," *Proc. International Conference on Military Communications and Information Systems*, May 2018.

[20] Riihonen, T., et al., "Tactical Communication Link Under Joint Jamming and Interception by Same-Frequency Simultaneous Transmit and Receive Radio," *Proc. IEEE Military Communications Conference*, October 2018.

[21] Riihonen, T., et al., "Military Full-Duplex Radio Shield for Protection Against Adversary Receivers," *Proc. International Conference on Military Communications and Information Systems*, May 2019.

[22] Saikanmäki, J., et al., "Simultaneous Jamming and RC System Detection by Using Full-Duplex Radio Technology," *Proc. International Conference on Military Communications and Information Systems*, May 2019.

[23] Pärlin, K., T. Riihonen, and M. Turunen, "Sweep Jamming Mitigation Using Adaptive Filtering for Detecting Frequency Agile Systems," *Proc. International Conference on Military Communications and Information Systems*, May 2019.

[24] Pärlin, K., and T. Riihonen, "Digitally Assisted Analog Mitigation of Narrowband Periodic Interference," *Proc. 16th International Symposium on Wireless Communication*, August 2019.

[25] Pärlin, K., and T. Riihonen, "Analog Mitigation of Frequency-Modulated Interference for Improved GNSS Reception," *Proc. 10th International Conference on Localization and GNSS*, June 2020.

[26] Pärlin, K., et al., "Jamming and Classification of Drones Using Full-Duplex Radios and Deep Learning," *Proc. 31st IEEE International Symposium on Personal, Indoor and Mobile Radio Communications*, September 2020.

[27] Riihonen, T., et al., "Full-Duplex Operation for Electronic Protection by Detecting Communication Jamming at Transmitter," *Proc. 31st IEEE International Symposium on Personal, Indoor and Mobile Radio Communications*, September 2020.

[28] Marin, J., et al., "Perfecting Jamming Signals Against RC Systems: An Experimental Case Study on FHSS with GFSK," *Proc. 31st IEEE International Symposium on Personal, Indoor and Mobile Radio Communications*, September 2020.

[29] Adrat, M., et al., "Full-Duplex Radio – Increasing the Spectral Efficiency for Military Applications," *Proc. International Conference on Military Communications and Information Systems*, May 2019.

[30] Research Task Group SET-285, "Multifunction RF Systems," activity details, NATO Science and Technology Organization, April 2020, https://www.sto.nato.int/Lists/test1/activitydetails.aspx?ID=16816.

[31] Research Task Group SET-287, "Characterization of Noise Radar," activity details, NATO Science and Technology Organization, April 2020, https://www.sto.nato.int/Lists/test1/activitydetails.aspx?ID=16858.

[32] "TrellisWare Awarded $15.7M to Lead the Military Full Duplex Radio Program," press release, TrellisWare Technologies, November 2017, https://www.trellisware.com/trellisware-awarded-15-7m-to-lead-the-military-full-duplex-radio-program/.

[33] Ranström, T., "Full Duplex in a Military Scenario: Feasibility of Practical Implementation," M.Sc. thesis, Linköping University, Sweden, December 2019.

[34] Ranström, T., and E. Axell, "Full Duplex Based Digital Out-of-Band Interference Cancellation for Collocated Radios," *Proc. IEEE Wireless Communications and Networking Conference*, May 2020.

[35] Samara, L., et al., "On the Performance of Tactical Communication Interception Using Military Full Duplex Radios," *Proc. 30th IEEE International Symposium on Personal, Indoor and Mobile Radio Communications*, September 2019.

[36] Mourougayane, K., and S. Srikanth, "A Tri-Band Full-Duplex Cognitive Radio Transceiver for Tactical Communications," *IEEE Communications Magazine*, Vol. 58, No. 2, February 2020, pp. 61–65.

Acronyms and Abbreviations

5G	Fifth generation
A	Analog
ACLR	Adjacent channel leakage ratio
ADC	Analog-to-digital converter
AGC	Automatic gain control
ALSTAR	Aperture-level simultaneous transmit and receive
AP	Access point
AR	Antenna reflection
AUX	Auxiliary
AWG	Arrayed waveguide grating
AWGN	Additive white Gaussian noise
BB	Baseband
BER	Bit error rate
BFN	Beamforming network
BM	Backscatter modulation
BON	Bed of nails
BPD	Balanced photodetector
BPF	Bandpass filter
BPSK	Binary phase-shift keying
BS	Base station
BW	Bandwidth
CAD	Computer-Aided Design
CCS	Cube coefficient subspaces
CDFE	Code-domain front-end
CDF	Cumulative distribution function
CDMA	Code-division multiple access
CG	Common gate
CMOS	Complementary metal-oxide semiconductor
COSMOS	Cloud Enhanced Open Software Defined Mobile Wireless Testbed for City-Scale Deployment
COTS	Commercial off-the-shelf
CP	Circular polarization
C-Rx	Circulator receiver
CS	Common source
CSF	Code-selective filter

CW	Continuous wave
D	Digital
DAC	Digital-to-analog converter
DeMOD	Demodulation
DFL	Distributed feedback laser
DoD	Department of Defense
DOF	Degrees of freedom
DP	Direct path
DPD	Digital predistortion
DR	Dynamic range
DSB	Double-sideband
DSP	Digital signal processing
DSSS	Direct-sequence spread spectrum
DTFT	Discrete-time Fourier transform
EAM	Electro-absorption modulator
EBD	Electrical balance duplexer
EII	Effective isotropic isolation
EIRP	Effective isotropic radiated power
ELSTAR	Element-level simultaneous transmit and receive
EM	Electromagnetic
E-O	Electro-optic
EOC	Electrical-to-optical converter
E-O-E	Electro-optic-electro
ERP	Effective radiated rower
ET	Exploratory team
EVM	Error vector magnitude
EW	Electronic warfare
EXT SIGNAL	External signal
F/B	Front-to-back ratio
FBAR	Film bulk acoustic resonator
FBG	Fiber Bragg grating
FBLMS	Fast block least mean squares
FBW	Fractional bandwidth
FD	Full-duplex
FDD	Frequency-division duplex
FDE	Frequency-domain equalization
FFT	Fast Fourier transform
FIR	Finite impulse response
FM	Phase modulator
FMCW	Frequency-modulated continuous wave
FOM	Figure of merit
FPGA	Field-programmable gate array
GaAs	Gallium arsenide
GaN	Gallium nitride
GC	Grating coupler
GD	Gradient descent
GMP	Generalized memory polynomial
HD	Half-duplex

HEMT	High-electron-mobility transistor
HF	High frequency
HFET	Heterojunction field-effect transistor
HIS	High-impedance surface
HPA	High-power amplifier
HWP	Half-wave plate
IBFD	In-band full-duplex
IC	Integrated circuit
ICMCIS	International Conference on Military Communications and Information Systems
IED	Improvised explosive device
IEEE	Institute of Electrical and Electronics Engineers
IF	Intermediate frequency
IIP2	Input second-order intercept point
IIP3	Input third-order intercept point
IL	Insertion loss
IM3	Third-order intermodulation distortion
INT	Interference signal
IoT	Internet of Things
IRR	Image rejection ratio
IRS	Interference reference signal
IS	Interference signal
ISI	Intersymbol interference
ISM	Industrial, Scientific, and Medical
IST	Information systems technology
JADIC	Joint Analog and Digital Interference Cancellation
LC	Inductor-capacitor
LFM	Linear frequency modulated
LHCP	Left-handed circular polarization
LMS	Least mean squares
LNA	Low-noise amplifier
LNTA	Low-noise transconductance amplifier
LO	Local oscillator
LP	Linearly polarized
LPTV	Linear periodically time-varying
LRA	Limited rate adaptation
LTE	Long term evolution
LTI	Linear time-invariant
LUT	Look-up table
M0	Mode 0
M1	Mode 1
M2	Mode 2
MATINE	Scientific Advisory Board for Defence
MCCN	Mutual Coupling Compensation Network
MCS	Modulation and coding scheme
MEMS	Microelectromechanical system
MFDR	Military full-duplex radio
MIMO	Multiple-input multiple-output

MOD	Modulator
MP	Memory polynomial
MPC	Microwave photonic canceller
M-PSK	*M*-ary phase-shift keying
MRR	Microring resonator
MSE	Mean square error
MU-MIMO	Multi-user multiple-input multiple-output
MZI	Mach-Zehnder interferometer
MZM	Mach-Zehnder modulator
NBWF	Narrowband waveform
NF	Noise figure
NN	Neural network
NR	New Radio
NRL	Naval Research Laboratory
NSC	National Spectrum Consortium
NSF	National Science Foundation
OA	Over-the-air
O-E	Optic-electro
OEC	Optical-to-electrical converter
OFDM	Orthogonal frequency-division multiplexing
OOB	Out-of band
OSR	Oversampling ratio
P	Propagation
PA	Power amplifier
PAPR	Peak-to-average power ratio
PAWR	Platforms for Advanced Wireless Research
PBS	Polarization beam splitter
PC	Polarization controller
PCB	Printed circuit board
PE	Processing element
PET	Pulse-encoded transitions
PG	Processing gain
PIC	Photonic integrated circuit
PN	Pseudorandom noise
PolM	Polarization modulator
PRR	Packet reception ratio
PRS	Printed reactive surfaces
PSD	Power spectral density
Q	Quality factor
QFN	Quad-flat no-leads
QPSK	Quadrature phase-shift keying
R&D	Research and development
RC	Radio control
RC	Resistor-capacitor
RCS	Radar cross-section
REF	Reference
RF	Radio frequency
RFIC	Radio frequency integrated circuit

RFID	Radio frequency identification
RHCP	Right-handed circular polarization
RIN	Relative intensity noise
RIS	Reactive-impedance surface
RLC	Resistor-inductor-capacitor
RLS	Recursive least squares
RP	Reflection path
RPS	Reflection phase shifter
RTG	Research Task Group
Rx	Receive
SAW	Surface acoustic wave
SBIR	Small Business Innovation Research
SDR	Software-defined radio
SET	Sensors and electronics technology
SFFD	Same-frequency full-duplex
SI	Self-interference
SIC	Self-interference cancellation
SINR	Signal-to-interference-plus-noise ratio
SISO	Single-input single-output
SNR	Signal-to-noise ratio
SOA	Semiconductor optical amplifier
SOI	Signal of interest
SOI	Silicon-on-insulator
SPH	Spline-based Hammerstein
SRW	Soldier radio waveform
SSB	Single-sideband
STAR	Simultaneous transmit and receive
STO	Science and technology organization
STTR	Small business technology transfer
SWaP	Size, weight, and power
TCDA	Tightly coupled dipole array
TDD	Time-division duplex
TDL	Tunable delay line
TIA	Transimpedance amplifier
TIPRx	Transmit-isolating photonic receiver
TRL	Technology readiness level
TRS	Tapered ribbed surface
TSA	Tapered slot antenna
TWT	Traveling wave tube
Tx	Transmit
Tx-Rx	Transmit-to-receive
UE	User equipment
UHF	Ultrahigh frequency
USRP	Universal Software Radio Peripheral
V2X	Vehicle-to-everything
VGA	Variable-gain transconductance amplifier
VHF	Very high frequency
VM	Vector modulator

VNA	Vector network analyzer
VOA	Variable optical attenuator
VST	Vector signal transceiver
VSWR	Voltage standing-wave ratio
WDM	Wavelength division multiplexing
XI	Crosstalk interference
XIC	Crosstalk-interference cancellation
ΣΔM	Sigma-delta modulation

About the Editor

Kenneth E. Kolodziej is a technical staff member in the RF Technology Group at Massachusetts Institute of Technology (MIT) Lincoln Laboratory, where he is a principal investigator on wideband electronic and in-band full-duplex (IBFD)—commonly known as simultaneous transmit and receive (STAR)—systems. Since joining Lincoln Laboratory, Mr. Kolodziej has conducted research on RF, microwave and photonic circuits, including antenna, radar and communications systems. He also teaches both an electromagnetics course and several "Build-a-Radar" courses on MIT campus. Previously, he worked at BAE Systems in Wayne, New Jersey, designing RF electronics for handheld communication devices. Mr. Kolodziej received his BE and ME degrees in electrical engineering from Stevens Institute of Technology in Hoboken, New Jersey.

He has published numerous journal articles and conference papers, and holds several patents in the field of antennas and RF systems as well as delivered invited lectures and organized technical workshops on IBFD. Mr. Kolodziej is a member of the IEEE Microwave Theory and Techniques, IEEE Antennas and Propagation and IEEE Communications Societies, where he reviews papers for several journals and conferences. He also serves on the technical program review committees for the IEEE International Microwave Symposium (IMS), the IEEE International Symposium on Antennas and Propagation (APS), the IEEE International Symposium on Phased Array Systems and Technology (PAST), and the IEEE International Conference on Communications (ICC).

About the Authors

Edward I. Ackerman cofounded Photonic Systems, Inc., and is its vice-president of research and development. His accomplishments in the field of analog photonics include the demonstration of the first amplifierless direct modulation analog optical link with gain, demonstration of the lowest noise figure ever measured for an amplifierless analog optical link, and invention of architectures for the linearization of such links across broader bandwidths than what previous linearization techniques can realize. In 2019, the IEEE Microwave Theory & Techniques Society (MTT-S) selected Dr. Ackerman as a Distinguished Microwave Lecturer. He is a fellow of the IEEE and holds 16 U.S. patents.

Lauri Anttila received an M.Sc. in 2004 and a D.Sc. (Hons.) in 2011 in electrical engineering from the Tampere University of Technology (TUT), Tampere, Finland. Since 2016, he has been a university researcher with the Department of Electrical Engineering, Tampere University (formerly TUT). From 2016 to 2017, he was a visiting research fellow with the Department of Electronics and Nanoengineering, Aalto University, Espoo, Finland. His research interests are in radio communications and signal processing, with a focus on the radio implementation challenges in systems, such as 5G, full-duplex radio, and large-scale antenna systems.

Mahmood Baraani Dastjerdi received an M.S. from Sharif University of Technology, Tehran, Iran, in 2014, and a Ph.D. in electrical engineering from Columbia University, New York, in 2020. His career is focused on the theory, design, and experimental validation of analog, radio frequency (RF), and millimeter-wave integrated circuits and systems. Dr. Baraani Dastjerdi joined MixComm Inc. as a member of the technical staff in 2019. He is a recipient of the 2020 Columbia University Electrical Engineering Department Jury Award, 2019 IEEE SSCS Predoctoral Achievement Award, 2018 ADI ISSCC outstanding student designer award and creative tech award in engineering at the 2018 NYC Media Lab.

Eric C. Blow is a sixth-year Ph.D. candidate in the Electrical Engineering Department of Princeton University. He is a senior member of the Lightwave Laboratory led by Professor Paul R. Prucnal. He has authored several journal papers in the fields of microwave photonics and neuromorphic photonics. His research focuses on developing microwave photonic systems to improve the efficiency of high-speed wireless communication, including interference cancellation, beamforming, and predistortion. He earned a B.S. from The College of New Jersey and an M.A. from Princeton University in electrical engineering. Mr. Blow is also a member of the engineering team at Bascom Hunter, Inc.

Luciano Boglione received a Ph.D. from the University of Leeds in 1998. In 2011, he joined the U.S. Naval Research Laboratory, where he has been leading a team to design subsystems in silicon technology for full-duplex applications and investigating new noise characterization techniques. He has authored and coauthored more than 30 papers in peer-reviewed publications, holds 1 patent, and has 2 patents pending. He was the 2007 Radio Frequency Integrated Circuits (RFIC) Conference General Chair. He was the past chair and is currently a member of the MTT-S Microwave Low-Noise Techniques Committee (MTT14), and an elected AdCom member of the MTT Society for 2 terms.

James F. Buckwalter is a professor of electrical and computer engineering at the University of California Santa Barbara (UCSB). His research interests are in wireless and optical communication systems covering RF to millimeter-wave frequency bands and device technologies, including silicon CMOS, silicon germanium, and III-V devices. Professor Buckwalter received a B.S. with Honors from California Institute of Technology (Caltech), an M.S. in electrical and computer engineering from UCSB, and a Ph.D. from Caltech in electrical engineering in 2006.

Yuhe Cao received a bachelor's degree from the Huazhong University of Science and Technology, Wuhan, China, and the University of Birmingham, Birmingham, United Kingdom, in 2013, and an M.S. in electrical engineering from Columbia University, New York, in 2015. He joined the University of Illinois at Urbana-Champaign in 2017 as a Ph.D. student. Mr. Cao was a recipient of the 2019 International Microwave Symposium (IMS) and RFIC Symposium Ph.D. Student Sponsorship.

Tingjun Chen received a Ph.D. in electrical engineering from Columbia University in 2020 and a B.Eng. in electronic engineering from Tsinghua University in 2014. He is a postdoctoral associate at Yale University and will join the Department of Electrical and Computer Engineering at Duke University as an assistant professor in Fall 2021. His research interests are in the areas of wireless networking and systems with focuses on algorithms, optimization, and system design and implementation. He received the Facebook Fellowship, Wei Family Private Foundation Fellowship, Columbia Electrical Engineering Armstrong Memorial Award, and ACM CoNEXT'16 Best Paper Award.

Charles H. Cox III is the cofounder and president of Photonic Systems and is one of the pioneers of the field that is now generally referred to as analog or RF photonics. He developed the theoretical basis for, and the first experimental demonstration of, amplifierless optical links with gain. Dr. Cox holds 27 issued patents, has served on the committees of 16 conferences, has given 90 invited talks, and has written 2 books, 7 book chapters, and 80 refereed papers. He is a fellow of the IEEE and the Optical Society of America (OSA). Dr. Cox served as a member of the Advisory Group on Electron Devices from 2003 to 2009.

Ian T. Cummings is a postdoctoral research associate in the Space and Remote Sensing Group in the Intelligence and Space Research Division (ISR-2) at the Los Alamos National Laboratory. During graduate school, he studied adaptive beamforming and self-interference cancellation techniques for simultaneous transmit and receive (STAR) phased arrays as a collaboration with the MIT Lincoln Laboratory. He received the National Science Foundation Graduate Research Fellowship, and earned his B.S. and M.S. in computer engineering and his Ph.D. in electrical engineering from Michigan Technological University.

Clayton G. Davis is a signals research scientist in the resilient communications group at The MITRE Corporation. Prior to joining MITRE, Dr. Davis researched in-band full-duplex (IBFD) and signal intelligence techniques in the tactical electronic warfare division at the Naval Research Laboratory. Dr. Davis received a Ph.D. in physics from the University of Maryland in College Park, Maryland, and an M.S. in mathematics and a B.S. in physics from the University of Virginia in Charlottesville, Virginia.

Jonathan Doane has been a member of the technical staff at the MIT Lincoln Laboratory since 2013. He was the principal investigator of the STAR Phased Array program, focused on isolation improvement technologies to enable full-duplex phased arrays and multifunctional converged RF applications. Dr. Doane received a Ph.D. from The Ohio State University and a B.S. and an M.S. from Michigan Technological University. He holds 16 U.S. patents and has published numerous articles on antenna and phased array technologies.

Melissa Duarte received a B.Sc. in electrical engineering from the Pontificia Universidad Javeriana, Colombia, in 2005. She received an M.Sc. in 2007 and a Ph.D. in 2012 in electrical and computer engineering from Rice University, Houston, Texas. From 2012 to 2013, she was a postdoctoral researcher at the School of Computer and Communication Sciences, EPFL Switzerland. In 2014, she joined the Mathematical and Algorithmic Sciences Lab, Huawei Technologies, France. She received the ACM MobiHoc 2013 Best Paper Award, the 2017 IEEE Jack Neubauer Memorial Award and the 2018 IEEE Communications Society Award for Advances in Communication.

Dejan S. Filipovic is the Hudson Moore Jr. Endowed Chair at the University of Colorado Boulder. His broader research interests are in applied electromagnetics, including antenna theory and design with an emphasis on frequency-independent and wideband antennas; development of passive millimeter-wave components, systems, and electronic warfare front ends; low-cost fabrication of RF systems; simultaneous transmit and receive; and multiphysics, multiscale modeling. He was a two-time recipient of the University of Colorado Provost's Faculty Achievement Award and Electrical and Computing Engineering (ECEN) Department Holland's Teaching Award. Professor Filipovic has directed more than 30 Ph.D. and M.Sc. dissertations. He is a Fellow of the IEEE.

Joel Goodman has been on the staff at the U.S. Naval Research Laboratory since 2010. He was a recipient of the Eastman Technical Achievement award for his work on magnetic imaging, an invited lecturer for an IEEE advanced signal processing symposium on the topic of nonlinear signal processing, a recipient of the MIT LL Technical Excellence award for his work on nonlinear equalization, recipient of three NRL Alan Berman Research Publication awards and an NRL Review award, and an invited lecturer at the Virginia Tech Summer School on Wireless Communications. Mr. Goodman received a B.S. and an M.S. in electrical engineering from Boston University.

Ahmed Hamza is a Ph.D. student at the University of California at Santa Barbara, California. His research interests include RF and millimeter-wave integrated circuits, and analog/mixed-signal circuit design for wireline and wireless applications. Mr. Hamza received a B.Sc. and an M.Sc. in electronics and communication systems engineering from Ain Shams University, Cairo, Egypt.

Cameron Hill is a postdoctoral researcher at the University of California, Santa Barbara. His research interests include RF and millimeter-wave integrated circuits, specifically in high-linearity CMOS design methodologies. Dr. Hill received a Ph.D. in electrical engineering from the University of California, Santa Barbara, in 2020 and an M.E. and a B.E. in electrical engineering from the Stevens Institute of Technology in Hoboken, New Jersey in 2015.

Seiran Khaledian is a staff antenna engineer at Motorola Mobility. She is working on smart phone antenna design for products at the edge of technology, such as the first foldable smartphone and 5G products. She received a Ph.D. in electrical engineering from the University of Illinois at Chicago (UIC) in 2019. She was a member of the Networks Information Communications and Engineering Systems Laboratory at UIC, where she was working on full-duplex communication systems, analog self-interference cancellation techniques, reconfigurable antenna, and RF/microwave components. She holds several patents and many published researches in the peer-reviewed IEEE journals.

Dani Korpi received an M.Sc. in 2014 and a D.Sc. (Hons.) in 2017 in communications engineering and electrical engineering from the Tampere University of Technology, Finland. He is currently a senior specialist with Nokia Bell Labs, Espoo, Finland. He received the Best Dissertation of the Year Award from the Tampere University of Technology, for his Ph.D. thesis, as well as the Finnish Technical Sector's Award for the best doctoral dissertation in 2017. His research interests include IBFD radios, machine learning for wireless communications, and beyond 5G radio systems.

Harish Krishnaswamy received a B.Tech. in electrical engineering from IIT Madras, Chennai, India, in 2001, and an M.S. in 2003 and a Ph.D. in 2009 in electrical engineering from the University of Southern California (USC), Los Angeles. In 2009, he joined the Electrical Engineering Department, Columbia University, New York, where he is currently an associate professor and the director of the Columbia High-Speed and Millimeter-Wave IC Laboratory (CoSMIC). In 2017, he cofounded MixComm Inc., Chatham, New Jersey, a venture-backed startup, to commercialize CoSMIC Laboratory's advanced wireless research.

Leo Laughlin is the cofounder and chief executive officer of Forefront RF Ltd., a company developing self-interference cancellation technologies for wireless applications. He was previously a research fellow at the University of Bristol, United Kingdom, where he developed novel transceiver architectures and cancellation algorithms for IBFD and frequency-division duplex systems, with a particular focus on electrical balance duplexers. He has a Ph.D. from the University of Bristol and an M.Eng. from the University of York, United Kingdom.

Carlos A. Mulero Hernández is a Ph.D. student with the Antenna Research Group at the University of Colorado Boulder, where his research has focused on lens-based high-directivity IBFD, also known as STAR, systems. Mr. Mulero Hernández received a B.S. and an M.S. in electrical engineering from the University of Puerto Rico, Mayagüez.

Pablo Pascual Campo received a B.Sc. in 2012 and an M.Sc. in 2014 in telecommunications and electrical engineering from Universidad Politécnica de Madrid, Madrid, Spain. He is currently pursuing a D.Sc. at Tampere University's Department of Electrical Engineering, Tampere, Finland. His research interests include digital predistortion, full-duplex systems and applications, and signal processing for wireless communications in the millimeter-wave spectrum.

Bradley T. Perry is an assistant leader of the RF Technology Group at the MIT Lincoln Laboratory, where he currently leads programs on the development of next-generation systems and technology related to isolation improvement. These programs include work in the areas of microwave circuit and antenna design, compact receiver and transmitter designs for electronic warfare systems and active decoys, and techniques for STAR applications. Dr. Perry received a B.S., an M.S., and a Ph.D. in electrical engineering from Michigan State University, where he focused on antenna design and electromagnetic propagation through layered media.

Prathap Valale Prasannakumar is an antenna engineer in the wireless connectivity team at Oculus AR/VR, Facebook Inc. His current research is focused on Wi-Fi (2.4 GHz, 5 GHz, and 6 GHz), Bluetooth, and ultra-wideband (UWB) antennas and improving the isolation between collocated antennas in augmented reality and virtual reality (AR/VR) consumer electronics devices. His Ph.D. thesis focused on the design of monostatic, quasi-monostatic, and bistatic STAR antenna systems and wideband isolation enhancement techniques. Dr. Prathap received an M.S. and a Ph.D. in electrical engineering from the University of Colorado, Boulder. Previously, he worked as deputy engineer at Bharat Electronics, India.

Paul R. Prucnal is a professor of electrical engineering at Princeton University. He is best known for his seminal work in neuromorphic photonics, optical code division multiple access (OCDMA), and the invention of the terahertz optical asymmetric demultiplexer (TOAD). He earned an M.S., an M.Phil., and a Ph.D. in electrical engineering from Columbia University. Professor Prucnal is the coauthor of the book, *Neuromorphic Photonics*. He is a fellow of the IEEE, the OSA, and the National Academy of Inventors (NAI).

Negar Reiskarimian is the X-Window Consortium Career Development Assistant Professor within the MIT Electrical Engineering and Computer Science Department. She earned a Ph.D. in 2020 and an M.Phil. in 2017 in electrical engineering from Columbia University. She received an M.Sc. and a B.Sc. from Sharif University of Technology, Tehran, Iran. Her research has been focused on developing RF and millimeter-wave integrated circuits and system technologies by exploring ideas from applied physics and electromagnetics.

Taneli Riihonen received a D.Sc. (with honors) in electrical engineering from Aalto University, Helsinki, Finland, in 2014. He was a visiting associate research scientist and an adjunct assistant professor at Columbia University, New York, in 2015. He is a tenure-track assistant professor at Tampere University, Finland, since 2017 and a lieutenant in the Finnish Defence Forces' reserve. His research activity is focused on physical-layer OFDM(A), multiantenna, multihop, and full-duplex wireless techniques with current interest in the evolution of beyond 5G radio systems.

Besma Smida is an associate professor of electrical and computer engineering at University of Illinois at Chicago. Before that, she held academic positions at Purdue University Northwest and Harvard University. Her research focuses on the areas of wireless communication theory. She has been the chair of the IEEE Communication Chapter, Chicago Section, since 2019. She was awarded the INSIGHT Into Diversity Magazine's 2015 "100 Inspiring Women in STEM." She was the recipient of the Academic Gold Medal of the Governor General of Canada in 2007 and the NSF CAREER award in 2015.

Vincent J. Urick is a program manager at the Defense Advanced Research Projects Agency, Arlington, Virginia. From 2001 to 2016, he was a research physicist and the head of the Applied RF Photonics Section at the Naval Research Laboratory, Washington, D.C. He has published more than 125 works in microwave photonics and advanced technology for defense applications. Dr. Urick received a B.S. (magna cum laude) in physics from Bloomsburg University and an M.S. and a Ph.D. in physics from George Mason University.

Mikko Valkama received an M.Sc. (with honors) in 2000 and a D.Sc. (with honors) in 2001 from Tampere University of Technology. Currently, he is a full professor and the head of the Unit of Electrical Engineering at Tampere University, Finland. His general research interests include radio communications, radio localization, and radio-based sensing, with a particular emphasis on 5G and beyond mobile radio networks.

Jin Zhou received a Ph.D. in electrical engineering from Columbia University in 2017. From 2011 to 2012, he was an RF integrated circuits design engineer with MediaTek Singapore. In 2017, he joined the Department of Electrical and Computer Engineering at the University of Illinois at Urbana-Champaign, where he is currently an assistant professor. Dr. Zhou is a recipient of the 2015-2016 Qualcomm Innovation Fellowship, the 2015-2016 IEEE Solid-State Circuits Society Predoctoral Achievement Award, and the 2016 Eli Jury Award from the Department of Electrical Engineering at Columbia University.

Gil Zussman received a Ph.D. in electrical engineering from the Technion in 2004 and was a postdoctoral associate at MIT from 2004 to 2007. He is a professor of electrical engineering at Columbia University. He has been an associate editor of the *IEEE/ACM Transactions on Networking*, *IEEE Transactions on Control of Network Systems*, and *IEEE Transactions on Wireless Communications*. He is a corecipient of 7 paper awards, including ACM SIGMETRICS'06 Best Paper Award, 2011 IEEE Communications Society Award for Advances in Communication, and ACM CoNEXT'16 Best Paper Award. He received the Fulbright Fellowship, the DTRA Young Investigator Award, and NSF CAREER Award.

Index

H

I

N

O

P

T

U

V

W

Recent Titles in the Artech House Mobile Communications Series

William Webb, Series Editor

3G CDMA2000 Wireless System Engineering, Samuel C. Yang

3G Multimedia Network Services, Accounting, and User Profiles, Freddy Ghys, Marcel Mampaey, Michel Smouts, and Arto Vaaraniemi

5G New Radio: Beyond Mobile Broadband, Amitav Mukherjee

5G Spectrum and Standards, Geoff Varrall

802.11 WLANs and IP Networking: Security, QoS, and Mobility, Anand R. Prasad and Neeli R. Prasad

Achieving Interoperability in Critical IT and Communications Systems, Robert I. Desourdis, Peter J. Rosamilia, Christopher P. Jacobson, James E. Sinclair, and James R. McClure

Advances in 3G Enhanced Technologies for Wireless Communications, Jiangzhou Wang and Tung-Sang Ng, editors

Advances in Mobile Information Systems, John Walker, editor

Advances in Mobile Radio Access Networks, Y. Jay Guo

Applied Satellite Navigation Using GPS, GALILEO, and Augmentation Systems, Ramjee Prasad and Marina Ruggieri

Artificial Intelligence in Wireless Communications, Thomas W. Rondeau and Charles W. Bostian

Broadband Wireless Access and Local Network: Mobile WiMax and WiFi, Byeong Gi Lee and Sunghyun Choi

CDMA for Wireless Personal Communications, Ramjee Prasad

CDMA Mobile Radio Design, John B. Groe and Lawrence E. Larson

CDMA RF System Engineering, Samuel C. Yang

CDMA Systems Capacity Engineering, Kiseon Kim and Insoo Koo

CDMA Systems Engineering Handbook, Jhong S. Lee and Leonard E. Miller

Cell Planning for Wireless Communications, Manuel F. Cátedra and Jesús Pérez-Arriaga

Cellular Communications: Worldwide Market Development, Garry A. Garrard

Cellular Mobile Systems Engineering, Saleh Faruque

Cognitive Radio Interoperability through Waveform Reconfiguration, Leszek Lechowicz and Mieczyslaw M. Kokar

Cognitive Radio Techniques: Spectrum Sensing, Interference Mitigation, and Localization, Kandeepan Sithamparanathan and Andrea Giorgetti

The Complete Wireless Communications Professional: A Guide for Engineers and Managers, William Webb

Digital Communication Systems Engineering with Software-Defined Radio, Di Pu and Alexander M. Wyglinski

EDGE for Mobile Internet, Emmanuel Seurre, Patrick Savelli, and Pierre-Jean Pietri

Emerging Public Safety Wireless Communication Systems, Robert I. Desourdis, Jr., et al.

From LTE to LTE-Advanced Pro and 5G, Moe Rahnema and Marcin Dryjanski

The Future of Wireless Communications, William Webb

Geographic Information Systems Demystified, Stephen R. Galati

Geospatial Computing in Mobile Devices, Ruizhi Chen and Robert Guinness

GPRS for Mobile Internet, Emmanuel Seurre, Patrick Savelli, and Pierre-Jean Pietri

GPRS: Gateway to Third Generation Mobile Networks, Gunnar Heine and Holger Sagkob

GSM and Personal Communications Handbook, Siegmund M. Redl, Matthias K. Weber, and Malcolm W. Oliphant

GSM Networks: Protocols, Terminology, and Implementation, Gunnar Heine

GSM System Engineering, Asha Mehrotra

Handbook of Land-Mobile Radio System Coverage, Garry C. Hess

Handbook of Mobile Radio Networks, Sami Tabbane

Handbook of Next-Generation Emergency Services, Barbara Kemp and Bart Lovett

High-Speed Wireless ATM and LANs, Benny Bing

Implementing Full Duplexing for 5G, David B. Cruickshank

In-Band Full-Duplex Wireless Systems Handbook, Kenneth E. Kolodziej, Editor

Inside Bluetooth Low Energy, Second Edition, Naresh Gupta

Interference Analysis and Reduction for Wireless Systems, Peter Stavroulakis

Interference and Resource Management in Heterogeneous Wireless Networks, Jiandong Li, Min Sheng, Xijun Wang, and Hongguang Sun

Internet Technologies for Fixed and Mobile Networks, Toni Janevski

Introduction to 3G Mobile Communications, Second Edition, Juha Korhonen

Introduction to 4G Mobile Communications, Juha Korhonen

Introduction to Communication Systems Simulation, Maurice Schiff

Introduction to Digital Professional Mobile Radio, Hans-Peter A. Ketterling

Introduction to GPS: The Global Positioning System, Ahmed El-Rabbany

An Introduction to GSM, Siegmund M. Redl, Matthias K. Weber, and Malcolm W. Oliphant

Introduction to Mobile Communications Engineering, José M. Hernando and F. Pérez-Fontán

Introduction to OFDM Receiver Design and Simulation, Y. J. Liu

Introduction to Radio Propagation for Fixed and Mobile Communications, John Doble

Introduction to Wireless Local Loop, Broadband and Narrowband, Systems, Second Edition, William Webb

IS-136 TDMA Technology, Economics, and Services, Lawrence Harte, Adrian Smith, and Charles A. Jacobs

Location Management and Routing in Mobile Wireless Networks, Amitava Mukherjee, Somprakash Bandyopadhyay, and Debashis Saha

LTE Air Interface Protocols, Mohammad T. Kawser

Metro Ethernet Services for LTE Backhaul, Roman Krzanowski

Mobile Data Communications Systems, Peter Wong and David Britland

Mobile IP Technology for M-Business, Mark Norris

Mobile Satellite Communications, Shingo Ohmori, Hiromitsu Wakana, and Seiichiro Kawase

Mobile Telecommunications Standards: GSM, UMTS, TETRA, and ERMES, Rudi Bekkers

Mobile-to-Mobile Wireless Channels, Alenka Zajić

Mobile Telecommunications: Standards, Regulation, and Applications, Rudi Bekkers and Jan Smits

Multiantenna Digital Radio Transmission, Massimiliano Martone

Multiantenna Wireless Communications Systems, Sergio Barbarossa

Multi-Gigabit Microwave and Millimeter-Wave Wireless Communications, Jonathan Wells

Multipath Phenomena in Cellular Networks, Nathan Blaunstein and Jørgen Bach Andersen

Multiuser Detection in CDMA Mobile Terminals, Piero Castoldi

OFDMA for Broadband Wireless Access, Slawomir Pietrzyk

Personal Wireless Communication with DECT and PWT, John Phillips and Gerard MacNamee

Practical Wireless Data Modem Design, Jonathon Y. C. Cheah

The Practitioner's Guide to Cellular IoT, Cameron Kelly Coursey

Prime Codes with Applications to CDMA Optical and Wireless Networks, Guu-Chang Yang and Wing C. Kwong

Quantitative Analysis of Cognitive Radio and Network Performance, Preston Marshall

QoS in Integrated 3G Networks, Robert Lloyd-Evans

Radio Engineering for Wireless Communication and Sensor Applications, Antti V. Räisänen and Arto Lehto

Radio Propagation in Cellular Networks, Nathan Blaunstein

Radio Resource Management for Wireless Networks, Jens Zander and Seong-Lyun Kim

Radiowave Propagation and Antennas for Personal Communications, Third Edition, Kazimierz Siwiak and Yasaman Bahreini

RDS: The Radio Data System, Dietmar Kopitz and Bev Marks

Resource Allocation in Hierarchical Cellular Systems, Lauro Ortigoza-Guerrero and A. Hamid Aghvami

RF and Baseband Techniques for Software-Defined Radio, Peter B. Kenington

RF and Microwave Circuit Design for Wireless Communications, Lawrence E. Larson, editor

RF Positioning: Fundamentals, Applications, and Tools, Rafael Saraiva Campos and Lisandro Lovisolo

Sample Rate Conversion in Software Configurable Radios, Tim Hentschel

Signal Processing Applications in CDMA Communications, Hui Liu

Signal Processing for RF Circuit Impairment Mitigation, Xinping Huang, Zhiwen Zhu, and Henry Leung

Smart Antenna Engineering, Ahmed El Zooghby

Software-Defined Radio for Engineers, Travis F. Collins, Robin Getz, Di Pu, and Alexander M. Wyglinski

Software Defined Radio for 3G, Paul Burns

Spread Spectrum CDMA Systems for Wireless Communications, Savo G. Glisic and Branka Vucetic

Technical Foundations of the IoT, Boris Adryan, Dominik Obermaier, and Paul Fremantle

Technologies and Systems for Access and Transport Networks, Jan A. Audestad

Third-Generation and Wideband HF Radio Communications, Eric E. Johnson, Eric Koski, William N. Furman, Mark Jorgenson, and John Nieto

Third Generation Wireless Systems, Volume 1: Post-Shannon Signal Architectures, George M. Calhoun

Traffic Analysis and Design of Wireless IP Networks, Toni Janevski

Transmission Systems Design Handbook for Wireless Networks, Harvey Lehpamer

UMTS and Mobile Computing, Alexander Joseph Huber and Josef Franz Huber

Understanding Cellular Radio, William Webb

Understanding Digital PCS: The TDMA Standard, Cameron Kelly Coursey

Understanding GPS: Principles and Applications, Second Edition, Elliott D. Kaplan and Christopher J. Hegarty, editors

Understanding WAP: Wireless Applications, Devices, and Services, Marcel van der Heijden and Marcus Taylor, editors

Universal Wireless Personal Communications, Ramjee Prasad

WCDMA: Towards IP Mobility and Mobile Internet, Tero Ojanperä and Ramjee Prasad, editors

Wireless Communications in Developing Countries: Cellular and Satellite Systems, Rachael E. Schwartz

Wireless Communications Evolution to 3G and Beyond, Saad Z. Asif

Wireless Intelligent Networking, Gerry Christensen, Paul G. Florack and Robert Duncan

Wireless LAN Standards and Applications, Asunción Santamaría and Francisco J. López-Hernández, editors

Wireless Sensor and Ad Hoc Networks Under Diversified Network Scenarios, Subir Kumar Sarkar

Wireless Technician's Handbook, Second Edition, Andrew Miceli

For further information on these and other Artech House titles,including previously considered out-of-print books now available through our In-Print-Forever® (IPF®) program, contact:

Artech House
685 Canton Street
Norwood, MA 02062
Phone: 781-769-9750
Fax: 781-769-6334
e-mail: artech@artechhouse.com

Artech House
16 Sussex Street
London SW1V 4RW UK
Phone: +44 (0)20 7596-8750
Fax: +44 (0)20 7630-0166
e-mail: artech-uk@artechhouse.com

Find us on the World Wide Web at: www.artechhouse.com